MW01291577

ASTRONOMICAL ALGORITHMS

Jean Meeus

Published by:
Willmann–Bell, Inc.
P.O. Box 35025
Richmond, Virginia 23235

Published by Willmann-Bell, Inc.
P.O. Box 35025, Richmond, Virginia 23235

First English Edition

Printed in the United States of America

Library of Congress Cataloging-in-Publication Data.
Meus, Jean.
Astronomical algorithms / Jean Meeus.
p. cm.
Includes bibliographical references and index.
ISBN 0-943396-35-2
1. Astronomy–Data processing. 2.Astronomy–Problems, exercises, etc. 3. Algorithms. I. Title.
QB51.3.E43M42 1991
520–dc20 91-23501
CIP

97 98 9 8 7 6 5 4

Foreword

People who write their own computer programs often wonder why the machine gives inaccurate planet positions, an unreal eclipse track, or a faulty Moon phase. Sometimes they insist, bewildered, "and I used double precision, too." Even commercial software is sometimes afflicted with gremlins, which comes as quite a shock to anyone caught up in the mystique and presumed infallibility of computers. Good techniques can help us avoid erroneous results from a flawed program or a simplistic procedure—and that's what this book is all about.

In the field of celestial calculations, Jean Meeus has enjoyed wide acclaim and respect since long before microcomputers and pocket calculators appeared on the market. When he brought out his *Astronomical Formulae for Calculators* in 1979, it was practically the only book of its genre. It quickly became the "source among sources," even for other writers in the field. Many of them have warmly acknowledged their debt (or should have), citing the unparalleled clarity of his instructions and the rigor of his methods.

And now this Belgian astronomer has outdone himself yet again! Virtually every previous handbook on celestial calculations (including his own earlier work) was forced to rely on formulae for the Sun, Moon, and planets that were developed in the last century—or at least before 1920. The past 10 years, however, have seen a stunning revolution in how the world's major observatories produce their almanacs. The Jet Propulsion Laboratory in California and the U.S. Naval Observatory in Washington, D.C., have perfected powerful new machine methods for modeling the motions and interactions of bodies within the solar system. At the same time in Paris, the Bureau des Longitudes has been a beehive of activity aimed at describing these motions analytically, in the form of explicit equations.

Yet until now the fruits of this exciting work have remained mostly out of reach of ordinary people. The details have existed mainly on reels of magnetic tape in a form comprehensible only to the largest brains, human or electronic. But *Astronomical Algorithms* changes all that. With his special knack for computations of all sorts, the author has made the essentials of these modern techniques available to us all.

We also stand at a confusing crossroads for astronomy. In just the last few years the International Astronomical Union has introduced subtle changes in the reference frame used for the coordinates of celestial objects, both within and far beyond our solar system. So sweeping are these revisions that a highly respected work for professional astronomers, the *Explanatory Supplement to the Astronomical Ephemeris*, published in 1961, is now seriously out of date. While the technical journals have seen a flurry of scientific papers on these issues, the book you're holding now is the first to offer succinct and practical methods for coping with the changeover. It will be many years before astronomical data bases and catalogues are fully converted to the new system, and anyone who needs a detailed understanding of what's going on will appreciate this book's many comments about the FK4 and FK5 reference frames, "equinox error," and the distinction between "J" and "B" when placed before an epoch like 2000.0.

Scarcely any formula is presented without a fully worked numerical example—so crucial to the debugging process. The emphasis throughout is on testing, on the proper arrangement of formulae, and on not pushing them beyond the time span over which they are valid. Chapter 2 contains much wisdom of this sort, growing out of the author's long experience with various computers and their languages. He alerts us to other pitfalls throughout the text. Anyone who tries to chart the path of a comet, for instance, soon encounters Kepler's equation. It has so vexed astronomers over the years that literally hundreds of solutions have been proposed; the striking graphs in Chapter 29 give a good idea why.

Whenever I read about interpolation techniques, as in Chapter 3, I'm reminded poignantly of Comet Kohoutek. News of its discovery caused a great stir in the spring of 1973, and then it let observers down with a lackluster performance. But this comet also taught me an important mathematical lesson. After preparing a chart of its motion from a list of ephemeris points, I noticed that it was going to pass very near the Sun and tried several interpolation schemes in hopes of finding out what the exact time and minimum distance would be. Much to my surprise, they all failed to give an answer matching what was perfectly obvious from my chart! Readers of this book can save themselves a similar frustration by paying close attention to the remarks on page 107.

When he's not busy writing or conducting seminars on computing techniques, Meeus likes to seize hold of an astronomical problem with great zeal, especially if he senses it is a calculation that has never been done before. Once I asked him about the dates in the past and future when the Moon reaches its most extreme near and far distances from the Earth. Within weeks he had created a table much like that given in Table 48.C of this book. He later confided that this calculation had taken 470 hours on his HP-85 computer, consuming 12 kilowatt-hours of electricity.

On another occasion I heard about a program that was much too large for the mainframe computer he was using at the time. So he devised a scheme to avoid storing the vast number of coefficients in the computer's limited memory; his Fortran program simply read and rewound the same magnetic tape 915 times in the course of generating the hour-by-hour lunar ephemeris he sought. No problem, except that the computer-room operators began to take notice, getting mildly perturbed!

Astronomical calculations have a variety of uses, some scarcely foreseen by the person making them. As long ago as 1962, for example, Meeus published an article in the British Astronomical Association *Journal* about a rare and remarkable forthcoming event. If any observers happened to be on Mars on 1984 May 11, he explained, they should be able to see the silhouette of Earth pass directly across the face of the Sun. Among his readers was the science-fiction writer Arthur C. Clarke, who later incorporated the calculations in a short story, *Transit of Earth.* The piece tells of an astronaut, stranded on the red planet, who barely manages to witness this event before his oxygen supply runs out.

Many of the topics in this book are targeted at serious observers of the sky. Thus, Chapter 51 can help in predicting the illumination at a specific spot on the Moon, for any date and time. Observers often want to know the exact moments when sunlight will just glance across a particular crater, sinuous rille, or gently sloping lunar dome, because oblique lighting is ideal for telescopic scrutiny, making subtle reliefs stand out better than in most of NASA's closeup spacecraft photographs. This chapter can also help us find when the Moon will undergo extreme librations, turning craters near the limb our way.

Chapter 43 holds a special treat for students of Jupiter. First there is a simple method for locating the four famous satellites, quite adequate for identifying them in your own telescope or on historical drawings back to the time of Galileo. Then comes a second set of formulae of the utmost accuracy. Here the computer hobbyist can have a field day, creating observing schedules not only for ordinary satellite eclipses and transits but also for the mutual events between one satellite and another. Astronomy journals have been lax in forecasting these dramatic events, so that many of them have gone unobserved except by accident. For handling the Jovian moons, the routines presented in this book rival or exceed in accuracy those used by the great national almanac offices.

Other unusual topics are offered, like the method in Chapter 50 for computing the dates when the Moon's declination becomes extreme. This is no frivolous calculation, for the very issue came up in recent findings about a century-old murder trial involving the Illinois lawyer and soon-to-be U.S. President Abraham Lincoln. Historians had long tried to reconcile conflicting testimony about the Moon and its role in allowing a witness to see the details of the murder. Some suggested that Lincoln, as lawyer for the defense, may have tampered with an almanac. Not until 1990 was this curious situation explained, and Lincoln's integrity upheld, when Donald W. Olson and Russell Doescher noticed something quite unusual about the Moon on the night in question: 1857 August 29. As any user of this book can confirm, the Moon had a far southerly declination that night, nearly the most extreme value possible in its 18.6-year cycle, and this circumstance made the time of moonset appear quite at odds with its phase. Here is a beautiful instance of astronomers stepping in, bringing their special knowledge and calculations to bear on a longstanding puzzle for historians.

We now live in a thrilling time for practitioners of the number-crunching art. The four-function pocket calculators that were so costly 20 years ago are now incorporated as a gimmick on certain wristwatches. The memory capacity of the 1K RAM board in the pioneering MITS Altair microcomputer is exceeded 500-fold by a single chip in some of today's laptop and notebook computers. Who knows what other marvels lie just ahead? By presenting these astronomical algorithms in standard mathematical notation, rather than in the form of program listings, the author has made them accessible to users of a wide variety of machines and computer languages—including those not yet invented.

Roger W. Sinnott
Sky & Telescope magazine

Introduction

When, in 1978, we wrote the first (Belgian) edition of our *Astronomical Formulae for Calculators*, the industry of microcomputers was just starting its worldwide expansion. Because these 'personal computers' were then not yet within reach of everybody, the aforesaid book was written mainly for the users of *pocket* calculating machines and therefore calculation methods requiring a large amount of computer memory, or many steps in a program, were avoided as far as possible, or kept to a minimum.

The present work is a greatly revised version of the former one. It is, in fact, a completely new book. The subjects have been expanded and the content has been improved. Changes were needed to take into account new resolutions of the International Astronomical Union, particularly the adoption of the new standard epoch J2000.0, while moreover we profited by the new planetary and lunar theories constructed at the Bureau des Longitudes of Paris.

As Gerard Bodifée wrote in the Preface of our previous work:

> "Anyone who endeavours to make astronomical calculations has to be very familiar with the essential astronomical conceptions and rules and he must have sufficient knowledge of elementary mathematical techniques. As a matter of fact he must have a perfect command of his calculating machine, knowing all possibilities it offers the competent user. However, all these necessities don't suffice. Creating useful, successful and beautiful programs requires much practice. Experience is the mother of all science. This general truth is certainly valid for the art of programming. Only by experience and practice can one learn the innumerable tricks and dodges that are so useful and often essential in a good program."

Astronomical Algorithms intends to be a guide for the (professional or amateur) astronomer who wants to do calculations. An algorithm (from the Arabic mathematician *Al-Khārezmi*) is a set of rules for getting something done; for us it is a mathematical procedure, a sequence of reasonings and operations which provides the solution to a given problem.

This book is not a general textbook on astronomy. The reader will find no theoretical derivations. Some definitions are kept to a minimum. Nor is this a textbook on mathematics or a manual for microcomputers. The reader is assumed to be able to use his machine properly.

Except in a few rare cases, no programs are given in this book. The reasons are clear. A program is useful only for one computer language. Even if we consider BASIC only, there are so many versions of this language that a given program cannot be used as such by everybody without making the necessary changes. Every calculator thus must learn to create his own programs. There is the added circumstance that the precise contents of a program usually depend on the specific goals of the computation, that are impossible to anticipate by anybody else.

The few programs we give are in standard BASIC. They can easily be converted into FORTRAN or any other computer language.

Of course, in the formulae we still use the classical mathematical symbols and notations, not the symbolism used in program languages. For example, we write $\sqrt{a}$ instead of SQR(A), or $a(1-e)$ instead of A*(1-E), or $\cos^2 x$ instead of COS(X)^2 or COS(X)**2.

The writing of a program to solve some astronomical problem will require a study of more than one chapter of this book. For instance, in order to create a program for the calculation of the altitude of the Sun for a given time on a given date at a given place, one must first convert the date and time to Julian Day (Chapter 7), then calculate the Sun's longitude for that instant (Chapter 24), its right ascension (Chapter 12), the sidereal time (Chapter 11), and finally the required altitude of the Sun (Chapter 12).

This book is restricted to the 'classical', mathematical astronomy, although a few astronomy oriented mathematical techniques are dealt with, such as interpolation, fitting curves and sorting data. But astrophysics is not considered at all. Moreover, it is clear that not all topics of mathematical astronomy could have been covered in this book. So nothing is said about orbit determination, occultations of stars by the Moon, meteor astronomy, or eclipsing binaries. For solar eclipses, the interested reader will find Besselian elements and many useful formulae in *Canon of Solar Eclipses -2003 to +2526* by H. Mucke and J. Meeus (Astronomisches Büro, Vienna, 1983), or in *Elements of Solar Eclipses 1951 to 2200* by the undersigned (1989). Elements and formulae about transits of Mercury and Venus across the Sun's disk are provided in our *Transits* (1989). The latter two books are published by Willmann-Bell, Inc.

The author wishes to express his gratitude to Dr. S. De Meis (Milan, Italy), to A. Dill (Germany), and to E. Goffin and C. Steyaert (Belgium), for their valuable advice and assistance.

Jean Meeus

Contents

Some Symbols and Abbreviations

e	Eccentricity (of an orbit)
h	Altitude above the horizon
r	Radius vector, or distance of a body to the Sun, in AU
v	True anomaly
A	Azimuth
H	Hour angle
M	Mean anomaly
R	Distance from Earth to Sun, in AU
T	Time in Julian centuries (36525 days) from J2000.0
α	Right ascension
δ	Declination
ε	Obliquity of the ecliptic (ε_0 is used for the mean obliquity)
θ	Sidereal time (θ_0 is the sidereal time at Greenwich)
π	Parallax
τ	Time in Julian millennia (365250 days) from J2000.0
ϕ	Geographical latitude
ϕ'	Geocentric latitude
Δ	Distance to the Earth, in AU
ΔT	Difference TD - UT
$\Delta\varepsilon$	Nutation in obliquity
$\Delta\psi$	Nutation in longitude
AU	Astronomical unit
INT	Integer part of a number
JD	Julian Day
JDE	Julian Ephemeris Day
TD	Dynamical Time
UT	Universal Time

Following an old, general astronomical practice, small superior symbols are placed immediately above the decimal point, not after the last decimal. For instance, 28°.5793 means 28.5793 degrees.

Moreover, note carefully the difference between hours with decimals, and hours-minutes-seconds. For example, $1^h.30$ is *not* 1 hour and 30 minutes, but 1.30 hours, that is 1 hour and 30 hundredths of an hour, or 1 hour and 18 minutes.

Do not use the symbols ' and " for minutes and seconds of time: they are used for minutes and seconds of a *degree* (or arcminutes and arcseconds, respectively). Minutes and seconds of time have the symbols *m* and *s*. For example,

the angle 23°26'44", but the instant $15^h 22^m 07^s$.

Indeed, we have

1' = one minute of *arc* = 1/60th of a *degree*
1^m = one minute of *time* = 1/60th of an *hour*

Do not use the symbol ± for 'approximately'. That symbol means: plus *or* minus (or both). For instance, the square root of 25 is ±5. Writing $\pi = \pm 3$ is incorrect, because π is equal to neither +3 nor −3; the correct symbol to be used here is ≈. For example, 1002 ≈ 1000.

In general, we shall use the 'scientific' form for calendar dates, which reads from the largest to the smallest unit of time, for example 1993 November 6. It contrasts with the common 'American' form (November 6, 1993), and with the 'European' form (6 November 1993). Anyway, it is recommended to spell out the month, because one person's '11/6/93' is another's '6/11/93'.

Chapter 1

Hints and Tips

To explain how to calculate or to program on a computer is out of the scope of this book. The reader should, instead, study carefully his instructions manual. However, even then writing good programs cannot be learned in the lapse of time of one day. It is an art which can be acquired only progressively. Only by practice can one learn to write better and shorter programs.

In this first Chapter, we will give some practical hints and tips, which may be of general interest.

Trigonometric functions of large angles

Large angles frequently appear in astronomical calculations. In Example 24.a we find that on 1992 October 13.0 the mean longitude of the Sun is -2318.19281 degrees. Even larger angles are found for rapidly moving objects, such as the Moon and the bright satellites of Jupiter, or the rotations of the planets (see, for instance, the angle W in Step 9 of Example 41.a).

It may be necessary to reduce the angles to the interval 0 – 360 degrees, because some pocket calculators or some program languages give incorrect values for the trigonometric functions of large angles. Try, for instance, to calculate the sine of 36 000 030 degrees. The result must be 0.5 exactly.

Angle modes

The calculating machines do not calculate directly the trigonometric functions of an angle which is given in degrees, minutes and seconds. Before performing the trigonometric functions, the angle should be converted to degrees and *decimals*. Thus, to calculate the cosine of 23°26'49", first convert this angle to 23.446 944 44 degrees, and *then* use the COS function.

There is the added complication that most computers can calculate only in radians, not in degrees. It is an infernal nuisance having to convert degrees to radians all the time, but on most computers this has to be done before calculating a trigonometric function of an angle given in degrees.

Right ascensions

Right ascensions are generally expressed in hours, minutes and seconds of time. If the trigonometric function of a right ascension must be calculated, it is necessary to convert that value to degrees (and then to radians, if necessary). Remember that one hour corresponds to 15 degrees.

Example 1.a — Calculate $\tan\alpha$, where $\alpha = 9^h14^m55^s.8$.

We first convert α to hours and decimals :

$$9^h14^m55^s.8 = 9 + 14/60 + 55.8/3600 = 9.248\,833\,333 \text{ hours.}$$

Then, multiplying by 15,

$$\alpha = 138^\circ.73250.$$

Dividing this value by $180/\pi = 57.295\,779\,513...$ gives α in radians. We then find $\tan\alpha = -0.877\,517$.

The correct quadrant

When the sine, the cosine or the tangent of an angle is known, the angle itself can be obtained by using the 'inverse' function arcsine (ASN or ASIN), arccosine (ACS or ACOS), or arctangent (ATN or ATAN). It should be noted that the functions arcsine and arccosine are absent one some machines or in some languages, principally on almost all early microcomputers.

The inverse trigonometric functions (arcsine, arccosine, arctangent) are not single valued. For instance, if $\sin\alpha = 0.5$, then $\alpha = 30°, 150°, 390°$, etc. For this reason, the electronic computers return inverse trigonometric functions correctly over only half the range of 0 to 360 degrees: arcsine and arctangent give an angle lying between -90 and +90 degrees, while arccosine gives a value between 0 and +180 degrees.

For example, try cos 147°. The answer is -0.8387, which reverts to 147° when you take the inverse function. But now try cos 213°. The answer is again -0.8387 which, when you take its arccosine, gives 147°.

Hence, whenever the inverse function of SIN, COS or TAN is taken, an ambiguity arises which has to be cleared up by one or other means *when it is necessary*. Each problem must be examined separately.

For instance, formulae (12.4) and (24.7) give the sine of the declination of celestial body. The function arcsine then will always give this declination in the correct quadrant, because all declinations lie between -90 and +90 degrees. So, no special test should be performed here.

This is also the case for the angular separation whose cosine is given by formula (16.1). Indeed, any angular separation is in the range of 0° to +180°, which matches the range of the inverse cosine function.

But consider the conversion from right ascension (α) and declination (δ) to celestial longitude (λ) and latitude (β) by means of the following formulae

$$\cos\beta \sin\lambda = \sin\delta \sin\varepsilon + \cos\delta \cos\varepsilon \sin\alpha$$
$$\cos\beta \cos\lambda = \cos\delta \cos\alpha$$

Call A and B the second members. Then, dividing the first equation by the second one, we obtain $\tan\lambda = A/B$. Applying the function arctangent to the quotient A/B will yield the angle λ between -90° and +90°, with an ambiguity of ±180°. This ambiguity can be removed with the following test: if $B < 0$, add 180° to the result. However, some computer languages contain the important 'second' arctangent function, ATN2 or ATAN2, which uses the *two* arguments A and B separately and returns the angle in the proper quadrant. For instance, suppose that A = -0.45, B = -0.72; then ATN(A/B) will give the angle 32°, while ATN2(A, B) will yield the correct value -148°, or +212°.

The input of negative angles

Angles expressed in degrees, minutes and seconds can be input as three different numbers (INPUT D, M, S). For instance, the angle 21°44'07" can be entered as the three numbers 21, 44, and 7. Then, in the program the angle H in degrees is calculated by means of the instruction H = D + M/60 + S/3600.

In such a case, care must be taken for negative angles. If the angle is, for example, -13°47'22", then this means -13° and -47' and -22". In this case, the three numbers are D = -13, M = -47, and S = -22. All three numbers have the same sign!

Mislead by the notation -13°47'22", one can have the tendency to input -13, +47 and +22 instead, and in that case the angle entered would actually be -12°12'38". It is possible to write the program in such a way that similar errors are corrected automatically:

```
200  INPUT D,M,S
210  IF D<0 THEN M=-ABS(M) : S = -ABS(S)
220  H=D+M/60+S/3600
```

In line 210, the minutes and seconds are made negative when the degrees are negative. The two ABS functions make sure that no error is made when M and S are actually entered as negative numbers.

This procedure does not work, however, when the angle is between 0° and -1°. If the angle is, for instance, equal to -0°32'41", then we have D = -0, which a computer automatically converts to 0, which is not negative, so the machine will conclude that the angle is +0°32'41" instead. One solution (in BASIC) is to enter the degrees as a 'string' instead of a numeric variable, hence by means of INPUT D$ instead of INPUT D. Then one can use the VAL function and test on the first character of the string D$.

Powers of time

Some quantities are calculated by means of a formula containing powers of the time (T, T^2, T^3, ...). It is important to note that such polynomial expressions are valid only for values of T that are not too large. For instance, the formula

$$e = 0.046\,295\,90 - 0.000\,027\,337\,T + 0.000\,000\,0790\,T^2 \qquad (1.1)$$

gives the eccentricity e of the orbit of Uranus; T is the time measured in Julian centuries (36525 days) from the beginning of the year 2000. It is evident that this formula is valid for only a limited number of centuries before and after A.D. 2000, for instance for T lying between -30 and +30. For $|T|$ much larger than 30, the above expression is no longer valid. For $T = -3305.8$ the formula would give $e = 1$, and an incompetent person, thinking that "the computer cannot make errors", would deduce that in the year -328580 the orbit of Uranus was parabolic and hence that this planet originates from outside our solar system — bringing us in the realm of pseudoscience.

In fact, the eccentricity e of a planet's orbit varies rather irregularly in the course of time, though it cannot exceed a well-defined upper limit. But for a time interval of a few millennia the eccentricity can be accurately represented by a polynomial of the second degree such as (1.1).

One should further carefully note the difference between periodic terms (terms in sine and/or cosine), which remains small throughout the centuries, and secular terms (terms in T, T^2, T^3, ...) which increase more and more rapidly with time. A term in T^2, which is very small when T is small, becomes increasingly important for larger values of $|T|$. Thus, for large values of $|T|$ it is meaningless to take into account small periodic terms if terms in T^2, etc., are not taken into account in the calculation.

Avoiding powers

Suppose that one wants to calculate the polynomial

$$y = A + Bx + Cx^2 + Dx^3 + Ex^4$$

with A, B, C, D and E constants, and x a variable. Now, one may program the machine to calculate this polynomial directly term after term and adding all terms, so that for each given x the machine obtains the value of the polynomial. However, instead of calculating all the powers of x, it appears to be wiser to write the polynomial as follows:

$$y = A + x(B + x(C + x(D + xE)))$$

In this expression all power functions have disappeared and only additions and multiplications are to be performed. This way of expressing a polynomial is called *Horner's method*, an approach especially well suited for automatic calculation because powers are avoided.

Also, it may be wise to calculate the square of a number A by means of A*A instead of using the power function. We calculated the squares of the first 200 positive integers on the HP-85 microcomputer. Using the procedure

```
FOR I = 1 TO 200
K = I^2
NEXT I
```

The complete calculation took 10.75 seconds. But when the second line was replaced by K = I*I, then the calculation time was only 0.96 second!

To shorten a program

To make a program as short as possible is not always an art for art's sake, but sometimes a necessity as long as the memory capacities of the calculating machine have their limits.

There exist many tricks to make a program shorter, even for simple calculations. Suppose that one wants to calculate the sum S of many terms:

$$\begin{aligned} S = {} & 0.0003233 \sin(2.6782 + 15.54204\,T) \\ & + 0.0000984 \sin(2.6351 + 79.62980\,T) \\ & + 0.0000721 \sin(1.5905 + 77.55226\,T) \\ & + 0.0000198 \sin(3.2588 + 21.32993\,T) \\ & + \ldots\ldots\ldots\ldots \end{aligned}$$

First, because the coefficients of all sines are small numbers, one can avoid typing in all those decimals by taking as unit the last decimal (10^{-7} in this case). So, instead of 0.0003233, etc., we use 3233, etc. Then, *after* the sum of the terms has been calculated, we divide the result by 10^7.

Secondly, it would be unwise to write all those many terms explicitely in the program. Instead, we could make use of a so-called *loop*. Each of the above terms is of the form $A \sin(B + CT)$, so we put all values *A*, *B*, *C* as DATA in the program. Suppose there are 50 terms. Then the program will look like this:

```
100  S = 0
110  RESTORE 170
120  FOR I = 1 TO 50
130  READ A, B, C
140  S = S + A*SIN(B+C*T)
150  NEXT I
160  S = S/10000000
170  DATA 3233, 2.6782, 15.54204, 984, etc....
```

Safety tests

Include a safety test in case an 'impossible' situation might occur, for example in order to stop the calculation when, after a specified number of iterations, the required accuracy has not been reached.

Or consider the case of the occultation of a star by the Moon. In a program for local circumstances, the times of disappearance and of reappearance of the star are calculated. It may happen, however, that the star is not occulted as seen from the given place; in such a case, of course, the times of ingress and egress do not exist, and trying to calculate them would correspond to calculating the square root of a negative number. To avoid this problem, the program should be written in such a way that first of all the value of the star's least distance to the center of the lunar disk (as seen from the given place) is calculated; if, and only if, this distance is smaller than the radius of the Moon's disk, can the times of ingress and egress be calculated.

Debugging

After a program has been written, it must be checked for errors, which are called *bugs*. The process of locating the bugs and correcting them is called *debugging*. Several types of errors can occur when programming in any language:

a. *syntax errors* violate the rules of the language, such as spelling, a forgotten parenthesis, or other conventions specific to each language. For instance, in BASIC,

   `A = SIM(B)` should be `A = SIN(B)`
   `P = SQR(ABS(A+B)` should be `P = SQR(ABS(A+B))`

b. *semantic errors*, such as a forgotten program line. For instance, `GOTO 800` when no line 800 exists in the program.

c. *run-time errors*, which occur during the execution of a program. For example:

A = SQR(B). The variable B is calculated during execution of the program, but its value happens to be negative;

ON X GOTO 1000,2000,3000, but X is larger than 3.

d. other programmer's errors. The following ones happen frequently :

- Typing the letter O ('oh') instead of the number zero (0 or Ø), or vice versa, or typing the number 1 instead of the letter I. A cross reference list, if available, can help here.
- The name of a variable is used twice in the program (with different meanings).
- Error in copying down a numerical constant (such as 127.3 instead of 127.03, or 15 instead of .15), typing an * instead of a +, etc.
- Incorrect units are used. For instance, an angle is expressed in degrees instead of radians, or a right ascension expressed in hours has not been converted to degrees or radians.
- The angle is in the wrong quadrant. See 'The correct quadrant' on page 8.
- Rounding errors. For example, the cosine of an angle *d* has been calculated, from which one wants to deduce that angle. This does not work well when the angle is very small. Indeed, if *d* is very small, its cosine is almost equal to 1 and varies quite slowly as a function of *d*. In that case, the value of *d* is ill-defined and cannot be calculated accurately.

  For instance, $\cos 15'' = 0.999\,999\,997$ but $\cos 0''$ is 1 exactly. If one expects that the angle *d* can be very small, then its value should be calculated by means of another method. See, for instance, Chapter 16.
- An iteration procedure which does not guarantee convergence in some cases. See Chapters 5 (Iteration) and 29 (Equation of Kepler).
- An incorrect method of calculation has been used. For example, to interchange two numbers X and Y, an extra variable A is needed (*):

(*) This is not quite exact. Theoretically, it *is* possible to interchange two numbers without using a third, auxiliary variable, as follows :

```
X = X + Y
Y = X - Y
X = X - Y
```

But, of course, this is rather a curiosity than a useful method, because the execution of these operations require extra computer time, and moreover because rounding errors can occur.

Incorrect procedure	*Correct procedure*
Y = X	A = Y
X = Y	Y = X
	X = A

In QUICKBASIC, GWBASIC and some other BASIC versions, there exists the SWAP function: SWAP(X,Y) interchanges the numbers X and Y.

Checking the results

Of course, a program should not only be 'grammatically' correct: it must give correct results. *Test* your program using a known solution. If, for instance, you wrote a program for the calculation of planetary positions or for the times of lunar phases, compare your results with the values given in an astronomical almanac.

Test your program for some 'special' cases. For instance, are the results still correct for a negative value of the declination? Or for a declination lying between 0° and −1°? Or if the observer's latitude is exactly zero? Or for negative values of the time T?

Chapter 2

About Accuracy

The following topics will be considered in this Chapter: the accuracy needed for a particular problem, the accuracy with which a given program language works, and finally the accuracy of the published results.

The accuracy needed for a given problem

The accuracy needed in a calculation depends on its aim. For example, if one wants to calculate the position of a planet with the goal of obtaining the times of rising or setting, an accuracy of 0.01 degree will be sufficient. The reason is evident: the apparent diurnal motion of the celestial sphere corresponds to a rotation over one degree during a time interval of four minutes, and so an error of 0.01 degree in the object's position will result in an error of only 0.04 minute (approximately) in its time of rising or setting. Taking hundreds of periodic terms into account in order to obtain the planet's position with an accuracy of $0''.01$ would just be a waste of effort and of computer time for *this* problem.

But if the position of the planet is needed to calculate the occultation of a star by that planet, then an accuracy of better than $1''$ will be necessary by reason of the small size of the planet's disk.

A program written for one aim may not be suitable for another application. Suppose that, for the calculation of the position of a star, a program uses the low-accuracy method for the precession (see Chapter 20). While the results will be good enough for the observer who wants to find celestial objects with a telescope on a parallactic mounting, that program will be completely worthless when *accurate* results are required, for instance in occultation work, or for the calculation of close conjunctions.

If a given accuracy is required, one has to use an algorithm that really provides this precision. John Mosley [1] mentions a commercially available program which calculates planetary positions; but because perturbations are not considered, the positions of Saturn, Uranus and Neptune can be up to 1 degree off, even though displayed to the nearest arcsecond!

To obtain a better accuracy it is often necessary to use another method of calculation, not just to keep more decimals in the result of an approximate calculation. For example, if one has to know the position of Mars with an accuracy of 0.1 degree, it suffices to use an unperturbed elliptical orbit (Keplerian motion). However, if the position of Mars is to be known with a precision of 10" or better, perturbations due to the other planets have to be calculated and the program will be a much longer one.

The programmer, who knows his formulae and the desired accuracy in a given problem, must himself consider which terms, if any, may be omitted in order to keep the program handsome and as short as possible. For instance, the mean geometric longitude of the Sun, referred to the mean equinox of the date, is given by

$$L = 280°27'59''.244 + 129\,602\,771''.380\,T + 1''.0915\,T^2$$

where T is the time in Julian centuries of 36525 ephemeris days from the epoch 2000 January 1.5 TD. In this expression, the last term (secular acceleration of the Sun) is smaller than 1" if $|T| < 0.95$, that is, between the years 1905 and 2095. If an accuracy of 1" is sufficient, the term in T^2 may thus be dropped for any instant in that period. But for the year +100 we have $T = -19$, so that the last term becomes 394", which is larger than 0.1 degree.

The computer's accuracy

This is a much more complex problem. The program language should work with a sufficient number of significant digits. (Note that this is not the same as the number of decimals! For instance, the number 0.0000183 has seven decimals, but only three significant digits. The significant digits of a number are those digits which are left over when the leading and trailing zeros are suppressed).

On a machine rounding operations to 6 significant figures, the result of 1 000 000 + 2 will just be 1 000 000.

There can be dangerous situations, for instance when the difference is made of two *nearly-equal* numbers. Suppose that the following subtraction is performed:

$$6.92736 - 6.92735 = 0.00001.$$

Each number is given to six figures, but subtracting them gives a number with just *one* significant figure! Moreover, the two given numbers perhaps have already been rounded. If such is the case, then the situation can even be worse. Suppose that the two numbers are actually 6.927 3649 and 6.927 3451. Then the correct result is

0.000 0198, which is almost twice the previous result!

Six or eight significant digits, as was the general rule for the early microcomputers, or is nowadays often the case in 'single-precision', are generally not sufficient for mathematical astronomy.

For many applications, it is necessary that the machine calculates with a larger number of significant digits than it is required in the final result. Let us consider, for example, the following formula giving the mean longitude L' of the Moon for any given instant, in degrees (Chapter 45):

$$L' = 218.316\,4591 + 481\,267.881\,342\,36\,T - 0.001\,3268\,T^2 + 0.000\,0019\,T^3$$

where T is the time measured in Julian centuries of 36525 days elapsed since the standard epoch 2000 January 1.5 TD (JDE 2451 545.0). Suppose now that we wish to obtain the Moon's mean longitude to an accuracy of 0.001 degree. Because longitudes are restricted to the interval 0 – 360 degrees, one might think that a machine calculating with only six significant digits internally will be just sufficient for our purpose (3 digits before, and 3 digits after the decimal point). This is not the case in the present problem, however, because L' can reach large values before to be reduced to less than 360 degrees.

For instance, let us calculate L' for $T = 0.4$, which corresponds to 2040 January 1 at 12^h TD. We find $L' = 192\,725^\circ.469$, which reduces to 125°.469, the correct answer. But if the machine works with only six significant digits, it will not find $L' = 192\,725^\circ.469$, but rather 192 725° (six digits!), which will reduce to 125°, so in this case the final result is only to the nearest degree, and the error is 0.469 degree or 28'; and this happens only 40 years after the starting epoch. Under such circumstances it is just impossible to calculate eclipses or occultations.

To find out with which internal accuracy a computer works, the following short program (in BASIC) can be used.

```
10  X=1
20  J=0
30  X=X*2
40  IF X+1 <> X THEN 60
50  GOTO 80
60  J = J + 1
70  GOTO 30
80  PRINT J, J*0.30103
90  END
```

Here, J is the number of significant bits in the mantissa of a floating number, while 0.30103 J is the number of significant digits in a *decimal* number. The constant 0.30103 is $\log_{10} 2$. For instance, the HP-85 computer gives J = 39, whence 11.7 digits. With the HP-UX Technical Basic 5.0, working on the HP-Integral microcomputer, we find J = 52, whence 15.6 internal digits. The QUICK-BASIC 4.5 gives J = 63, whence 19.0 digits.

However, this accuracy refers only to simple arithmetics, not to the trigonometric functions. Although the Toshiba with GWBASIC has J = 55, that is 16.6 internal digits, it gives the sines with only 7 correct decimals; the last 9 figures are completely wrong!

One rapid manner to check the accuracy of trigonometric functions is PRINT 4*ATN(1). If the computer works in radians, this must give the famous number π = 3.14 15 92 65 35 89 79... Or one may calculate the sine of an angle whose value is accurately known, for example

$$\text{SIN}(0.61 \text{ rad}) = 0.572\,867\,460\,100\,48...$$

Rounding is inevitable in a computer. Consider for instance the value 1/3 = 0.33333333... Because the machine cannot handle an infinite number of decimals, such a number must necessarily be truncated somewhere.

Rounding errors can *accumulate* from one calculation to the next. In most cases this is of no importance because the errors almost cancel each other, but in some arithmetical applications the accumulated error can increase beyond any limit. Although this topic is outside of the scope of this book, we shall mention two cases.

Consider the following program.

```
10  X = 1/3
20  FOR I = 1 TO 30
30  X = (9*X + 1)*X - 1
40  PRINT I, X
50  NEXT I
60  END
```

The operation on line 30 actually replaces X by itself. Yet on most computers the results diverge. The above-mentioned HP-UX Technical Basic yields

```
  0.333 333 333 333 308   after  4 steps
  0.333 326 162 117 054   after 14 steps
  0.215 899 338 763 055   after 19 steps
286.423 ....              after 24 steps
```

and a value of the order of 10^{217} after 30 steps!

The difference in accuracy between microcomputers or even handheld calculators can be demonstrated by a simple test [2]: repeatedly squaring the number 1.000 0001. After 27 times, the result to 10 significant figures must be 674 530.4707. The results for some machines or programming languages are as follows:

```
674 494.06     on the HP-67 calculator
674 514.87     on the HP-85
674 520.61     on the TI-58 calculator
674 530.4755   on the HP-Integral (HP-UX Techn. Basic)
674 530.4755   in QUICKBASIC 4.5
```

But that is still not the end of the story. There are two basically different ways for the internal representation of numerical information into a computer. Some machines, such as the older HP-85, use the BCD (Binary Coded Decimal) scheme for representing numbers internally, but in most other cases the binary representation is used.

BCD is a scheme where the actual value of *each digit* of a number is stored individually. This allows numbers to be represented exactly, to the specified digits of precision of the given machine or program language. Binary, on the other hand, represents all numbers as some combination of powers of 2. In binary, fractions are also represented as being powers of 2, so it is impossible to represent numbers which are not exact combinations of negative powers of 2 in a binary system. For instance, 1/10 is not rationally expressed as combinations of negative powers of 2, because 1/10 = 1/16 + 1/32 + 1/128 ....

Binary arithmetic functions are usually faster in their execution than BCD counterparts, but the inconvenience is that some numbers, even with a small number of decimals, are not represented exactly.

As a consequence, the result of an arithmetic operation may be incorrect, even when numbers with only a few decimals are involved. Suppose that X = 4.34. Then the correct result of the operation H = INT(100*(X - INT(X))) is 34. However, many computer languages give H = 33 here. The reason is that in this case the value of X is represented internally as 4.3399999998, or something like that.

Another surprising example is

$$2 + 0.2 + 0.2 + 0.2 + 0.2 + 0.2 - 3.$$

On many computers, the result is *not* zero! On the HP-Integral, using the HP-UX Technical Basic 5.0, the result is 8.88×10^{-16}. But on the same machine 0.2 + 0.2 + 0.2 + 0.2 + 0.2 + 2 - 3 does give zero, so the order in which the operations are performed can be of importance here!

Surprisingly, 2 + (5*0.2) - 3 gives exactly zero on the HP-Integral, and so does the following:

```
A = 0.2 + 0.2 + 0.2 + 0.2 + 0.2
B = 2 + A
C = B - 3
PRINT C
```

Consider the following program:

```
10  FOR I = 0 TO 100 STEP 0.1
20  U = I
30  NEXT I
40  PRINT U
50  END
```

Here, I and U take the successive values from 0 to 100 with steps of

0.1, and the last value of U must be exactly 100. The HP-85 does give 100 indeed, but the HP-Integral gives 99.999 999 999 9986, which can have a disastrous consequence in some applications. The error is due to the fact that the step value of 0.1 is translated into binary as 0.0999999.... The difference with 0.1 is very small but, because there are 1000 steps, the final error is 1000 times as large as that small difference. In this case, one remedy may consist in taking an integer value for the step :

```
10  FOR J = 0 TO 1000
20  I = J/10
30  U = I
40  NEXT J
50  PRINT U
60  END
```

We may find other surprises with

```
A = 3 * (1/3)
PRINT INT(A)
```

whose result is correctly 1 on some computers, but zero on others. Or try, for instance, A = 0.1, PRINT INT(1000 * A).

Another interesting test is

```
INPUT A
B = A/10
C = 10 * B
PRINT A - C
```

The result must be zero. But for *some* numbers A the answer can be different.

One easy way to find out if a computer language works in BCD or not, consists of looking at the largest possible integer value (that is, a number defined as an INTEGER). If this is a 'nice, round' number, then this indicates that the machine works in BCD. For example, on the HP-85 that largest integer is 99 999 (or $10^5 - 1$). But if the largest possible integer is a 'strange' number (in fact, a power of 2 minus one), then this means that the computer does not work in BCD. On the old TRS-80, that largest integer is 32767 (or $2^{15} - 1$), while for the HP-UX Technical Basic 5.0 on the HP-Integral it is 2 147 483 647 (or $2^{31} - 1$).

Rounding by inexact arithmetics can yield other surprising results. On most microcomputers, the result of SQR(25) - 5 is *not* zero! This can be a problem if testing on the result. Is 25 a perfect square? One might think the answer is no, since the computer tells us that SQR(25) - INT(SQR(25)) is not zero!

Rounding the final results

Results should be rounded correctly and meaningfully, where it is necessary.

Rounding should be made to the *nearest* value. For instance, 15.88 is to be rounded to 15.9, or to 16, not to 15. However, calendar dates and years are exceptions. For example, March 15.88 denotes an instant belonging to March 15 : it means 0.88 day after March 15, 0^h. So, if we read that an event occurs on March 15.88, it takes place on March 15, not on March 16. Similarly, 1977.69 denotes an instant belonging to the year 1977, not to 1978.

Only meaningful digits should be retained. For example, Müller's formula for calculating the visual magnitude of Jupiter is

$$m = -8.93 + 5 \log r\Delta$$

where r is Jupiter's distance to the Sun, Δ its distance to the Earth (both in astronomical units), and the logarithm is to the base 10. Now, on 1992 May 14, at 0^h TD, we have

$$r = 5.417\,149$$
$$\Delta = 5.125\,382$$

whence $m = -1.712\,514\,898$. But giving all these decimals, under the pretext that they were given like this by the computer, would be ridiculous and would give the reader a false impression of high accuracy. Since the constant -8.93 in Müller's formula is given to 0.01 magnitude, no higher accuracy can be expected in the result. And, in any case, the meteorological phenomena in the atmosphere of Jupiter are such that the magnitude of that giant planet cannot be predicted with an accuracy better than 0.01 or even 0.1.

As another example, John Mosley [3] mentions a commercially available program giving rising and setting times of heavenly bodies to the nearest 0.1 second, which is impossibly precise.

Some 'feeling' and sufficient astronomical knowledge are necessary here. For instance, it would be completely irrelevant to give the illuminated fraction of the Moon's disk accurate to 0.000 000 001.

The rounding should be performed *after* the whole calculation has been made, not before the start or before the input of the data into the computer.

Example: Calculate 1.4 + 1.4 to the nearest integer. If we first round the given numbers, we obtain 1 + 1 = 2. In fact, 1.4 + 1.4 = 2.8, which is to be rounded to 3.

Here is another example. At its opposition date, 1996 July 18, the declination of Neptune is $\delta = -20°24'$. What is the planet's altitude h_m at the transit through the southern meridian, at Sonneberg Observatory, Germany, to the nearest degree? The observatory's latitude is $\phi = +50°23'$. The formula to be used is

$$h_m = 90° - \phi + \delta$$

The answer is $h_m = 90° - 50°23' - 20°24' = 19°13'$, whence 19°.

Rounding ϕ and δ to the nearest degree *before* the calculation would yield the incorrect result $90° - 50° - 20° = 20°$.

A similar error occurs when distances, already rounded to the nearest mile, are converted to kilometers. In this case the value of 17 km, for instance, will never be reached, because

10 miles will give 16.09 km, which is rounded to 16 km,
11 miles will give 17.70 km, which is rounded to 18 km.

Right ascensions and declinations. — Since 24 hours correspond to 360 degrees, one hour corresponds to 15°, one minute of *time* corresponds to 15 minutes of *arc*, and one second of time to 15 seconds of arc: during a time interval of one second the Earth rotates over an arc of 15".

For this reason, if the declination of a celestial body is given, for instance, to 1", then the right ascension should be given to the nearest *tenth* of a second of time, since otherwise the declination would be given with a much greater accuracy than the right ascension. The following table gives the approximate correspondence between the accuracies in right ascension (α) and in declination (δ). For example, if δ is given with an accuracy of 1', then α must be given to the nearest 0.1 minute of time. As examples, we give the position of Nova Cygni 1975 with different accuracies.

in α	in δ	Example (Nova Cygni 1975)	
1^m	$0°.1$	$\alpha = 21^h10^m$	$\delta = +47°.9$
$0^m.1$	$1'$	$21^h09^m.9$	$+47°57'$
1^s	$0'.1$	$21^h09^m53^s$	$+47°56'.7$
$0^s.1$	$1''$	$21^h09^m52^s.8$	$+47°56'41''$

As a final remark, let us mention that trailing zeros can be important. For instance, 18.0 is not the same as 18. The former value means that the actual number lies between 17.95 and 18.05, while the second value has been rounded to the nearest integer and can actually be equal to any number between 17.5 and 18.5. For this reason, trailing zeros *must* be given in the result to indicate the accuracy: a star of magnitude 7 is not the same as a star of magnitude 7.00.

References

1. John Mosley, *Sky and Telescope*, Vol. 78, p. 300 (September 1989).

2. F. Gruenberger, 'Computer Recreation', *Scientific American*, Vol. 250, p. 10 (April 1984).

3. John Mosley, *Sky and Telescope*, Vol. 81, p. 201 (February 1991).

Chapter 3

Interpolation

The astronomical almanacs or other publications contain numerical tables giving some quantities y for *equidistant* values of an argument x. For example, y is the right ascension of the Sun, and the values x are the different days of the year at 0^h TD.

Interpolation is the process of finding values for instants, quantities, etc., intermediate to those given in a table.

Of course, the 'table' should not necessarily be taken from a book, but may have been calculated in a computer program. Suppose that the position of the Sun is to be calculated for many (>3) instants of the *same* day. Then one may calculate the Sun's position for 0^h, 12^h and 24^h of that day, and then use these values to perform the interpolation for every given instant. This will require less computer time than calculating the position of the Sun directly for every instant.

In this Chapter we will consider two cases: interpolation from three or from five tabular values. In both cases we will also show how an extremum or a zero of the function can be found. The case of only two tabular values will not be considered here, for in that case the interpolation can but be linear, and this will give no difficulty at all.

Three tabular values

Three tabular values y_1, y_2, y_3 of the function y are given, corresponding to the values x_1, x_2, x_3 of the argument x. Let us form the table of differences

$$\begin{array}{llll} x_1 & y_1 & & \\ & & a & \\ x_2 & y_2 & & c \\ & & b & \\ x_3 & y_3 & & \end{array} \qquad (3.1)$$

where $a = y_2 - y_1$ and $b = y_3 - y_2$ are called the *first differences*. The *second* difference c is equal to $b - a$, that is

$$c = y_1 + y_3 - 2y_2$$

Generally, the differences of the successive orders are gradually smaller. Interpolation from three tabular values is permitted when the second differences are almost constant in that part of the table, that is, when the third differences are almost zero. Some good sense and experience are needed here. For example, the Moon's position can be interpolated accurately from three positions given at hourly intervals, but not when the interval is one day.

Let us consider, for instance, the distance of Mars to the Earth from 5 to 9 November 1992, at 0^h TD. The values are given in astronomical units, and the differences are in units of the 6th decimal:

1992 November 5	0.898 013			
		-6904		
6	0.891 109		+21	
		-6883		+2
7	0.884 226		+23	
		-6860		+2
8	0.877 366		+25	
		-6835		
9	0.870 531			

Since the third differences are almost zero, we may interpolate from only three tabular values.

The central value x_2 must be chosen in such a manner that it is that value of x that is closest to the value of x for which we want to perform the interpolation. For example, if from the table above we must deduce the value of the function for November 7 at 22^h14^m, then y_2 is the value for November 8.00. In that case, we should consider the tabular values for November 7, 8 and 9, namely the table

November 7	$y_1 = 0.884\,226$	
8	$y_2 = 0.877\,366$	(3.2)
9	$y_3 = 0.870\,531$	

and the differences are

$$a = -0.006\,860 \qquad c = +0.000\,025 \qquad b = -0.006\,835$$

Let n be the interpolating factor. That is, if the value y of the function is required for the value x of the argument, we have $n = x - x_2$ in units of the tabular interval. The value n is positive if $x > x_2$, that is for a value 'later' than x_2, or from x_2 towards the bottom of the table. If x precedes x_2, then $n < 0$.

If y_2 has been correctly chosen, then n will be between -0.5 and +0.5, although the following formulae will also give correct results for all values of n between -1 and +1.

The interpolation formula is

$$y = y_2 + \frac{n}{2}(a + b + nc) \tag{3.3}$$

Example 3.a — From the table (3.2), calculate the distance of Mars to the Earth on 1992 November 8, at 4^h21^m TD.

We have 4^h21^m = 4.35 hours and, since the tabular interval is 1 day or 24 hours, n = 4.35/24 = +0.18125.

Formula (3.3) then gives y = 0.876 125, the required value.

If the tabulated function reaches an *extremum* (that is, a maximum or a minimum value), this extremum can be found as follows. Let us again form the difference table (3.1) for the appropriate part of the ephemeris. The extreme value of the function then is

$$y_m = y_2 - \frac{(a+b)^2}{8c} \tag{3.4}$$

and the corresponding value of the argument x is given by

$$n_m = -\frac{a+b}{2c} \tag{3.5}$$

in units of the tabular interval, and again measured from the central value x_2.

Example 3.b — Calculate the time of passage of Mars through the perihelion in May 1992, and the value of its radius vector at that instant.

The following values for the distance Sun–Mars have been calculated at intervals of four days :

1992 May 12.0 TD	1.381 4294
16.0	1.381 2213
20.0	1.381 2453

The differences are

a = -0.000 2081

c = +0.000 2321

b = +0.000 0240

from which we deduce

$$y_m = 1.381\,2030 \qquad \text{and} \qquad n_m = +0.39\,660$$

Hence, the least distance from Mars to the Sun is 1.381 2030 AU. The corresponding time is found by multiplying 4 days (the tabular interval) by +0.39 660. This gives 1.58 640 days, or 1 day and 14 hours later than the central time, that is 1992 May 17, at 14^h TD.

[Of course, if n_m were negative, the extremum would take place *earlier* than the central time.]

The value of the argument x for which the function y becomes zero can be found by again forming the difference table (3.1) for the appropriate part of the ephemeris. The interpolating factor corresponding to a zero of the function is then given by

$$n_o = \frac{-2y_2}{a + b + cn_o} \tag{3.6}$$

Equation (3.6) can be solved by first putting $n_o = 0$ in the second member. Then the formula gives an approximate value for n_o. This value is then used to calculate the right hand side again, which gives a still better value for n_o. This process, called *iteration* (Latin: *iterare* = to repeat), can be continued until the value found for n_o no longer varies, to the precision of the computer.

Example 3.c — Given the following values for the declination of Mercury,

1973 February 26.0 TD	−0° 28′ 13″.4
27.0	+0 06 46.3
28.0	+0 38 23.2

calculate when the planet's declination was zero.

Firstly, we convert the tabulated values into seconds of a degree and then form the differences :

$$y_1 = -1693.4$$
$$a = +2099.7$$
$$y_2 = +406.3 \qquad c = -202.8$$
$$b = +1896.9$$
$$y_3 = +2303.2$$

Formula (3.6) then becomes

$$n_o = \frac{-812.6}{+3996.6 - 202.8\,n_o}$$

Putting $n_o = 0$ in the second member, we find $n_o = -0.20332$.

Repeating the calculation, we find successively -0.20125 and -0.20127. Hence, n_o = -0.20127 and therefore, the tabular interval being one day, Mercury crossed the celestial equator on

1973 February 27.0 - 0.20127 = February 26.79873
= February 26, at 19^h10^m TD.

For the calculation of the value of the interpolating factor n_o for which the function is zero, formula (3.6) is excellent when, as in Example 3.*c*, the function is 'almost a straight line' in the interval considered. If, however, the curvature of the function is important, use of the formula may require a large number of iterations; moreover, it can lead to divergence even when starting from an almost correct value for n_o. In this case, a better method for calculating n_o is as follows : the correction to the assumed value of n_o is

$$\Delta n_o = -\frac{2y_2 + n_o(a + b + cn_o)}{a + b + 2cn_o} \qquad (3.7)$$

The calculation should be repeated, using the new value of n_o, until n_o no longer varies.

Example 3.d — Consider the following values of a function :

$x_1 = -1 \qquad y_1 = -2$
$x_2 = 0 \qquad y_2 = +3$
$x_3 = +1 \qquad y_3 = +2$

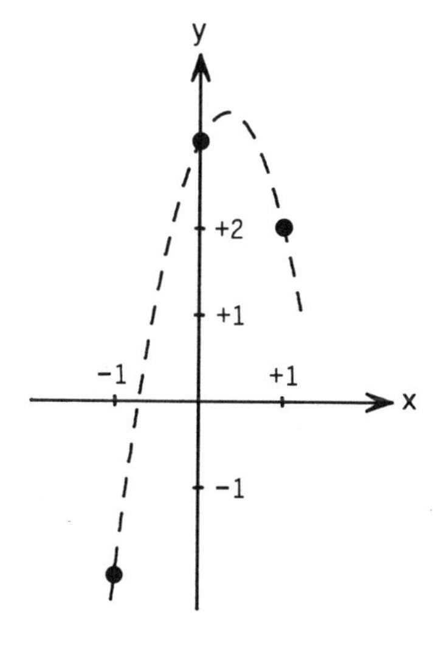

These three points actually define the parabola $y = 3 + 2x - 3x^2$, which has a strong curvature between $x = -1$ and $x = +1$ (see the Figure at left).

Starting with $n_o = 0$, formula (3.6) gives successively

-1.5
-0.461538...
-0.886363...
-0.643902...
-0.763027...
-0.699450...

and so on. The correct value of the *sixth* decimal is obtained after not less than 24 iterations. But if we use formula (3.7), again starting with n_o = zero, we find successively

-1.5
-0.886 363 636 364
-0.732 001 693 959
-0.720 818 540 935
-0.720 759 221 726
-0.720 759 220 056
-0.720 759 220 056

so the twelfth decimal is correctly obtained with only six iterations in this case.

Five tabular values

When the third differences may not be neglected, more than three tabular values must be used. Taking five consecutive tabular values, y_1 to y_5, we form, as before, the table of differences

$$\begin{array}{cccccc}
 & y_1 & & & & \\
 & & A & & & \\
 & y_2 & & E & & \\
 & & B & & H & \\
n \downarrow & y_3 & & F & & K \\
 & & C & & J & \\
 & y_4 & & G & & \\
 & & D & & & \\
 & y_5 & & & &
\end{array}$$

where $A = y_2 - y_1$, $H = F - E$, etc. If n is the interpolating factor, measured from the central value y_3 in units of the tabular interval, positively towards y_4, the interpolation formula is

$$y = y_3 + \frac{n}{2}(B + C) + \frac{n^2}{2}F + \frac{n(n^2 - 1)}{12}(H + J) + \frac{n^2(n^2 - 1)}{24}K$$

which may also be written (3.8)

$$y = y_3 + n\left(\frac{B + C}{2} - \frac{H + J}{12}\right) + n^2\left(\frac{F}{2} - \frac{K}{24}\right) + n^3\left(\frac{H + J}{12}\right) + n^4\left(\frac{K}{24}\right)$$

Example 3.e — Consider the following values of the equatorial horizontal parallax of the Moon :

1992 February 27.0 TD	54'36".125
27.5	54 24.606
28.0	54 15.486
28.5	54 08.694
29.0	54 04.133

The differences (in ") are

$$A = -11.519 \quad B = -9.120 \quad C = -6.792 \quad D = -4.561$$

$$E = +2.399 \quad F = +2.328 \quad G = +2.231$$

$$H = -0.071 \quad J = -0.097$$

$$K = -0.026$$

We see that the third differences (H and J) may not be neglected, unless an accuracy of about 0".1 is sufficient.

Let us now calculate the Moon's parallax on February 28 at 3^h20^m TD. The tabular interval being 12 hours, we have

$$n = \frac{3^h20^m}{12^h} = \frac{3.333\,333}{12} = +0.277\,7778.$$

Formula (3.8) then gives

$$y = 54'15''.486 - 2''.117 = 54'13''.369.$$

The interpolating factor n_m corresponding to an extremum of the function can be obtained by solving the equation

$$n_m = \frac{6B + 6C - H - J + 3n_m^2(H+J) + 2n_m^3 K}{K - 12F} \tag{3.9}$$

As before, this may be performed by iteration, firstly putting $n_m = 0$ in the second member. Once n_m is found, the corresponding value of the function can be calculated by means of formula (3.8).

Finally, the interpolating factor n_o corresponding to a zero of the function may be found from

$$n_o = \frac{-24y_3 + n_o^2(K - 12F) - 2n_o^3(H+J) - n_o^4 K}{2\,(6B + 6C - H - J)} \tag{3.10}$$

where, again, n_o can be found by iteration, starting by putting $n_o = 0$ in the second member.

The remark made on p. 27 about formula (3.6) holds here too. If the curvature of the function in the considered interval is important, a better method for calculating n_o is as follows. Calculate

$$M = \frac{K}{24} \qquad N = \frac{H+J}{12} \qquad P = \frac{F}{2} - M \qquad Q = \frac{B+C}{2} - N$$

Then the correction to the assumed value of n_0 is

$$\Delta n_0 = -\frac{Mn_0^4 + Nn_0^3 + Pn_0^2 + Qn_0 + y_3}{4Mn_0^3 + 3Nn_0^2 + 2Pn_0 + Q} \tag{3.11}$$

and, again, the calculation should be repeated with the new value of n_0 until n_0 no longer varies.

Exercise. — From the following values of the heliocentric latitude of Mercury, find the instant when the latitude was zero, by using formula (3.10).

1988 January 25.0 TD	−1° 11′ 21″.23
26.0	−0 28 12.31
27.0	+0 16 07.02
28.0	+1 01 00.13
29.0	+1 45 46.33

Answer : Mercury reached the ascending node of its orbit for n_0 = −0.361 413, that is on 1988 January 26.638 587, or January 26 at 15^h20^m TD.

Using only the three central values and formula (3.6), one would find n_0 = −0.362 166, a difference of 0.000 753 day, or 1.1 minute, with the previous result.

Important remarks

1. Interpolation cannot be performed on complex (*) quantities directly. These quantities should be converted, in advance, into a single, suitable unit. For instance, angles expressed in degrees, minutes and seconds should be converted either to degrees and decimals, or to arcseconds, before they can be used for interpolation.

2. *Interpolating times and right ascensions.* — We draw attention to the fact that times and right ascensions jump to zero when the value of 24 hours is reached. This should be taken into account when interpolation is performed on tabulated values. Suppose, for example, that we wish to calculate the right ascension of Mercury for the instant 1992 April 6.2743 TD, using the three following values :

1992 April 5.0 TD	α = $23^h 51^m 56^s.04$
6.0	23 56 28.49
7.0	0 01 00.71

(*) By definition, a *complex* number is a number composed of different units, having among them a ratio different from a power of 10. Examples of 'complex' quantities: $10^h29^m55^s$; 23° 26′ 44″; £, shillings, pence; yd, ft, inch; $a + bi$.

Not only is it necessary to convert these values to hours and decimals, but the last value should be written as $24^h01^m00^s.71$, otherwise the machine will consider that, from April 6.0 to 7.0, the value of α *decreases* from 23^h56^m.... to 0^h01^m....

We find a similar situation in some other cases. For instance, here is the longitude of the central meridian of the Sun for a few dates :

1992 June 14.0 UT	37°.96
15.0	24.72
16.0	11.48
17.0	358.25

It is evident that the variation is approximately -13.24 degrees per day. Hence, one should *not* interpolate directly between 11.48 and 358.25. Either the first value should be written as 371°.48, or the second value should be considered as being -1.75 degrees.

3. As much as possible, avoid making an interpolation for $|n| > 0.5$. In any case, the interpolating factor n should be restricted between the limits -1 and +1. This same rule applies to the calculation of an extremum (n_m) or a zero (n_0) of the function. Choose the central value of y in such a manner that this is the tabular value which is closest to the extremum or to the zero. Of course, the exact value of n_m or n_0 is not known in advance, but an approximate value can be calculated first, after which the choice of the central value (y_3 or y_2) of the function can be changed accordingly.

If the chosen value is too far from the zero or from the extremum, the formulae given in this Chapter for calculating these points will give incorrect, or even absurd results. Let us give an example. We know that $\sin x$ reaches a maximum for $x = 90°$. But let us consider the following sines, with ten decimals :

sin 29°	0.484 809 6202
sin 30°	0.500 000 0000
sin 31°	0.515 038 0749
sin 32°	0.529 919 2642
sin 33°	0.544 639 0350

Using the *three* central values, formula (3.4) gives $y_m = 1.22827$ (instead of 1 exactly), and (3.5) yields $n_m = +95.35$, or the maximum taking place at 31° + 95°.35 = 126°.35 instead of 90°.

Using all *five* values, formula (3.9) gives $n_m = +57.30$, whence the maximum taking place at 88°.30, from which the value of 0.99348 is found for that maximum. Although these results are much better than those obtained with only three points, they are still unsatisfactory !

Interpolation to halves

If the values y_1, y_2, y_3, y_4 of the function are given for four equally-spaced abscissae x_1, x_2, x_3 and x_4, then the value of the function for the point exactly half-way between x_2 and x_3 is easily calculated :

$$y = \frac{9\,(y_2 + y_3) - y_1 - y_4}{16} \tag{3.12}$$

This formula is valid when the fourth differences of the tabulated values are negligible.

Example 3.f — Given the following values for the apparent right ascension of the Moon,

1994 March 25	8^h TD	$\alpha = 10^h18^m48^s.732$
	10	10 23 22.835
	12	10 27 57.247
	14	10 32 31.983

calculate the right ascension for 11^h00^m TD.

Converting the minutes and seconds, after 10^h, into seconds, we change the four given data into

y_1 = 1128.732 seconds
y_2 = 1402.835
y_3 = 1677.247
y_4 = 1951.983

Formula (3.12) then gives $y = 1540.001$ seconds $= 25^m40^s.001$, so that the required right ascension is $\alpha = 10^h25^m40^s.001$.

Interpolation with unequally-spaced abscissae : Lagrange's interpolation formula

When the abscissae (the values of the independent x coordinate) of the given points are not equally spaced, the interpolation formula of Lagrange may be used. (Of course, this formula may also be used when the points are evenly spaced).

This simple formula, developed by the French mathematician J. L. Lagrange (1736 - 1813), determines a polynomial of degree $n-1$ matching n given points exactly. If the given points are x_i, y_i (i = 1 to n), the formula is, for a given x,

$$y = y_1 L_1 + y_2 L_2 + \ldots.. + y_n L_n \tag{3.13}$$

where

$$L_i = \prod_{\substack{j=1 \\ j \neq i}}^{n} \frac{x - x_j}{x_i - x_j}$$

The $\prod$ means that the product of the fractions should be calculated for all values $j = 1$ to n, except for $j = i$. That is,

$$L_i = \frac{(x - x_1)\ (x - x_2)\ \ldots (x - x_{i-1})\ (x - x_{i+1})\ \ldots\ (x - x_n)}{(x_i - x_1)\ (x_i - x_2)\ \ldots (x_i - x_{i-1})\ (x_i - x_{i+1})\ \ldots\ (x_i - x_n)}$$

It should be noted that the values x_i of the given points must all be different.

The following program in BASIC can be used.

```
10  DIM X(50), Y(50)
20  PRINT "NUMBER OF GIVEN POINTS = ";
30  INPUT N
40  IF N < 2 OR N > 50 THEN 20
50  PRINT
60  FOR I = 1 TO N
70  PRINT "X, Y FOR POINT No."; I
80  INPUT X(I), Y(I)
90  IF I = 1 THEN 130
100 FOR J = 1 TO I-1
110 IF X(I) = X(J) THEN PRINT "THIS VALUE OF X HAS ALREADY
      BEEN USED !" : GOTO 70
120 NEXT J
130 NEXT I
140 PRINT : PRINT "POINT X FOR INTERPOLATION = ";
150 INPUT Z
160 V = 0
170 FOR I = 1 TO N
180 C = 1
190 FOR J = 1 TO N
200 IF J = I THEN 220
210 C = C*(Z-X(J))/(X(I)-X(J))
220 NEXT J
230 V = V + C*Y(I)
240 NEXT I
250 PRINT : PRINT "INTERPOLATED VALUE = "; V
260 PRINT : PRINT "STOP (0) OR INTERPOLATION AGAIN (1) ";
270 INPUT A
280 IF A = 0 THEN STOP
290 IF A = 1 THEN 140
300 GOTO 260
```

The program first asks how many known values you are going to enter from a table and allows you to input these one at a time. Then it asks you repeatedly for intermediate values of interest, returning the interpolated value for each.

A remarkable feature of Lagrange interpolation is that the values entered initially do not have to be in order, or evenly spaced. Accuracy is usually better with uniform spacing, however.

As an exercise, try the program on the following six given points.

x = *angle in degrees*	y = *sine*
29.43	0.491 359 8528
30.97	0.514 589 1926
27.69	0.464 687 5083
28.11	0.471 165 8342
31.58	0.523 688 5653
33.05	0.545 370 7057

Asking for the sine of 30°; you should obtain 0.5 exactly. It is remarkable that even for the remote values $x = 0°$ and $x = 90°$, the Lagrange interpolation formula performed with these six data points yields the still rather good values +0.000 0482 and +1.00 007, respectively, the correct values being 0 and 1 exactly.

The expression (3.13) is a polynomial of degree $n - 1$, and it is the *unique* polynomial of that degree which takes the values y_1, y_2, ... , y_n for $x = x_1, x_2, \ldots, x_n$. But Lagrange's formula has the disadvantage that in itself it gives no indication of the number of points required to secure a desired degree of accuracy. However, when we wish to express the interpolating polynomial explicitely as a function of the variable x rather than making an actual interpolation, the use of Lagrange's formula is advantageous.

Example 3.g — Construct the (unique) 3rd-order polynomial passing through the following values :

x :	1	3	4	6
y :	−6	6	9	15

By substituting the given values of x and y into (3.13), we obtain

$$y = (-6)\frac{(x-3)(x-4)(x-6)}{(1-3)(1-4)(1-6)} + (6)\frac{(x-1)(x-4)(x-6)}{(3-1)(3-4)(3-6)}$$

$$+ (9)\frac{(x-1)(x-3)(x-6)}{(4-1)(4-3)(4-6)} + (15)\frac{(x-1)(x-3)(x-4)}{(6-1)(6-3)(6-4)}$$

which upon simplification reduces to

$$y = \frac{1}{5}(x^3 - 13x^2 + 69x - 87)$$

Chapter 4

Curve Fitting

In many cases, the result of a large number of observations is a series of points in a graph, each point being defined by an x-value and an y-value. It may be necessary to draw through the points the 'best' fitting curve.

Several curves can be fitted through a series of points: a straight line, an exponential, a polynomial, a logarithmic curve, etc.

To avoid individual judgment, it is necessary to agree on a definition of a 'best fitting' curve. Consider Figure 1 in which the N data points are given by (X_1, Y_1), (X_2, Y_2), ..., (X_N, Y_N). The values of X are supposed to be rigorously exact, while the Y-values are measured quantities, hence subject to an error.

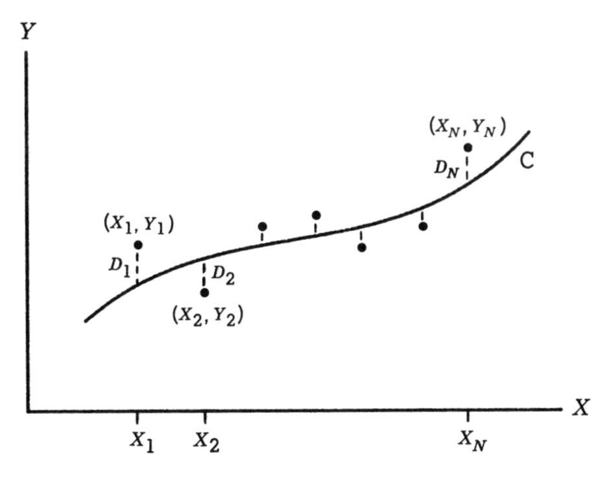

Figure 1

For a given value of X, say X_1, there will be a difference between the value Y_1 and the corresponding value as determined from the curve C. As indicated in the figure, we denote this difference by D_1, which is sometimes referred to as a *deviation*, *error* or *residual* and may be positive, negative or zero. Similarly, corresponding to the values X_2, ..., X_N we obtain the deviations D_2, ..., D_N.

A measure of the 'goodness of fit' of the curve C to the given data is provided by the quantity $D_1^2 + D_2^2 + \ldots + D_N^2$. If this is small the fit is good; if it is large the fit is bad. We therefore make the following definition: of all curves approximating a given set of data points, the curve having the property that ΣD_i^2 is a minimum, is called a best fitting curve. The Σ means 'sum of'.

A curve having this property is said to fit the data in the *least square sense* and is called a *least square curve*.

As has been said above, all values of the independent variable X are supposed to be exact. Of course, it is possible to define another least square curve by considering perpendicular distances from each of the data points to the curve instead of vertical distances; however, this is not used too often.

In this Chapter we will consider principally the case where the best fitting curve is a straight line, a problem called *linear regression*.

The name 'regression' may seem strange, because in the calculation of the best curve nothing 'regresses'! Alt [1] writes :

> "Die Benennung Regression wurde von *Galton* (1822-1911) eingeführt, der die Körperlängen von Eltern und Kindern verglich und dabei beobachtete, daß zwar im allgemeinen große Väter große Söhne haben, daß diese Beziehung jedoch nicht immer stimmt, da die Körpergröße der Söhne im Mittel etwas kleiner ist, als die der Väter, umgekehrt aber kleine Eltern im Mittel etwas größere Kinder haben. Diesen 'Rückschlag' in Richtung auf die Durchschnittsgröße der Bevölkerung bezeichnete er als Regression.„

A better term is *curve fitting*, and in the case of a straight line it is a linear curve fitting.

Linear curve fitting (linear regression)

We wish to calculate the coefficients of the linear equation

$$y = ax + b \tag{4.1}$$

using the least-squares method. The slope a and the y-intercept b can be calculated by means of the formulae

$$a = \frac{N\,\Sigma xy - \Sigma x\,\Sigma y}{N\,\Sigma x^2 - (\Sigma x)^2}$$

$$b = \frac{\Sigma y\,\Sigma x^2 - \Sigma x\,\Sigma xy}{N\,\Sigma x^2 - (\Sigma x)^2} \tag{4.2}$$

where N is the number of points. Note that both fractions have the same denominator. The sign Σ indicates the summation. Thus, Σx is the sum of all x-values, Σy the sum of all the y-values, Σx^2 the sum of the squares of all the x-values, Σxy the sum of the products xy of all the couples of values, etc. It should be noted that Σxy is not the same as $\Sigma x \times \Sigma y$ (the sum of the products is not the same as the product of the sums), and that $(\Sigma x)^2$ is not the same as Σx^2 (the square of the sum is not the same as the sum of the squares)!

An interesting astronomical application is to find the relation between the intrinsic brightness of a comet and its distance to the Sun. The apparent magnitude m of a comet can generally be represented by a formula of the form

$$m = g + 5 \log \Delta + \kappa \log r$$

Here, Δ and r are the distances in astronomical units of the comet to the Earth and to the Sun, respectively. The logarithms are to the base 10. The absolute magnitude g and the coefficient κ must be deduced from the observations. This can be performed when the magnitude m has been measured during a sufficiently long period. More precisely, the range of r should be sufficiently large. For each value of m, the values of Δ and r must be deduced from an ephemeris, or calculated from the orbital elements.

In this case, the unknowns are g and κ. The formula above can be written

$$m - 5 \log \Delta = \kappa \log r + g$$

which is of the form (4.1), when we write $y = m - 5 \log \Delta$, and $x = \log r$. The quantity y may be called the 'heliocentric' magnitude, because the effect of the variable distance to the Earth has been removed.

Example 4.a — Table 4.A contains visual magnitude estimates m of the periodic comet Wild 2 (1978*b*), made by John Bortle. The corresponding values of r and Δ have been calculated from the orbital elements [2].

The quantities x and y are used to calculate the sums Σx, Σy, Σx^2, and Σxy. We find

$$N = 19 \qquad \Sigma x = 4.2805 \qquad \Sigma x^2 = 1.0031$$
$$\Sigma y = 192.0400 \qquad \Sigma xy = 43.7943$$

whence, by formula (4.2),

$$a = 13.67 \qquad b = 7.03$$

Consequently, the 'best' straight line fitting the observations is

$$y = 13.67x + 7.03$$

or $\quad m - 5 \log \Delta = 13.67 \log r + 7.03$

Hence, for the periodic comet Wild 2 in 1978, we have

$$m = 7.03 + 5 \log \Delta + 13.67 \log r$$

TABLE 4.A

1978, UT		m	r	Δ	$x = \log r$	$y = m - 5 \log \Delta$
Febr.	4.01	11.4	1.987	1.249	0.2982	10.92
	5.00	11.5	1.981	1.252	0.2969	11.01
	9.02	11.5	1.958	1.266	0.2918	10.99
	10.02	11.3	1.952	1.270	0.2905	10.78
	25.03	11.5	1.865	1.335	0.2707	10.87
March	7.07	11.5	1.809	1.382	0.2574	10.80
	14.03	11.5	1.772	1.415	0.2485	10.75
	30.05	11.0	1.693	1.487	0.2287	10.14
April	3.05	11.1	1.674	1.504	0.2238	10.21
	10.06	10.9	1.643	1.532	0.2156	9.97
	26.07	10.7	1.582	1.592	0.1992	9.69
May	1.08	10.6	1.566	1.610	0.1948	9.57
	3.07	10.7	1.560	1.617	0.1931	9.66
	8.07	10.7	1.545	1.634	0.1889	9.63
	26.09	10.8	1.507	1.696	0.1781	9.65
	28.09	10.6	1.504	1.703	0.1772	9.44
	29.09	10.6	1.503	1.707	0.1770	9.44
June	2.10	10.5	1.498	1.721	0.1755	9.32
	6.09	10.4	1.495	1.736	0.1746	9.20

Coefficient of Correlation

A correlation coefficient is a statistical measure of the degree to which two variables are related to each other. In the case of a linear equation, the coefficient of correlation is

$$r = \frac{N\,\Sigma xy - \Sigma x\,\Sigma y}{\sqrt{N\,\Sigma x^2 - (\Sigma x)^2}\ \sqrt{N\,\Sigma y^2 - (\Sigma y)^2}} \qquad (4.3)$$

This coefficient is always between +1 and −1. A value of +1 or −1 would indicate that the two variables are totally correlated; it would denote a perfect linear relationship, all the points representing paired values of x and y falling exactly on the straight line representing this relationship. If $r = +1$, an increase of x corresponds to an increase of y (Figure 2). If $r = -1$, there is again a perfect linear relationship, but y decreases when x increases (see Figure 3).

When r is zero, there is no relationship between x and y (Figure 4). In practice, however, when there is no relationship, one may find that r is not exactly zero, due to fortuitous coincidences that generally occur except for an infinite number of points.

When $|r|$ is between 0 and 1, there is a trend between x and y,

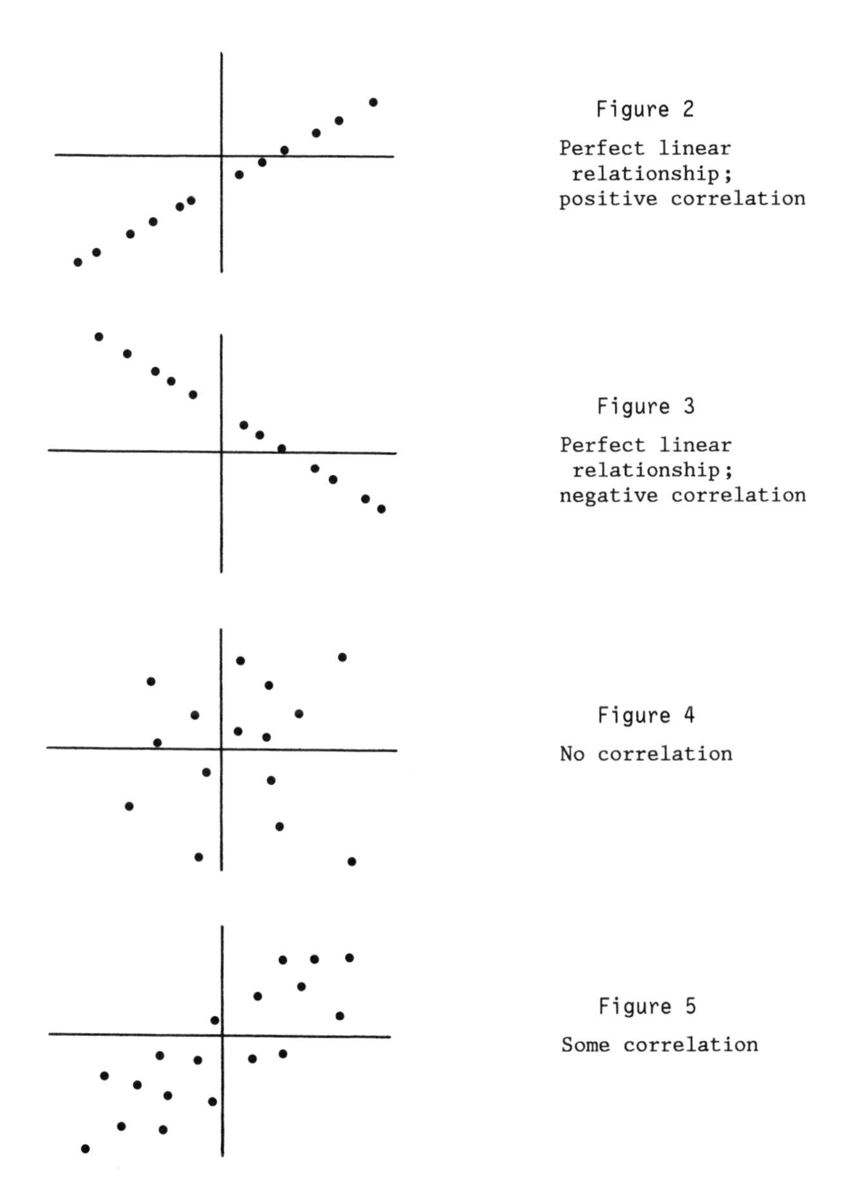

Figure 2
Perfect linear relationship; positive correlation

Figure 3
Perfect linear relationship; negative correlation

Figure 4
No correlation

Figure 5
Some correlation

although there is no strict relationship (Figure 5). Here, again, it should be noted that, *if* there is actually a strict relationship between the two variables, the calculation may give a value of r not exactly equal to +1 or to −1, by reason of inaccuracies inherent to all measures.

It should be noted that r is a dimensionless quantity; that is, it does not depend on the units employed.

The sign of r only tells us whether y is increasing or decreasing when x increases. The important fact is not the sign, but the magnitude of r because it is this magnitude which indicates how well the linear approximation is.

It must be emphasized that the computed value of r in any case measures the degree of relationship relative to the assumed type of function, namely the linear equation. Thus, if the value of r appears to be nearly zero, it means that there is almost no *linear* correlation between the variables. However, it does not necessarily mean that there is no correlation at all, since there may actually be a high *non-linear* correlation between the variables. As an example, consider the seven points

x	-4	-3	-2	-1	0	+1	+2
y	-6	-1	+2	+3	+2	-1	-6

Formula (4.3) yields r = zero, although the points lie *exactly* on the parabola $y = 2 - 2x - x^2$ (Figure 6).

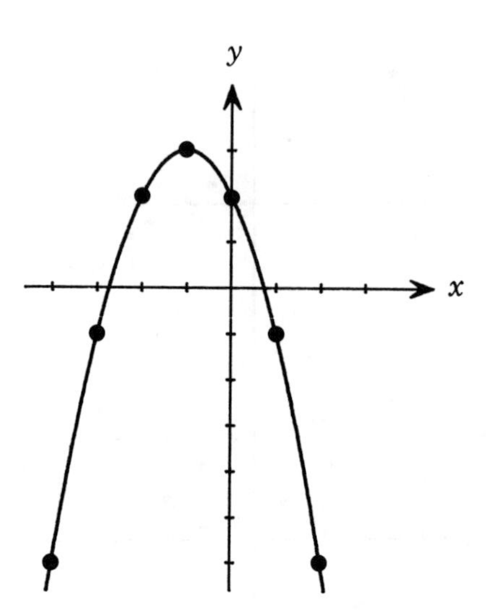

Figure 6

It should also be pointed out that a high correlation coefficient (that is, near +1 or -1) does not necessarily indicate a direct, physical dependence of the variables. Thus, if we consider a sufficiently large number of administrative territories, one can find a high correlation between the number of beds in the psychiatric hospitals and the number of television receivers of each territory. A high *mathematical* correlation, indeed, but a physical nonsense.

Example 4.b — Table 4.B gives, for each of the twenty-two sunspot maxima which have occurred from 1761 to 1989, the time interval x, in months, since the previous sunspot minimum, and the height y of the maximum (highest smoothed monthly mean).

We find

$$\Sigma x = 1120 \qquad \Sigma x^2 = 60608 \qquad \Sigma xy = 122\,337.1$$
$$\Sigma y = 2578.9 \qquad \Sigma y^2 = 340\,225.91 \qquad N = 22$$

and then, by formulae (4.2) and (4.1),

$$y = 244.18 - 2.49x \tag{4.4}$$

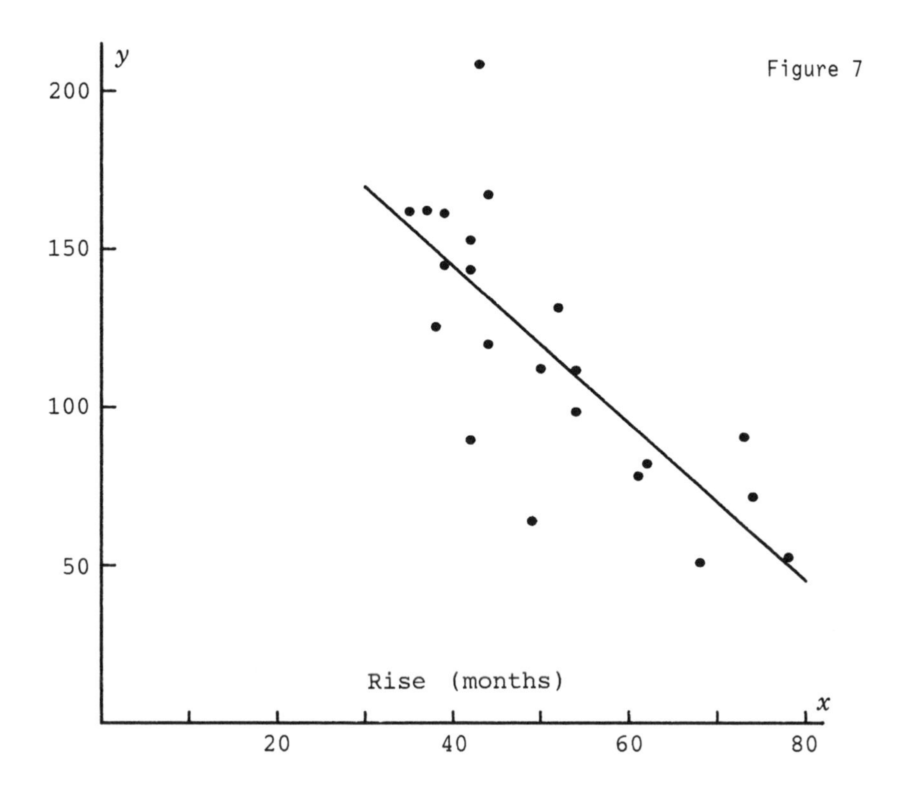

TABLE 4.B

Epoch of maximum	*x*	*y*	*Epoch of maximum*	*x*	*y*
1761 June	73	90.4	1884 Jan.	61	78.1
1769 Oct.	38	125.3	1893 Aug.	42	89.5
1778 May	35	161.8	1905 Oct.	49	63.9
1787 Nov.	42	143.4	1917 Aug.	50	112.1
1804 Dec.	78	52.5	1928 June	62	82.0
1816 March	68	50.8	1937 May	44	119.8
1829 June	74	71.5	1947 July	39	161.2
1837 Febr.	42	152.8	1957 Nov.	43	208.4
1847 Nov.	52	131.3	1969 Febr.	54	111.6
1860 July	54	98.5	1979 Nov.	44	167.1
1870 July	39	144.8	1989 Oct.	37	162.1

Equation (4.4) represents the best straight line fitting the given 22 points. These points and the line are shown in Figure 7.

From formula (4.3) we find $r = -0.767$. This shows that there exists an evident trend to connection, and the negative sign of r indicates that the correlation between x and y is negative: the *longer* the duration of the rise from a minimum to the next maximum of the sunspot activity, the *lower* this maximum generally is.

It should be noted that here, as in all statistical studies, the sample must be sufficiently large in order to give a meaningful result. A correlation coefficient close to +1 or to -1 has no physical meaning if it is based on too small a number of cases. With too few cases the correlation coefficient can accidentally be quite large.

TABLE 4.C

year	*x*	*y*	*year*	*x*	*y*	*year*	*x*	*y*
1901	2.7	700	1931	21.2	858	1961	53.9	903
1902	5.0	762	1932	11.1	858	1962	37.5	862
1903	24.4	854	1933	5.7	738	1963	27.9	713
1904	42.0	663	1934	8.7	707	1964	10.2	785
1905	63.5	912	1935	36.1	916	1965	15.1	1073
1906	53.8	821	1936	79.7	763	1966	47.0	1054
1907	62.0	622	1937	114.4	900	1967	93.8	707
1908	48.5	678	1938	109.6	711	1968	105.9	776
1909	43.9	842	1939	88.8	928	1969	105.5	776
1910	18.6	990	1940	67.8	837	1970	104.5	727
1911	5.7	741	1941	47.5	744	1971	66.6	691
1912	3.6	941	1942	30.6	841	1972	68.9	710
1913	1.4	801	1943	16.3	738	1973	38.0	690
1914	9.6	877	1944	9.6	766	1974	34.5	1039
1915	47.4	910	1945	33.2	745	1975	15.5	734
1916	57.1	1054	1946	92.6	861	1976	12.6	541
1917	103.9	851	1947	151.6	640	1977	27.5	855
1918	80.6	848	1948	136.3	792	1978	92.5	767
1919	63.6	980	1949	134.7	521	1979	155.4	839
1920	37.6	760	1950	83.9	951	1980	154.6	913
1921	26.1	417	1951	69.4	878	1981	140.5	1016
1922	14.2	938	1952	31.5	926	1982	115.9	800
1923	5.8	917	1953	13.9	557	1983	66.6	689
1924	16.7	849	1954	4.4	741	1984	45.9	931
1925	44.3	1075	1955	38.0	616	1985	17.9	758
1926	63.9	896	1956	141.7	795	1986	13.4	946
1927	69.0	837	1957	190.2	801	1987	29.2	908
1928	77.8	882	1958	184.8	834	1988	100.2	1005
1929	64.9	688	1959	159.0	560	1989	157.6	639
1930	35.7	953	1960	112.3	962	1990	142.6	759

As an exercise, show that there is no correlation between the rainfall at the Uccle Observatory, Belgium, and the sunspot activity, using the data of Table 4.C, where

x = yearly mean of the definitive Zürich sunspot numbers,
y = total annual rainfall at Uccle, in millimeters.

Answer : The correlation coefficient is $r = -0.064$, which shows that there is no significant correlation between x and y.

If we drop the last two points, the correlation (for the years 1901 to 1988) even drops to -0.027.

Quadratic curve fitting

Suppose that we wish to draw, through a set of N given points (x, y), the best quadratic function

$$y = ax^2 + bx + c$$

This is a parabola with vertical axis.

Let

$$P = \Sigma x$$
$$Q = \Sigma x^2$$
$$R = \Sigma x^3$$
$$S = \Sigma x^4$$
$$T = \Sigma y$$
$$U = \Sigma xy$$
$$V = \Sigma x^2 y$$

$$D = NQS + 2PQR - Q^3 - P^2 S - NR^2 \qquad (4.5)$$

Then we have

$$\left.\begin{aligned} a &= \frac{NQV + PRT + PQU - Q^2 T - P^2 V - NRU}{D} \\ b &= \frac{NSU + PQV + QRT - Q^2 U - PST - NRV}{D} \\ c &= \frac{QST + QRU + PRV - Q^2 V - PSU - R^2 T}{D} \end{aligned}\right\} \qquad (4.6)$$

General curve fitting (multiple linear regression)

The principle of the best fitting straight line can be extended to other functions and with more than two unknown linear coefficients.

Let us consider the case of a linear combination of *three* functions. Suppose that we know that

$$y = af_0(x) + bf_1(x) + cf_2(x)$$

where f_0, f_1 and f_2 are three known functions of x, but that the coefficients a, b and c are not known. Suppose, moreover, that the value of y is known for at least three values of x. Then the coefficients a, b, c can be found as follows.

Calculate the sums

$$\begin{aligned} M &= \Sigma f_0^2 & \qquad U &= \Sigma y f_0 \\ P &= \Sigma f_0 f_1 & \qquad V &= \Sigma y f_1 \\ Q &= \Sigma f_0 f_2 & \qquad W &= \Sigma y f_2 \\ R &= \Sigma f_1^2 \\ S &= \Sigma f_1 f_2 \\ T &= \Sigma f_2^2 \end{aligned}$$

Then

$$\left.\begin{aligned} D &= MRT + 2PQS - MS^2 - RQ^2 - TP^2 \\ a &= \frac{U(RT - S^2) + V(QS - PT) + W(PS - QR)}{D} \\ b &= \frac{U(SQ - PT) + V(MT - Q^2) + W(PQ - MS)}{D} \\ c &= \frac{U(PS - RQ) + V(PQ - MS) + W(MR - P^2)}{D} \end{aligned}\right\} \quad (4.7)$$

Example 4.c — We know that y is of the form

$$y = a \sin x + b \sin 2x + c \sin 3x$$

and that y takes the following values :

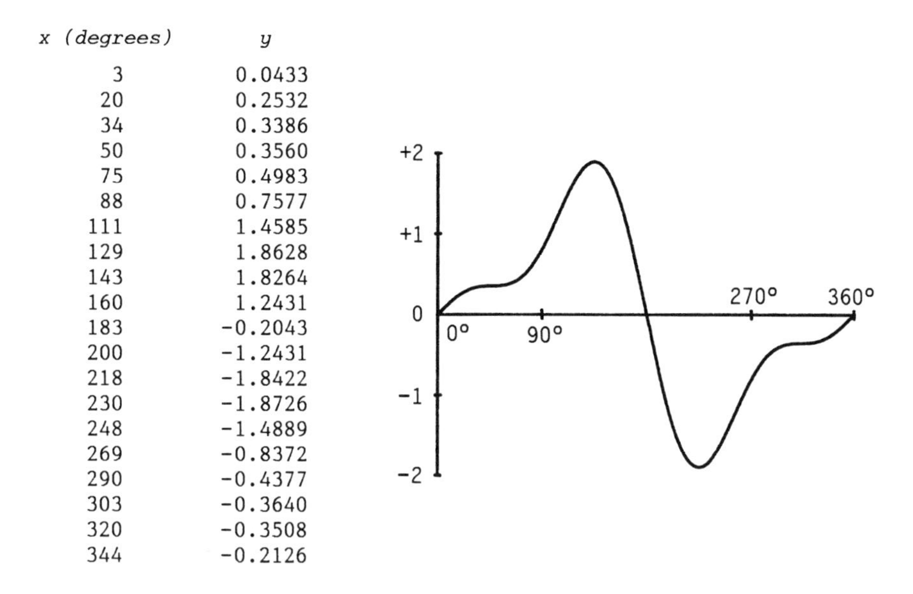

x (degrees)	*y*
3	0.0433
20	0.2532
34	0.3386
50	0.3560
75	0.4983
88	0.7577
111	1.4585
129	1.8628
143	1.8264
160	1.2431
183	-0.2043
200	-1.2431
218	-1.8422
230	-1.8726
248	-1.4889
269	-0.8372
290	-0.4377
303	-0.3640
320	-0.3508
344	-0.2126

Find the values of the coefficients a, b, c.

We leave it as an exercise to the reader. The function is

$$y = 1.2 \sin x - 0.77 \sin 2x + 0.39 \sin 3x$$

and is illustrated in the Figure above.

The reader will *not* find 1.2, -0.77 and +0.39 *exactly*, because in the table the values of y are given with only four decimals.

Let us consider the special case $y = ax^2 + bx + c$. Here we have

$$f_0 = x^2$$
$$f_1 = x$$
$$f_2 = 1$$

resulting in $T = N$ (the number of given points) and $Q = R$. The formulae (4.7) then reduce to (4.5) and (4.6), with other notations.

As another special case, consider $y = af(x)$ with only one unknown coefficient. The latter is easily found from

$$a = \frac{\Sigma y.f}{\Sigma f^2} \tag{4.8}$$

Example 4.d — $y = a\sqrt{x} \quad (x \geqslant 0)$

Find a for the best fitting curve through the following points :

x :	0	1	2	3	4	5
y :	0	1.2	1.4	1.7	2.1	2.2

Here, $f(x) = \sqrt{x}$, so Σf^2 is simply the sum of the x-values. Formula (4.8) gives

$$a = \frac{15.2437}{15}$$

so the required function is

$$y = 1.016 \sqrt{x}$$

References

1. Helmut Alt, *Angewandte Mathematik, Finanz-Mathematik, Statistik, Informatik für UPN-Rechner*, p. 125 (Vieweg, Braunschweig, 1979).

2. International Astronomical Union *Circular* No. 3177 (1978 Feb. 24).

Chapter 5

Iteration

Iteration (from the Latin *iterare* = to repeat) is a method consisting of repeating a calculation several times, until the value of an unknown quantity is obtained. Generally, after each repetition of the calculation, one obtains a result that is closer to the exact solution. We have already seen the use of iteration in Chapter 3, for solving equations (3.6), (3.7), (3.9), (3.10) and (3.11).

Iteration is used, for instance, when there is no method for calculating the unknown quantity directly in an easy way. Examples are:

- the equation of the fifth degree $x^5 + 17x - 8 = 0$;
- the calculation of the times of beginning and end of a solar eclipse, or of an occultation of a star by the Moon, for a given place at the Earth's surface;
- the equation of Kepler $E = M + e \sin E$ (see Chapter 29), where E is the unknown quantity.

To perform an iteration, one must start with an *approximate* value for the unknown quantity, and use must be made of a formula, or of a set of formulae, in order to obtain a *better* value for the unknown. This process is then *repeated* (iteration) until the required accuracy is reached.

A classical example is the calculation of the square root of a number. Of course, this method has nowadays lost its interest (except in special cases), because all pocket calculators and all program languages already possess the function $\sqrt{\ }$ or SQR. The calculation proceeds as follows.

Let N be the number whose square root is requested. Start with an approximate value n for this root; if none is known, the value 1 can be used. Divide N by n, and take the arithmetic mean of the quotient and n. The result is a better value for the square root. In other words, a better value is given by $(n + N/n)/2$. Then the calculation must be repeated.

Example 5.a — Calculate $\sqrt{159}$ to eight decimals.

We know that $12 \times 12 = 144$, so that 12 is an approximate value of the square root of 159. We divide 159 by 12, and find the quotient 13.2500. The arithmetic mean of 12 and 13.2500 is 12.6250, which is a better value for the required square root.

We now divide 159 by 12.6250; the quotient is 12.59406. The mean of 12.6250 (the previous result) and 12.59406 is 12.60953, which is a still better value for the square root.

In that way, we find successively

12 = starting value
12.625 000 00
12.609 529 71
12.609 520 22
12.609 520 22

As you see, 12.609 520 22 yields 12.609 520 22 again, so this is the required square root of 159.

Example 5.b — Calculate the (only) real root of the equation

$$x^5 + 17x - 8 = 0 \qquad (5.1)$$

Because there is no method or formula for the direct calculation of the roots of an equation of the fifth degree, we will have recourse to the iteration procedure. In equation (5.1) we put x^5 in the second member and solve for x; this gives

$$x = \frac{8 - x^5}{17} \qquad (5.2)$$

The unknown quantity is now present in the right-hand member too, but that does not matter, as we shall see. We start by letting $x = 0$ in the right-hand member. Formula (5.2) then yields

$$x = 8/17 = 0.470\,588\,235.$$

This is already a better value than $x = 0$. We now put the value $x = 0.470\,588\,235$ in the right-hand member, and now the formula gives $x = 0.469\,230\,684$. After four more iterations, we obtain the definitive value, namely $x = 0.469\,249\,878$.

The iteration process is not always without problems, however, as is shown in the following example.

Example 5.c — Consider the equation $x^5 + 3x - 8 = 0$.

As in the preceding example, we put x^5 in the right-hand member, and we obtain

$$x = \frac{8 - x^5}{3}$$

If we start, here again, with $x = 0$, we obtain successively

$$\begin{array}{r} 0.0000 \text{ (starting value)} \\ 2.6667 \\ -42.2826 \\ 45\,049\,099 \\ -6.18 \times 10^{37} \\ \text{etc...} \end{array}$$

and so the method does not work in this case! The successive results diverge; in absolute value they grow bigger and bigger. They go 'in the wrong direction'.

Why did the method work in Example 5.b, but not in Example 5.c ? When x lies between 0 and 1, then x^5 too is between 0 and 1. Moreover, x^5 is then smaller than x. That is the reason why in Example 5.b the results of the successive iterations *converge* to a well-defined value, the root of the equation. This root lies between 0 and 1.

But, as we shall see, the root of the equation in Example 5.c is larger than 1. When $x > 1$, then $x^5 > x > 1$ (for $x = 2$, we have already $x^5 = 32$), and a small increase of x gives rise to a larger increase of x^5.

Consequently, the iteration procedure, performed in the same way as in Example 5.b, cannot converge to the required result: the successive values diverge. However, it *is* possible to get the answer, on the condition that we write the iteration formula in another form.

Example 5.d — Let us again consider the equation $x^5 + 3x - 8 = 0$, but now we take into account the fact that the root is larger than 1, and hence that $x^5 > x$. For this reason, we do *not* put x^5 in the right-hand member here. Instead, we keep x^5 in the first member, so that the equation becomes

$$x^5 = 8 - 3x \qquad \text{or} \qquad x = \sqrt[5]{8 - 3x}$$

Starting again with $x = 0$, we obtain the required root after 14 iterations, namely $x = 1.321\,785\,627$.

In Example 5.b, we searched for the root of the equation

$$x^5 + 17x - 8 = 0.$$

However, we can write this equation as

$$x\,(x^4 + 17) = 8, \qquad \text{whence} \qquad x = \frac{8}{x^4 + 17}$$

We now can use this latter formula instead of (5.2). As an exercise, solve this equation by iteration; you should obtain the same result as in Example 5.b.

If we wish to work similarly for the equation of Example 5.c, we obtain the iteration formula

$$x = \frac{8}{x^4 + 3}$$

If we again start by putting the value $x = 0$ in the right-hand member, we obtain $x = 8/3 = 2.666....$ But then comes the surprise: after a few iterations, the successive results jump unceasingly from 2.666 223 459 to 0.149 436 927, and back. As you see, the iteration method does not succeed in all cases; much depends on the form of the iteration formula.

As another example, consider the equation $\sin\phi = 3\cos\phi$. Putting $\phi = 0°$ in the right member yields $\sin\phi = 3$, an impossibility. Putting, instead, $\phi = 90°$ in the second member gives $\sin\phi = 0$, whence $\phi = 0°$, which brings us back to the first case.

But if we write the equation as $\cos\phi = (\sin\phi)/3$ then, starting with $\phi = 0°$, we reach the solution $\phi = 71°.565\,051$ after a few iterations.

Or consider the equation $\sin\phi = \cos 2\phi$. The solution is $\phi = 30°$, since $\sin 30° = \cos 60°$. If we start by putting $\phi = 29°$ in the second member of that equation, the results of the successive iterations diverge. If, however, we write the equation the other way, namely $\cos 2\phi = \sin\phi$, then the successive results converge!

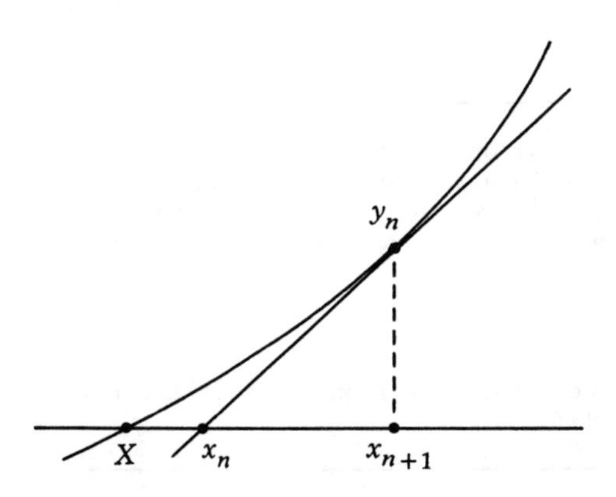

As a further illustration of the iteration procedure, let us consider Newton's method for searching the solution of an equation with one unknown by successive approximations.

Let $f(x)$ be a function of x, and we want to find for what value of x that function is zero. Let $f'(x)$ be the derivative function of $f(x)$. If x_n is an assumed value for the root X, then calculate the value y_n of the function $f(x)$, and the value y'_n of the derivative $f'(x)$, for that value of x. The value y'_n is the slope of the tangent to the curve at the point x_n, y_n — see the Figure on the preceding page. Then, a better value for the unknown quantity is given by

$$x_{n+1} = x_n - \frac{y_n}{y'_n}$$

The calculation is then repeated using this new value of x, until the final value X is reached.

In this procedure, the choice of a good starting value for x can be a problem. For example, for the equation $x^5 - 3x - 8 = 0$, the derivative function is $5x^4 - 3$ and, if we start with $x = 0$, we obtain oscillating results :

0.000 000 000
-2.666 666 667
-2.126 929 222
-1.672 392 941
-1.227 532 073
-0.376 965 299
-2.749 036 974
-2.194 266 642
-1.731 201 846
-1.293 218 530
-0.588 844 800
-3.216 865 068
-2.572 967 057
-2.049 930 313
-1.603 831 482
-1.145 086 797

The reason is that the function reaches a maximum for $x = -0.88$, so that the tangents on both sides of that point have slopes in opposite directions.

But if we start with $x = 1$, then the correct value (to 9 decimal places) is reached after 11 iterations :

+1.000 000 000
+6.000 000 000
+4.803 458 391
+3.850 111 311
+3.095 824 107
+2.510 476 381
+2.080 081 724
+1.807 461 730
+1.690 945 284
+1.671 102 262
+1.670 579 511
+1.670 579 156
+1.670 579 156

Test on "smaller than"

When an iteration procedure is used, one should — as has been mentioned above — repeat the calculation until the result no longer varies. In other words, as long as the last result differs from the previous one, a new iteration must be performed. But here we are faced with a small problem, due to the fact that the computer does not calculate 'exactly'.

Consider the following equation of the third degree

$$s^3 + 3s - W = 0$$

which occurs in the calculation of the motion in a parabolic orbit (see Chapter 33). W is a given constant, while s is the unknown quantity. This equation can very easily be solved by iteration. Start from *any* value; a good choice is $s = 0$. Then a better value for s is

$$\frac{2s^3 + W}{3(s^2 + 1)}$$

After some iterations the correct value of s is obtained. Take, for instance, the case $W = 0.9$. The calculation performed on the HP-85 microcomputer gives the following successive results :

0.000 000 000 000
0.300 000 000 000
0.291 743 119 266
0.291 724 443 641
0.291 724 443 546
0.291 724 443 548
0.291 724 443 548

and hence the exact value (with twelve significant digits) is 0.291 724 443 548. But if we repeat the calculation, on the same machine, for $W = 1.5$, we have a surprise : the machine does not stop and finds successively :

0.000 000 000 000
0.500 000 000 000
0.466 666 666 667
0.466 220 600 162
0.466 220 523 909
0.466 220 523 911
0.466 220 523 910
0.466 220 523 908
0.466 220 523 911
0.466 220 523 910
0.466 220 523 908

and forever again ...911, ...910, ...908. However, we tried this calculation (for $W = 1.5$) on two other machines, and the iteration procedure *did* converge; but then it did not converge for *other* values of W.

A remedy for this trouble consists of testing on 'smaller than' instead of on 'equal to'. In other words, let the iteration process stop when the difference between the new value of s and the previous one is, *in absolute value*, less than a given quantity, for instance 10^{-10}.

The binary search

There is a procedure which is absolutely foolproof, because it can neither stall nor diverge, and always converges in a fixed amount of time to the most exact value of the root the machine is capable. The method does not try to find successively better values of the root. Instead, it just uses a *binary search* to locate the correct value of the root.

Let us explain the procedure by reconsidering the equation of Example 5.b, namely $x^5 + 17x - 8 = 0$.

For $x = 0$ and $x = 1$, the first member of this equation takes the values -8 and +10, respectively. So we know that the root lies between 0 and 1 (*).

Let us now try $x = 0.5$, which is the arithmetical mean of 0 and 1. For $x = 0.5$, the function takes the value +0.53125, which has the opposite sign of the function's value for $x = 0$. So we now know that the root is between 0 and 0.5.

We now try $x = 0.25$, which is the arithmetical mean of 0 and 0.5. And so on.

After each step, the interval in which the root necessarily must be is halved. After 32 steps the value of the root is known with nine exact decimals. (In Example 5.b, the same accuracy was obtained after only 6 steps. But, as we already pointed out, the binary search is a method which is absolutely safe, and it can be used when the 'ordinary' iteration procedure is likely to fail).

With the binary search, one knows in advance the accuracy after n steps: it is the initial interval divided by 2^n.

For the example given above, the program in BASIC can be written as follows (see next page). Line 60 is not actually needed; it has been included to show the successively better values of x.

(*) This is true only if the function is continuous in the interval considered. From the fact that $\tan 86° > 0$ and $\tan 93° < 0$, we may *not* conclude that $\tan x$ becomes zero for a value of x between 86° and 93°.

```
 10  DEF FNA(X) = X*(X^4 + 17) - 8
 20  X1 = 0  :  Y1 = FNA(X1)
 30  X2 = 1  :  Y2 = FNA(X2)
 40  FOR J = 1 TO 33
 50  X = (X1 + X2)/2
 60  PRINT J, X
 70  Y = FNA(X)
 80  IF Y = 0 THEN PRINT J, X  :  END
 90  IF Y*Y1 > 0 THEN 120
100  X2 = X  :  Y2 = Y
110  GOTO 130
120  X1 = X  :  Y1 = Y
130  NEXT J
140  END
```

Chapter 6

Sorting Numbers

Computers are more than calculating machines. They can store and handle data. One example of handling is to rearrange or sort data. Sorting is a function with almost universal application for all users of computers. In astronomy, examples are: sorting stars by right ascension, or by declination; sorting times chronologically; sorting minor planets by increasing semimajor axis, or sorting their names alphabetically. Different algorithms are available to perform sorting. In this Chapter we shall give three methods, provide the BASIC programs, and compare the calculation times.

One of the simplest sorting algorithms is given in Table 6.A under the name 'SIMPLE SORT'. We start from N numbers $X(1)$, $X(2)$, ..., $X(N)$. The values of these elements are arbitrary, and the same value may occur more than once.

After the execution of the routine the numbers $X(I)$ are sorted in increasing order. If one wants them in decreasing order, one should, on line 120, replace >= by <= ; or, alternatively, one may replace $X(I)$ by $-X(I)$.

At each step, two elements are permuted. Successively, the smallest element is placed in front (for $I = 1$), then the second, and so on, to $N - 1$. Note that on line 100 the index I should go till $N - 1$, not till N.

This method is also called 'straight insertion'. The time needed to sort N numbers depends, of course, on the type of computer and on the program language, but in any case the sorting time will approximately be proportional to N^2. This means that the method is unsuitable for large N.

The method called 'BETTER' is somewhat faster, but again the sorting time is approximately proportional to N^2. Its principle is simple: find the smallest element, and place it in front by permuting two elements.

When the set of data to be sorted is large, a much better method

TABLE 6.A : Three sorting programs in BASIC

SIMPLE SORT

```
100  FOR I = 1 TO N-1
110  FOR J = I+1 TO N
120  IF X(J)>= X(I) THEN 160
130  A = X(I)
140  X(I) = X(J)
150  X(J) = A
160  NEXT J
170  NEXT I
```

BETTER

```
100  FOR I = 1 TO N-1
110  M = X(I)
120  K = I
130  FOR J = I+1 TO N
140  IF X(J)<M THEN M=X(J) : K=J
150  NEXT J
160  A=X(I) : X(I)=M : X(K)=A
170  NEXT I
```

QUICKSORT

```
100  DIM L(30), R(30)
110  S = 1 : L(1) = 1 : R(1) = N
120  L = L(S) : R = R(S)
130  S = S-1
140  I = L : J = R
150  V = X(INT((L+R)/2))
160  IF X(I) >= V THEN 190
170  I = I+1
180  GOTO 160
190  IF V >= X(J) THEN 220
200  J = J-1
210  GOTO 190
220  IF I > J THEN 250
230  W = X(I) : X(I) = X(J) :
       X(J) = W
240  I = I+1 : J = J-1
250  IF I <= J THEN 160
260  IF J-L < R-I THEN 320
270  IF L >= J THEN 300
280  S = S+1
290  L(S) = L : R(S) = J
300  L = I
310  GOTO 360
320  IF I >= R THEN 350
330  S = S+1
340  L(S) = I : R(S) = R
350  R = J
360  IF L < R THEN 140
370  IF S <> 0 THEN 120
```

is 'QUICKSORT', which was invented by C. A. R. Hoare. The program itself is longer, but the computer time is considerably shorter. Moreover, when N is sufficiently large, the computing time is approximately proportional to N, not to N^2. (In fact, it is nearly proportional to $N \log N$).

The QUICKSORT sorting technique needs two small auxiliary one-dimensional arrays: L(M) and R(M). M is at least the smallest integer larger than $\log_2 N$. A value of M = 30 is certainly sufficient for all practical purposes.

In Table 6.B we mention the calculation times for some values of N on the HP-85 microcomputer for the three programs mentioned in Table 6.A. As we already said, the times will be different on other computers, but in any case we find that these times increase rapidly for larger values of N, except for the QUICKSORT algorithm.

TABLE 6.B

Calculation times (in seconds) of the three sorting algorithms on the HP-85 microcomputer

N	SIMPLE SORT	BETTER	QUICKSORT
10	0.73	0.51	0.70
20	3.92	2.11	1.84
40	15.4	7.81	4.43
60	38.0	17.0	8.63
80	63.8	29.1	11.3
100	104.3	44.6	14.6
150	254	98.6	24.1
200	453	174	32.9
300	1002	387	56.7
500			97.7
1000			218
1500			342
2000			472

To gain some idea of the calculation speeds for larger values of *N*, we did appeal to a faster computer; the programs were written in FORTRAN and were compiled. The results are given in Table 6.C. The superiority of QUICKSORT is conspicuous here. For $N = 300$, the calculation time with QUICKSORT is still 15% of that with BETTER (Table 6.B); but for 15 000 numbers it is only one third of 1 per cent!

TABLE 6.C

Calculation times (in seconds) of the three sorting algorithms on a 'big' computer

N	SIMPLE SORT	BETTER	QUICKSORT
1 000	13	10	< 1
2 000	51	40	1
3 000	114	90	1
4 000	206	159	2
5 000	321	249	2
10 000	1272	994	5
15 000		2236	7
20 000			10
25 000			12
30 000			15

In some cases there is even no need to write a program. For instance, the TRS-80 Model I contains a built-in function which sorts 1000 numbers in 9 seconds, and 8000 numbers in 83 seconds. It appears that the sorting time is approximately proportional to N here, not to N^2, so probably the QUICKSORT method is used.

To conclude, we can recommend the 'straight insertion' (SIMPLE SORT) if the set of data to be sorted is not too large, for example for $N < 200$. For larger sets it is well worth while to use QUICKSORT.

Besides numerical data, often strings (names) are to be sorted, such as X$(1) = "Ceres", X$(2) = "Pallas", etc. Each character has its own value. The complete list with all signs constitutes the so-called ASCII table, a part of which is given in Table 6.D. [ASCII = 'American Standard Code for Information Interchange.]

TABLE 6.D : Visible ASCII Characters

After each character its decimal code is given

space	32	8	56	P	80	h	104
!	33	9	57	Q	81	i	105
"	34	:	58	R	82	j	106
#	35	;	59	S	83	k	107
$	36	<	60	T	84	l	108
%	37	=	61	U	85	m	109
&	38	>	62	V	86	n	110
'	39	?	63	W	87	o	111
(	40	@	64	X	88	p	112
)	41	A	65	Y	89	q	113
*	42	B	66	Z	90	r	114
+	43	C	67	[	91	s	115
,	44	D	68	\	92	t	116
-	45	E	69	]	93	u	117
.	46	F	70	^	94	v	118
/	47	G	71	_	95	w	119
0	48	H	72	`	96	x	120
1	49	I	73	a	97	y	121
2	50	J	74	b	98	z	122
3	51	K	75	c	99	{	123
4	52	L	76	d	100	|	124
5	53	M	77	e	101	}	125
6	54	N	78	f	102	~	126
7	55	O	79	g	103		

Chapter 7

Julian Day

In this Chapter we give a method for converting a date, given in the Julian or in the Gregorian calendar, into the corresponding Julian Day number (JD), or vice versa.

General remarks

The Julian Day number or, more simply, the *Julian Day* (*) (JD) is a continuous count of days and fractions thereof from the beginning of the year -4712. By tradition, the Julian Day begins at Greenwich mean *noon*, that is, at 12^h Universal Time. If the JD corresponds to an instant measured in the scale of Dynamical Time (or Ephemeris Time), the expression *Julian Ephemeris Day* (JDE) (**) is generally used. For example,

1977 April 26.4 UT = JD 2443 259.9
1977 April 26.4 TD = JDE 2443 259.9

In the methods described below, the Gregorian calendar reform is taken into account; thus, the day following 1582 October 4 (Julian calendar) is 1582 October 15 (Gregorian calendar).

The Gregorian calendar was not at once officially adopted by all countries. This should be kept in mind when making historical research. In Great Britain, for instance, the change was made as late as in A.D. 1752, and in Turkey not before 1927.

(*) In many books we read 'Julian *Date*' instead of 'Julian Day'. For us, a Julian date is a date in the Julian calendar, just as a Gregorian date refers to the Gregorian calendar. The JD has nothing to do with the Julian calendar.

(**) Not JED as it is sometimes written. The 'E' is a sort of index appended to 'JD'.

The Julian calendar was established in the Roman Empire by Julius Caesar in the year -45 and reached its final form about the year +8. Nevertheless, we shall follow the astronomers' practice consisting of extrapolating the Julian calendar indefinitely to the past. In this system, we can speak, for instance, of the solar eclipse of August 28 of the year -1203, although at that remote time the Roman Empire was not yet founded and the month of August was still to be conceived!

There is a disagreement between astronomers and historians about how to count the years preceding the year 1. In this book, the 'B.C.' years are counted astronomically. Thus, the year before the year +1 is the year zero, and the year preceding the latter is the year -1. The year which the historians call 585 B.C. is actually the year -584.

The astronomical counting of the negative years is the only one suitable for arithmetical purpose. For example, in the historical practice of counting, the rule of divisibility by 4 revealing the Julian leap-years no longer exists; these years are, indeed, 1, 5, 9, 13, ... B.C. In the astronomical sequence, however, these leap-years are called 0, -4, -8, -12 ..., and the rule of divisibility by 4 subsists.

We will indicate by INT(x) the integer part of the number x, that is the integer which precedes its decimal point. Examples :

INT(7/4) = 1 INT(5.02) = 5
INT(8/4) = 2 INT(5.9999) = 5

There may be a problem with negative numbers. On some computers or in some program languages, INT(x) is the greatest integer less than or equal to x. In that case we have, for instance, INT(-7.83) = -8, because -7 is indeed larger than -7.83.

But in other languages, INT is the integer part of the *written* number, that is, the part of the number that precedes the decimal point. In that case, INT(-7.83) = -7. This is called *truncation*, and some program languages have both functions: INT(x) having the first of the above-mentioned meanings, and TRUNC(x).

Hence, take care when using the INT function for negative numbers. (For positive numbers, both meanings yield the same result). In the formulae given in this book, the argument of the INT function is always positive.

Calculation of the JD

The following method is valid for positive as well as for negative years, but not for negative JD.

Let Y be the year, M the month number (1 for January, 2 for February, etc., to 12 for December), and D the day of the month (with decimals, if any) of the given calendar date.

- If $M > 2$, leave Y and M unchanged.
  If $M = 1$ or 2, replace Y by $Y - 1$, and M by $M + 12$.
  In other words, if the date is in January or February, it is considered to be in the 13th or 14th month of the preceding year.

- In the *Gregorian* calendar, calculate

$$A = \mathrm{INT}\left(\frac{Y}{100}\right) \qquad B = 2 - A + \mathrm{INT}\left(\frac{A}{4}\right)$$

  In the *Julian* calendar, take $B = 0$.

- The required Julian Day is then

$$\mathrm{JD} = \mathrm{INT}(365.25\,(Y + 4716)) + \mathrm{INT}(30.6001\,(M+1)) + D + B - 1524.5 \qquad (7.1)$$

The number 30.6 (instead of 30.6001) will give the correct result but 30.6001 is used so that the proper integer will always be obtained. [In fact, instead of 30.6001, one may use 30.601, or even 30.61.] For instance, 5 times 30.6 gives 153 exactly. However, most computers would not represent 30.6 exactly — see in Chapter 2 what we said about BCD — and might give a result of 152.999 9998 instead, whose integer part is 152. The calculated JD would then be incorrect.

Example 7.a — Calculate the JD corresponding to 1957 October 4.81, the time of launch of Sputnik 1.

Here we have $Y = 1957$, $M = 10$, $D = 4.81$.
Because $M > 2$, we leave Y and M unchanged.
The date is in the Gregorian calendar, so we calculate

$$A = \mathrm{INT}\left(\frac{1957}{100}\right) = \mathrm{INT}(19.57) = 19$$

$$B = 2 - 19 + \mathrm{INT}\left(\frac{19}{4}\right) = 2 - 19 + 4 = -13$$

$$\mathrm{JD} = \mathrm{INT}(365.25 \times 6673) + \mathrm{INT}(30.6001 \times 11) + 4.81 - 13 - 1524.5$$
$$\mathrm{JD} = 2436\,116.31$$

Example 7.b — Calculate the JD corresponding to January 27 at 12^h of the year 333.

Because $M = 1$, we have $Y = 333 - 1 = 332$ and $M = 1 + 12 = 13$.
Because the date is in the Julian calendar, we have $B = 0$.

$$\mathrm{JD} = \mathrm{INT}(365.25 \times 5048) + \mathrm{INT}(30.6001 \times 14) + 27.5 + 0 - 1524.5$$
$$\mathrm{JD} = 1842\,713.0$$

The following list gives the JD corresponding to some calendar dates. These data may be useful for testing a program.

Date	JD
2000 Jan. 1.5	2451 545.0
1987 Jan. 27.0	2446 822.5
1987 June 19.5	2446 966.0
1988 Jan. 27.0	2447 187.5
1988 June 19.5	2447 332.0
1900 Jan. 1.0	2415 020.5
1600 Jan. 1.0	2305 447.5
1600 Dec. 31.0	2305 812.5
837 Apr. 10.3	2026 871.8
-1000 July 12.5	1356 001.0
-1000 Feb. 29.0	1355 866.5
-1001 Aug. 17.9	1355 671.4
-4712 Jan. 1.5	0.0

If one is interested only in dates between 1900 March 1 and 2100 February 28, then in formula (7.1) we have $B = -13$.

In some applications it is needed to know the Julian Day JD_0 corresponding to January 0.0 of a given year. This is the same as December 31.0 of the preceding year. For a year in the *Gregorian* calendar, this can be calculated as follows.

$$Y = \text{year} - 1 \qquad A = \text{INT}\left(\frac{Y}{100}\right)$$

$$JD_0 = \text{INT}(365.25\,Y) - A + \text{INT}\left(\frac{A}{4}\right) + 1721\,424.5$$

For the years 1901 to 2099 inclusively, this reduces to

$$JD_0 = 1721\,409.5 + \text{INT}\,(365.25 \times (\text{year} - 1))$$

When is a given year a leap year ?

In the **Julian calendar**, a year is a leap (or bissextile) year of 366 days if its numerical designation is divisible by 4. All other years are common years (365 days).

For instance, the years 900 and 1236 were bissextile years, while 750 and 1429 were common years.

The same rule holds in the **Gregorian calendar**, with the following exception: the centurial years that are *not* divisible by 400, such as 1700, 1800, 1900, and 2100, are common years. The other century years, which *are* divisible by 400, are leap years, for instance 1600, 2000, and 2400.

The *Modified Julian Day* (MJD) sometimes appears in modern work, for instance when mentioning orbital elements of artificial satellites. Contrary to the JD, the Modified Julian Day begins at Greenwich mean *midnight*. It is equal to

$$\text{MJD} = \text{JD} - 2400\,000.5$$

and therefore MJD = 0.0 corresponds to 1858 November 17 at 0^h UT.

Calculation of the Calendar Date from the JD

The following method is valid for positive as well as for negative years, but not for negative Julian Day numbers.

Add 0.5 to the JD, and let Z be the integer part, and F the fractional (decimal) part of the result.

If $Z < 2299\,161$, take $A = Z$.

If Z is equal to or larger than 2299 161, calculate

$$\alpha = \text{INT}\left(\frac{Z - 1867\,216.25}{36524.25}\right)$$

$$A = Z + 1 + \alpha - \text{INT}\left(\frac{\alpha}{4}\right)$$

Then calculate

$$B = A + 1524$$

$$C = \text{INT}\left(\frac{B - 122.1}{365.25}\right)$$

$$D = \text{INT}(365.25\,C)$$

$$E = \text{INT}\left(\frac{B - D}{30.6001}\right)$$

The day of the month (with decimals) is then

$$B - D - \text{INT}(30.6001\,E) + F$$

The month number m is

$$m = \begin{cases} E - 1 & \text{if } E < 14 \\ E - 13 & \text{if } E = 14 \text{ or } 15 \end{cases}$$

The year is

$$\begin{cases} C - 4716 & \text{if } m > 2 \\ C - 4715 & \text{if } m = 1 \text{ or } 2 \end{cases}$$

Contrary to what has been said earlier about formula (7.1), in the formula for E the number 30.6001 may *not* be replaced by 30.6, even if the computer calculates exactly. Otherwise, one would obtain February 0 instead of January 31, or April 0 instead of March 31.

Example 7.c — Calculate the calendar date corresponding to JD 2436 116.31.

2436 116.31 + 0.5 = 2436 116.81

Z = 2436 116 and F = 0.81

Because $Z > 2299\,161$, we have

$$\alpha = \text{INT}\left(\frac{2436\,116 - 1867\,216.25}{36524.25}\right) = 15$$

$$A = 2436\,116 + 1 + 15 - \text{INT}\left(\frac{15}{4}\right) = 2436\,129$$

Then we find

B = 2437 653 C = 6673 D = 2437 313 E = 11

day of month = 4.81

month m = $E - 1$ = 10 (because $E < 14$)

year = $C - 4716$ = 1957 (because $m > 2$)

Hence, the required date is 1957 October 4.81.

Exercise : Calculate the calendar dates corresponding to JD = 1842 713.0 and JD = 1507 900.13.

Answers : 333 January 27.5 and -584 May 28.63.

Time interval in days

The number of days between two calendar dates can be found by calculating the difference between their corresponding Julian Days.

Example 7.d — The periodic comet Halley passed through perihelion on 1910 April 20 and on 1986 February 9. What is the time interval between these two passages ?

1910 April 20.0 corresponds to JD 2418 781.5
1986 Febr. 9.0 corresponds to JD 2446 470.5

The difference is 27 689 days.

Exercise : Find the date exactly 10 000 days after 1991 July 11.

Answer : 2018 November 26.

Day of the Week

The day of the week corresponding to a given date can be obtained as follows. Compute the JD for that date at 0^h, add 1.5, and divide the result by 7. The remainder of this division will indicate the weekday, as follows: if the remainder is 0, it is a Sunday, 1 a Monday, 2 a Tuesday, 3 a Wednesday, 4 a Thursday, 5 a Friday, and 6 a Saturday.

The week was not modified in any way by the Gregorian reform of the Julian calendar. Thus, in 1582, Thursday October 4 was followed by Friday October 15.

Example 7.e — Find the weekday of 1954 June 30.

1954 June 30.0 corresponds to JD 2434 923.5
2434 923.5 + 1.5 = 2434 925
The remainder of the division of 2434 925 by 7 is 3.
Hence it was a Wednesday.

Day of the Year

The number N of a day in the year can be computed by means of the following formula [1].

$$N = \mathrm{INT}\left(\frac{275\,M}{9}\right) - K \times \mathrm{INT}\left(\frac{M+9}{12}\right) + D - 30$$

where M is the month number, D the day of the month, and

$K = 1$ for a leap (bissextile) year,
$K = 2$ for a common year.

N takes integer values, from 1 on January 1, to 365 (or 366 in leap years) on December 31.

Example 7.f — 1978 November 14.

Common year, $M = 11$, $D = 14$, $K = 2$.
One finds $N = 318$.

Example 7.g — 1988 April 22.

Leap year, $M = 4$, $D = 22$, $K = 1$.
One finds $N = 113$.

Let us now consider the reverse problem: the day number N in the year is known, and the corresponding date is required, namely the month number M and the day D of that month. The following algorithm was found by A. Pouplier, of the Société Astronomique de Liège, Belgium [2].

As above, take

$K = 1$ in the case of a leap year,
$K = 2$ in the case of a common year.

$$M = \mathrm{INT}\left[\frac{9(K+N)}{275} + 0.98\right]$$

If $N < 32$, then $M = 1$

$$D = N - \mathrm{INT}\left(\frac{275M}{9}\right) + K \times \mathrm{INT}\left(\frac{M+9}{12}\right) + 30$$

References

1. Nautical Almanac Office, U.S. Naval Observatory, Washington, D.C., *Almanac for Computers for the Year 1978*, page B2.

2. A. Pouplier, letter to Jean Meeus, 1987 April 10.

Chapter 8

Date of Easter

In this Chapter we give a method for calculating the date of the Christian Easter Sunday of a given year — *not* the Jewish Pesah.

Gregorian Easter

The following method has been given by Spencer Jones in his book *General Astronomy* (pages 73-74 of the edition of 1922). It has been published again in the *Journal of the British Astronomical Association*, Vol. 88, page 91 (December 1977) where it is said that it was devised in 1876 and appeared in Butcher's *Ecclesiastical Calendar*.

Unlike the formula given by Gauss, this method has no exception and is valid for all years in the Gregorian calendar, hence from the year 1583 on. The procedure for finding the date of Easter is as follows :

Divide	by	Quotient	Remainder
the year x	19	—	a
the year x	100	b	c
b	4	d	e
$b + 8$	25	f	—
$b - f + 1$	3	g	—
$19a + b - d - g + 15$	30	—	h
c	4	i	k
$32 + 2e + 2i - h - k$	7	—	l
$a + 11h + 22l$	451	m	—
$h + l - 7m + 114$	31	n	p

Then n = number of the month (3 = March, 4 = April),
$p+1$ = day of that month upon which Easter Sunday falls.

If the computer language has no 'modulo' function or no 'remainder' function, the calculation of the remainder of a division must be programmed carefully. Suppose that the remainder of the division of 34 by 30 should be found. On the HP-67 and HP-41C pocket calculators, for instance, we find

$$34/30 = 1.133\,333\,333$$

the fractional part of which is 0.133 333 333. When multiplied by 30, this gives 3.999 999 990. This result differs from 4, the correct value, and may give a wrong date for Easter at the end of the calculation.

Try your program on the following years :

1991 → March 31	1954 → April 18
1992 → April 19	2000 → April 23
1993 → April 11	1818 → March 22

The extreme dates of Easter are March 22 (as in 1818 and 2285) and April 25 (as in 1886, 1943, 2038).

The rule for finding the date of Easter Sunday is well known : Easter is the first Sunday *after* the Full Moon that happens on or next after the March equinox. Actually, the rules for finding the Easter date were fixed long ago by the Christian clergy. For the purposes of these rules, the Full Moon is reckoned according to an ecclesiastical computation and is not the real, astronomical Full Moon. Likewise, the equinox is always assumed to fall on March 21; actually, it can occur a day or two sooner.

In 1967, for instance, the equinox was on March 21, and the Full Moon on March 26 (UT date). The first Sunday after March 26 was April 2. Nevertheless, Easter Sunday was March 26.

During the period 1900 - 2100, the purely astronomical rule yields another date for Easter Sunday than the ecclesiastical rule for the following years: 1900, 1903, 1923, 1924, 1927, 1943, 1954, 1962, 1967, 1974, 1981, 2038, 2049, 2069, 2076, 2089, 2095, and 2096.

A period of 5 700 000 years is required for the cyclical recurrence of the Gregorian Easter dates. It has been found that the most frequent Gregorian Easter date is April 19.

Julian Easter

In the Julian calendar, the date of Easter can be found as follows.

Divide	by	Quotient	Remainder
the year x	4	–	a
the year x	7	–	b
the year x	19	–	c
$19c + 15$	30	–	d
$2a + 4b - d + 34$	7	–	e
$d + e + 114$	31	f	g

Then f = number of the month (3 = March, 4 = April),
$g + 1$ = day of that month upon which Easter Sunday falls.

The date of the *Julian* Easter has a periodicity of 532 years. For instance, we find April 12 for the years 179, 711 and 1243.

Chapter 9

Dynamical Time and Universal Time

The Universal Time (UT), or Greenwich Civil Time, is based on the rotation of the Earth. The UT is necessary for civil life and for the astronomical calculations where local hour angles are involved.

However, the Earth's rotation is generally slowing down and, moreover, this occurs with unpredictable irregularities. For this reason, the UT is not a uniform time.

But the astronomers need a uniform time scale for their accurate calculations (celestial mechanics, orbits, ephemerides). From 1960 to 1983, in the great astronomical almanacs such as the *Astronomical Ephemeris*, use was made of a uniform time scale called the *Ephemeris Time* (ET) and defined by the laws of dynamics: it was based on the planetary motions. In 1984, the ET was replaced by the *Dynamical Time*, which is defined by atomic clocks. The Dynamical Time is, in fact, a prolongation of the Ephemeris Time.

One distinguishes a Barycentric Dynamical Time (TDB) and a Terrestrial Dynamical Time (TDT). These times differ by at most 0.0017 second, the difference being related to the motion of the Earth on its elliptical orbit around the Sun (relativistic effect). Because this very small difference can be neglected for most practical purposes, we will not make the distinction between TDB and TDT, and we will name both simply TD (Dynamical Time).

The exact value of the difference $\Delta T = \mathrm{TD} - \mathrm{UT}$ can be deduced only from observations. Table 9.A gives the value of ΔT for the *beginning* of some years. Except for the two last values, they are taken from the *Astronomical Almanac* for 1988 [1].

For epochs in the *near* future, one may extrapolate the values of Table 9.A. For instance, we can use the provisional values

$$\Delta T = +60 \text{ seconds in } 1993$$
$$\Delta T = +67 \text{ seconds in } 2000$$
$$\Delta T = +80 \text{ seconds in } 2010$$

TABLE 9.A

ΔT = TD − UT (in seconds) for the beginning of some years

Year	ΔT	*Year*	ΔT	*Year*	ΔT	*Year*	ΔT	*Year*	ΔT
1620	+124	1700	+ 9	1780	+17	1860	+ 7.9	1940	+24.3
1622	115	1702	9	1782	17	1862	7.5	1942	25.3
1624	106	1704	9	1784	17	1864	6.4	1944	26.2
1626	98	1706	9	1786	17	1866	5.4	1946	27.3
1628	91	1708	10	1788	17	1868	2.9	1948	28.2
1630	+85	1710	+10	1790	+17	1870	+ 1.6	1950	+29.1
1632	79	1712	10	1792	16	1872	− 1.0	1952	30.0
1634	74	1714	10	1794	16	1874	− 2.7	1954	30.7
1636	70	1716	10	1796	15	1876	− 3.6	1956	31.4
1638	65	1718	11	1798	14	1878	− 4.7	1958	32.2
1640	+62	1720	+11	1800	+13.7	1880	− 5.4	1960	+33.1
1642	58	1722	11	1802	13.1	1882	− 5.2	1962	34.0
1644	55	1724	11	1804	12.7	1884	− 5.5	1964	35.0
1646	53	1726	11	1806	12.5	1886	− 5.6	1966	36.5
1648	50	1728	11	1808	12.5	1888	− 5.8	1968	38.3
1650	+48	1730	+11	1810	+12.5	1890	− 5.9	1970	+40.2
1652	46	1732	11	1812	12.5	1892	− 6.2	1972	42.2
1654	44	1734	12	1814	12.5	1894	− 6.4	1974	44.5
1656	42	1736	12	1816	12.5	1896	− 6.1	1976	46.5
1658	40	1738	12	1818	12.3	1898	− 4.7	1978	48.5
1660	+37	1740	+12	1820	+12.0	1900	− 2.7	1980	+50.5
1662	35	1742	12	1822	11.4	1902	− 0.0	1982	52.2
1664	33	1744	13	1824	10.6	1904	+ 2.6	1984	53.8
1666	31	1746	13	1826	9.6	1906	5.4	1986	54.9
1668	28	1748	13	1828	8.6	1908	7.7	1988	55.8
1670	+26	1750	+13	1830	+ 7.5	1910	+10.5	1990	+56.9
1672	24	1752	14	1832	6.6	1912	13.4	1992	58.3
1674	22	1754	14	1834	6.0	1914	16.0		
1676	20	1756	14	1836	5.7	1916	18.2		
1678	18	1758	15	1838	5.6	1918	20.2		
1680	+16	1760	+15	1840	+ 5.7	1920	+21.2		
1682	14	1762	15	1842	5.9	1922	22.4		
1684	13	1764	15	1844	6.2	1924	23.5		
1686	12	1766	16	1846	6.5	1926	23.9		
1688	11	1768	16	1848	6.8	1928	24.3		
1690	+10	1770	+16	1850	+ 7.1	1930	+24.0		
1692	9	1772	16	1852	7.3	1932	23.9		
1694	9	1774	16	1854	7.5	1934	23.9		
1696	9	1776	17	1856	7.7	1936	23.7		
1698	9	1778	17	1858	7.8	1938	24.0		

For other epochs outside the time interval of Table 9.A, an *approximate* value of ΔT (in *seconds*) can be deduced from the following relation due to Morrison and Stephenson [2]:

$$\Delta T = -15 + 0.00325\,(\text{year} - 1810)^2$$

where 'year' can be taken with decimals, if needed. This formula can be written as

$$\Delta T = 102.3 + 123.5\,T + 32.5\,T^2 \qquad (9.1)$$

where T is measured in *centuries* from the epoch 2000.0, or, if the Julian Day is used, as

$$\Delta T = -15 + \frac{(\text{JD} - 2382\,148)^2}{41\,048\,480}$$

With these expressions, the uncertainty of UT can reach as much as two hours back to 4000 B.C. Future improvements of the formula will benefit the user when converting from TD to UT, but will not change algorithms, programs, ephemerides or tables given with the uniform time scale of TD.

In 1984, Stephenson and Morrison [3] published *two* other parabolic expressions for ΔT in the past. The period from 390 B.C. to A.D. 1600 was covered by two separate parabolic fits:

from −390 to +948 : $\Delta T = 1360 + 320\,T + 44.3\,T^2$
from +948 to +1600 : $\Delta T = 25.5\,T^2$

where T is the time difference in centuries from A.D. 1800, and ΔT is obtained in seconds.

Two years later, Stephenson and Houlden [4] gave yet two other expressions for ΔT in the past:

(i) at any time before A.D. 948 : $\Delta T = 1830 - 405\,E + 46.5\,E^2$

(ii) from A.D. 948 to 1600 : $\Delta T = 22.5\,t^2$

where E is the number of centuries from A.D. 948, and t is the number of centuries from A.D. 1850.

Formulae (i) and (ii) are equivalent to the following expressions, where T is the time in centuries from J2000.0 ($T < 0$):

before A.D. 948 : $2715.6 + 573.36\,T + 46.5\,T^2$
from 948 to 1600 : $50.6 + 67.5\,T + 22.5\,T^2$

The quantity ΔT was negative from A.D. 1871 to 1901. It should be noted that ΔT is positive *both* for the remote past and for the distant future.

Except for the years 1871–1901, an instant given in UT is *later* than the instant in TD having the same numerical value. For example, the instant 1990 January 27, 0^h UT is an instant 57 seconds later than 1990 January 27, 0^h TD. We have UT = TD − ΔT.

Example 9.a — New Moon took place on 1977 February 18 at $3^h 37^m 40^s$ Dynamical Time (see Example 47.a).

At that instant, ΔT was equal to +48 seconds. Consequently, the corresponding Universal Time of that lunar phase was

$$3^h 37^m 40^s - 48^s = 3^h 36^m 52^s .$$

Example 9.b — Suppose that the position of Mercury should be calculated for February 6 at 6^h Universal Time of the year +333.

Here we have

$$T = \frac{333.1 - 2000}{100} = -16.669$$

for which formula (9.1) gives the value $\Delta T = +7074$ seconds or 118 minutes. Hence, $\text{TD} = 6^h + 118$ minutes $= 7^h 58^m$, and the calculation must be performed for 333 February 6 at $7^h 58^m$ TD.

Schmadel and Zech [5] have constructed the following approximation for ΔT, valid for the entire time span 1800-1988. It represents the values given in Table 9.A with a maximum error of 1.9 seconds.

$$\begin{aligned}\Delta T = & -0.000\,014 + 0.001\,148\,\theta + 0.003\,357\,\theta^2 - 0.012\,462\,\theta^3 \\ & - 0.022\,542\,\theta^4 + 0.062\,971\,\theta^5 + 0.079\,441\,\theta^6 \\ & - 0.146\,960\,\theta^7 - 0.149\,279\,\theta^8 + 0.161\,416\,\theta^9 \\ & + 0.145\,932\,\theta^{10} - 0.067\,471\,\theta^{11} - 0.058\,091\,\theta^{12}\end{aligned}$$

In this formula, ΔT is expressed in *days*, and θ is the time elapsed since 1900.0 and expressed in Julian centuries.

Schmadel and Zech also provide expressions for shorter time spans. For the years 1800-1899, the following expression gives ΔT (in days) with a maximum error of 1.0 second :

$$\begin{aligned}\Delta T = & -0.000\,009 + 0.003\,844\,\theta + 0.083\,563\,\theta^2 + 0.865\,736\,\theta^3 \\ & + 4.867\,575\,\theta^4 + 15.845\,535\,\theta^5 + 31.332\,267\,\theta^6 \\ & + 38.291\,999\,\theta^7 + 28.316\,289\,\theta^8 + 11.636\,204\,\theta^9 \\ & + 2.043\,794\,\theta^{10}\end{aligned}$$

For the years 1900 to 1987, the following expression gives ΔT (in days) with a maximum error of 1.0 second :

$$\begin{aligned}\Delta T = & -0.000\,020 + 0.000\,297\,\theta + 0.025\,184\,\theta^2 - 0.181\,133\,\theta^3 \\ & + 0.553\,040\,\theta^4 - 0.861\,938\,\theta^5 + 0.677\,066\,\theta^6 - 0.212\,591\,\theta^7\end{aligned}$$

where θ has the same meaning as for the first formula.

It should be noted that these three expressions are empirical formulae. *Their use is prohibited outside of their defined validity range !*

References

1. *Astronomical Almanac* for 1988 (Washington, D.C.), pages K8 and K9.

2. L.V. Morrison and F.R. Stephenson, *Sun and Planetary System*, Vol. 96, page 73 (Reidel, Dordrecht, 1982). — Cited by P. Bretagnon and J.-L. Simon, *Planetary Programs and Tables from -4000 to +2800* (Willmann-Bell, Richmond, 1986), page 5.

3. F.R. Stephenson and L.V. Morrison, 'Long-term changes in the rotation of the Earth', *Phil. Trans. Royal Soc.*, A, Vol. 313, pages 47-70 (1984).

4. F.R. Stephenson and M.A. Houlden, *Atlas of Historical Eclipse Maps*, Cambridge University Press, England (1986), page x.

5. L.D. Schmadel and G. Zech, 'Empirical transformations from U.T. to E.T. for the period 1800-1988', *Astronomische Nachrichten*, Vol. 309, pages 219-221 (1988).

Chapter 10

The Earth's Globe

The actual figure of the Earth's surface, including all the inequalities of mountains and valleys, is incapable of geometric definition. Therefore, the ideal figure used in geodesy is that of the mean sea level, extended through the continents. This is the *geoid*, whose surface at every point is perpendicular to the local plumb-line.

However, the heterogeneity of the Earth's interior and the attraction of mountains are such that the surface of the geoid is not rigorously represented by any definable solid. An approximation sufficient for most geographical and astronomical purposes is obtained by considering it to be an ellipsoid of revolution.

Geocentric rectangular coordinates of an observer

The figure represents a meridian cross section of the Earth. C is the Earth's center, N its north pole, S its south pole, EF the equator, HK the horizontal plane of the observer O, and OP the perpendicular to HK. The direction OM, parallel to SN, makes with OH an angle ϕ which is the *geographical latitude* of O. The angle OPF too is equal to ϕ.

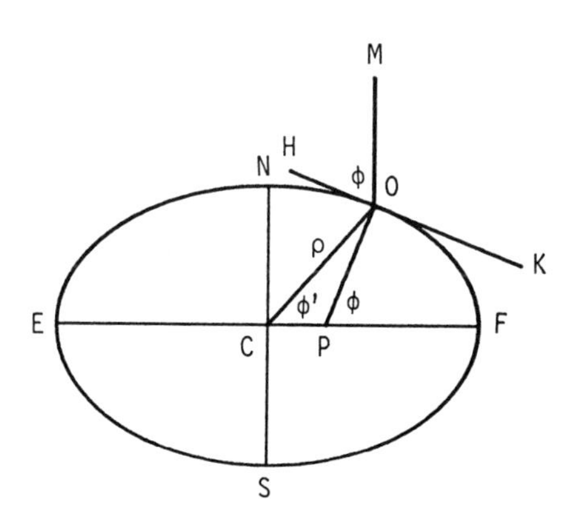

The radius vector OC, joining the observer to the center of the Earth, makes with the equator CF an angle ϕ' which is the *geocentric latitude* of O. We have $\phi = \phi'$ at the poles and at the equator; for all other latitudes

$$|\phi'| < |\phi|$$

Let f be the Earth's flattening, and b/a the ratio NC/CF of the polar radius $NC = b$ to the equatorial radius $CF = a$. In 1976 the International Astronomical Union adopted the values

$$a = 6378.14 \text{ km}, \qquad f = \frac{1}{298.257}$$

from which we have

$$b = a(1-f) = 6356.755 \text{ km}$$

$$\frac{b}{a} = 1 - f = 0.996\,647\,19$$

The eccentricity e of the Earth's meridian is

$$e = \sqrt{2f - f^2} = 0.081\,819\,22$$

We have the relations

$$f = \frac{a-b}{a} \qquad 1 - e^2 = (1-f)^2$$

For a place at sea level,

$$\tan \phi' = \frac{b^2}{a^2} \tan \phi$$

If H is the observer's height above sea level in *meters*, the quantities $\rho \sin \phi'$ and $\rho \cos \phi'$, needed in the calculation of diurnal parallaxes, eclipses and occultations, may be calculated as follows :

$$\tan u = \frac{b}{a} \tan \phi$$

$$\begin{cases} \rho \sin \phi' = \dfrac{b}{a} \sin u + \dfrac{H}{6\,378\,140} \sin \phi \\[2ex] \rho \cos \phi' = \cos u + \dfrac{H}{6\,378\,140} \cos \phi \end{cases}$$

The quantity $\rho \sin \phi'$ is positive in the northern hemisphere, negative in the southern one, while $\rho \cos \phi'$ is always positive.

The quantity ρ denotes the observer's distance to the center of the Earth (OC in the Figure), the Earth's equatorial radius being taken as unity.

Example 10.a — Calculate $\rho \sin \phi'$ and $\rho \cos \phi'$ for the Palomar Observatory, for which

$$\phi = +33°21'22'', \qquad H = 1706 \text{ meters}.$$

We obtain

$$\phi = 33^\circ.356\,111$$
$$u = 33^\circ.267\,796$$
$$\rho \sin \phi' = +0.546\,861$$
$$\rho \cos \phi' = +0.836\,339$$

Other formulae concerning the Earth's ellipsoid

For a given point on the ellipsoid, the difference between the geographic latitude and the geocentric latitude can be found from

$$\phi - \phi' = 692''.73 \sin 2\phi - 1''.16 \sin 4\phi$$

The difference $\phi - \phi'$ reaches a maximum value for $u = 45^\circ$. If ϕ_o and ϕ'_o are the corresponding geographic and geocentric latitudes, we have

$$\tan \phi_o = \frac{a}{b} \qquad \tan \phi'_o = \frac{b}{a} \qquad \phi_o + \phi'_o = 90^\circ$$

whence, for the IAU 1976 ellipsoid,

$$\phi_o = 45^\circ 05' 46''.36 \qquad \phi'_o = 44^\circ 54' 13''.64$$
$$\phi_o - \phi'_o = 11' 32''.73$$

The quantity ρ (for sea level) can be found from

$$\rho = 0.998\,3271 + 0.001\,6764 \cos 2\phi - 0.000\,0035 \cos 4\phi$$

The parallel of latitude ϕ is a circle whose radius is

$$R_p = \frac{a \cos \phi}{\sqrt{1 - e^2 \sin^2 \phi}}$$

where, as above, e is the eccentricity of the meridian ellipse.

Hence, one degree of longitude, at latitude ϕ, corresponds to a length of

$$\frac{\pi}{180} R_p$$

The rotational angular velocity of the Earth (with respect to the stars, *not* with respect to the vernal equinox) is

$$\omega = 7.292\,115\,018 \times 10^{-5} \text{ radian/second.}$$

Strictly speaking, this is the value at the epoch 1989.5 [1]. It decreases slowly with time because the rotation of the Earth is slowing down — see Chapter 9.

The linear velocity of a point at latitude ϕ, due to the rotation of the Earth, is ωR_p per second.

The radius of curvature of the Earth's meridian, at latitude ϕ, is

$$R_m = \frac{a\,(1 - e^2)}{(1 - e^2 \sin^2 \phi)^{3/2}}$$

and one degree of latitude corresponds to a length of $\frac{\pi}{180} R_m$.

R_m reaches a minimum value at the equator, namely $a\,(1 - e^2)$ = 6335.44 km, and a maximum value at the poles, $a / \sqrt{1 - e^2}$ = 6399.60 kilometers.

Example 10.b — For $\phi = +42°$, which is the latitude of Chicago, we find

R_p = 4747.001 km

1° of longitude = 82.8508 km

linear velocity = ωR_p = 0.34616 km/s

R_m = 6364.033 km

1° of latitude = 111.0733 km

Distance between two points on the Earth's surface

If the geographic coordinates of two points on the surface of the Earth are known, the shortest distance s between these points, measured along the Earth's surface, can be calculated.

Let the first point having longitude and latitude L_1 and ϕ_1, respectively. Let L_2 and ϕ_2 be the coordinates of the second point. We will suppose that these two points are *at sea level*.

If no great accuracy is needed, then we may consider the Earth as being spherical with a mean radius of 6371 kilometers. Find the angular distance d between the two points by means of the formula

$$\cos d = \sin \phi_1 \sin \phi_2 + \cos \phi_1 \cos \phi_2 \cos (L_1 - L_2) \qquad (10.1)$$

which is similar to formula (16.1) for the angular separation between two celestial bodies. Formula (10.1) does not work well when d is very small — see Chapter 16.

Then the required linear distance is

$$s = \frac{6371\,\pi\,d}{180} \text{ kilometers} \qquad (10.2)$$

where d is expressed in degrees.

Higher accuracy is obtained by the following method, due to H. Andoyer [2]; the relative error of the result is of the order of the square of the Earth's flattening.

As before, let a be the equatorial radius of the Earth, and f the flattening. Then calculate

$$F = \frac{\phi_1 + \phi_2}{2} \qquad G = \frac{\phi_1 - \phi_2}{2} \qquad \lambda = \frac{L_1 - L_2}{2}$$

$$S = \sin^2 G \cos^2 \lambda + \cos^2 F \sin^2 \lambda$$
$$C = \cos^2 G \cos^2 \lambda + \sin^2 F \sin^2 \lambda$$

$$\tan \omega = \sqrt{\frac{S}{C}}$$

$$R = \frac{\sqrt{SC}}{\omega} \qquad \text{where } \omega \text{ is expressed in radians}$$

$$D = 2\,\omega\,a \qquad H_1 = \frac{3R - 1}{2C} \qquad H_2 = \frac{3R + 1}{2S}$$

and the required distance will be

$$s = D\,(1 + fH_1 \sin^2 F \cos^2 G - fH_2 \cos^2 F \sin^2 G)$$

Example 10.c — Calculate the geodesic distance between the Observatoire de Paris (France) and the U.S. Naval Observatory at Washington (D.C.), adopting the following coordinates :

Paris : L_1 = 2°20'14" East = −2°20'14"
ϕ_1 = 48°50'11" North = +48°50'11"

Washington : L_2 = 77°03'56" West = +77°03'56"
ϕ_2 = 38°55'17" North = +38°55'17"

We find successively

F = +43°.878 8889
G = + 4°.957 5000
λ = −39°.701 3889

S = 0.216 426 96
C = 0.783 573 04

ω = 27°.724 274 = 0.483 879 87 radian

R = 0.851 0555

D = 6172.507 km

and finally s = 6181.63 km with a possible error of the order of

50 meters.

If we use the approximate expressions (10.1) and (10.2), we obtain

$$\cos d = 0.567\,146$$
$$d = 55^\circ.448\,55$$
$$s = 6166 \text{ km}$$

References

1. International Earth Rotation Service, *Annual Report* for 1989 (Observatoire de Paris, 1990).

2. *Annuaire du Bureau des Longitudes* pour 1950 (Paris), page 145.

Chapter 11

Sidereal Time at Greenwich

The sidereal time at the meridian of Greenwich, at 0^h Universal Time of a given date, can be obtained as follows.

Calculate the JD corresponding to that date at 0^h UT (Chapter 7). Thus, this is a number ending on .5. Then find T by

$$T = \frac{\text{JD} - 2451\,545.0}{36525} \tag{11.1}$$

The *mean* sidereal time at Greenwich at 0^h UT is then given by the following expression which was adopted in 1982 by the International Astronomical Union :

$$\begin{aligned}\theta_0 = 6^h41^m50^s.54841 &+ 8640\,184^s.812\,866\,T \\ &+ 0^s.093\,104\,T^2 - 0^s.000\,0062\,T^3\end{aligned} \tag{11.2}$$

Expressed in *degrees* and decimals, this formula can be written

$$\begin{aligned}\theta_0 = 100.460\,618\,37 &+ 36\,000.770\,053\,608\,T \\ &+ 0.000\,387\,933\,T^2 - T^3/38\,710\,000\end{aligned} \tag{11.3}$$

It is important to note that the formulae (11.2) and (11.3) are valid only for those values of T which correspond to 0^h UT of a date.

To find the sidereal time at Greenwich for any instant UT of a given date, multiply that instant by 1.002 737 909 35, and add the result to the sidereal time at 0^h UT.

The mean sidereal time at Greenwich, expressed in *degrees*, can also be found directly for any instant as follows. If JD is the Julian Day corresponding to that instant (not necessarily 0^h UT), find T by formula (11.1), and then

$$\theta_0 = 280.460\,618\,37 + 360.985\,647\,366\,29\,(\mathrm{JD} - 2451\,545.0) + 0.000\,387\,933\,T^2 - T^3/38\,710\,000 \qquad (11.4)$$

If high accuracy is needed, this formula requires the use of a computer working with a sufficient number of significant digits.

The sidereal time obtained by formulae (11.2), (11.3) or (11.4) is the *mean* sidereal time, that is, the Greenwich hour angle of the mean vernal point (the intersection of the ecliptic of the date with the mean equator of the date).

The *apparent* sidereal time, or the Greenwich hour angle of the true vernal equinox, is obtained by adding the correction $\Delta\psi \cos \varepsilon$, where $\Delta\psi$ is the nutation in longitude, and ε the true obliquity of the ecliptic (see Chapter 21). This correction for nutation is called the *nutation in right ascension* or *equation of the equinoxes*. Because $\Delta\psi$ is a small quantity, the value of ε may be taken to the nearest 10" here.

If $\Delta\psi$ is expressed in arcseconds (seconds of a degree), the correction in seconds of time is

$$\frac{\Delta\psi \cos \varepsilon}{15}$$

Example 11.a — Find the mean and the apparent sidereal time at Greenwich on 1987 April 10 at 0^h UT.

This date corresponds to JD 2446 895.5, and formula (11.1) gives

$$T = -0.127\,296\,372\,348$$

We then find by means of formula (11.2)

$$\theta_0 = 6^h41^m50^s.54841 - 1099\,864.18158 \text{ seconds}$$

or, by adding a convenient multiple of 86 400 seconds (the number of seconds in one day),

$$\begin{aligned}\theta_0 &= 6^h41^m50^s.54841 + 23\,335^s.81842 \\ &= 6^h41^m50^s.54841 + 6^h28^m55^s.81842 \\ &= 13^h10^m46^s.3668\end{aligned}$$

which is the required mean sidereal time.

From Example 21.a we have, for the same instant, $\Delta\psi = -3''.788$ and $\varepsilon = 23°26'36''.85$. [In fact, these values are for 0^h TD, not for 0^h UT, but here we will neglect the very small variation of $\Delta\psi$ during the time interval ΔT = TD − UT.]

Hence the nutation in right ascension is $\dfrac{-3.788}{15} \cos 23°.44357$ $= -0^s.2317$, and the required apparent sidereal time is

$$13^h10^m46^s.3668 - 0^s.2317 = 13^h10^m46^s.1351$$

Example 11.b — Find the mean sidereal time at Greenwich on 1987 April 10 at $19^h21^m00^s$ UT.

First, we calculate the mean sidereal time for that date at 0^h Universal Time. We find $13^h10^m46^s.3668$ (see the previous Example). Then

$$\begin{aligned} & 1.002\,737\,909\,35 \times 19^h21^m00^s \\ &= 1.002\,737\,909\,35 \times 69\,660 \text{ seconds} \\ &= 69\,850.7228 \text{ seconds} \\ &= 19^h24^m10^s.7228 \end{aligned}$$

and the required sidereal time is

$$\begin{aligned} 13^h10^m46^s.3668 + 19^h24^m10^s.7228 &= 32^h34^m57^s.0896 \\ &= 8^h34^m57^s.0896 \end{aligned}$$

Alternatively, we may use formula (11.4). The Julian Day corresponding to 1987 April 10 at $19^h21^m00^s$ UT is

$$\text{JD} = 2446\,896.30625$$

and, by (11.1), the corresponding value of T is $-0.127\,274\,30$. Formula (11.4) then gives

$$\theta_o = -1677\,831^\circ.262\,1266$$

or, by adding a convenient multiple of 360°,

$$\theta_o = 128^\circ.737\,8734$$

This is the required mean sidereal time in degrees. We obtain it in hours by dividing it by 15 (since one hour corresponds to 15°) :

$$\theta_o = 8^h.582\,524\,89 = 8^h34^m57^s.0896,$$

the same result as above.

Chapter 12

Transformation of Coordinates

We will use the following symbols :

α = right ascension. This quantity is generally expressed in hours, minutes and seconds of time, and hence should first be converted into degrees (and decimals) and then, if necessary, into radians, before it is used in a formula. Conversely, if α has been obtained by means of a formula and a calculating machine, it is expressed in radians or in degrees; it may be converted to hours by division of the degrees by 15, and then, if necessary, be converted into hours, minutes and seconds;

δ = declination, positive if north of the celestial equator, negative if south;

α_{1950} = right ascension referred to the standard equinox of B1950.0;

δ_{1950} = declination referred to the standard equinox of B1950.0;

α_{2000} = right ascension referred to the standard equinox of J2000.0;

δ_{2000} = declination referred to the standard equinox of J2000.0;

λ = ecliptical (or celestial) longitude, measured from the vernal equinox along the ecliptic;

β = ecliptical (or celestial) latitude, positive if north of the ecliptic, negative if south;

l = galactic longitude;

b = galactic latitude;

h = altitude, positive above the horizon, negative below;

A = azimuth, measured westwards from the *South*. It should be noted that the navigators and the meteorologists count the compass direction, or azimuth, from the North (0°), through East (90°), South (180°) and West (270°). But astronomers disagree (*) and we shall measure the azimuth from the South, because the hour angles too are measured from the South. Hence, a celestial body which is exactly in the southern meridian has $A = H = 0°$;

ε = obliquity of the ecliptic; this is the angle between the ecliptic and the celestial equator. The mean obliquity of the ecliptic is given by formula (21.2). If, however, the *apparent* right ascension and declination are used (that is, affected by the aberration and the nutation), the true obliquity $\varepsilon + \Delta\varepsilon$ should be used (see Chapter 21). If α and δ are referred to the standard equinox of J2000.0, then the value of ε for that epoch should be used, namely $\varepsilon_{2000} = 23°26'21''.448 = 23°.439\,2911$. For the standard equinox of B1950.0, we have $\varepsilon_{1950} = 23°.445\,7889$;

ϕ = the observer's latitude, positive if in the northern hemisphere, negative in the southern one;

H = the local hour angle, measured westwards from the South.

If θ is the local sidereal time, θ_0 the sidereal time at Greenwich, and L the observer's longitude (positive west, negative east from Greenwich), then the local hour angle can be calculated from

$$H = \theta - \alpha \qquad \text{or} \qquad H = \theta_0 - L - \alpha$$

If α is affected by the nutation, then the sidereal time too must be affected by it (see Chapter 11).

For the transformation from equatorial into ecliptical coordinates, the following formulae can be used:

$$\tan\lambda = \frac{\sin\alpha\cos\varepsilon + \tan\delta\sin\varepsilon}{\cos\alpha} \tag{12.1}$$

$$\sin\beta = \sin\delta\cos\varepsilon - \cos\delta\sin\varepsilon\sin\alpha \tag{12.2}$$

(*) William Chauvenet, on page 20 of his *A Manual of Spherical and Practical Astronomy* (5th edition, 1891), Vol. I: "The origin from which azimuths are reckoned is arbitrary; so also is the direction in which they are reckoned; but astronomers usually take the *south* point of the horizon as the origin, ... Navigators, however, usually reckon the azimuth from the north or south points, according as they are in north or south latitude."

S. Newcomb, on p. 95 of his *Compendium of Spherical Astronomy*: "in practice it is measured either from the north or the south point, and in either direction, east or west." — so this great American astronomer had no specific preference.

A. Danjon, on p. 39 of his excellent *Astronomie Générale* (Paris, 1959): "Le point S, origine des azimuts, (...) est l'intersection du méridien et de l'horizon, au sud."

Note on the geographical longitudes

In this work, the geographical longitudes are measured **positively westwards** from the meridian of Greenwich, and negatively to the east. This convention has been followed by most astronomers during more than one century — see for instance References 1 to 6. For example, the longitude of Washington, D.C., is +77°04'; that of Vienna, Austria, is -16°23'.

We cannot understand why the International Astronomical Union, having first decided to measure all planetographic longitudes in the direction opposite to that of rotation, then alters the system for the Earth (1982). We shall **not** follow this IAU resolution, and we shall continue to consider **west** longitudes as positive. This is in conformity with the longitude systems on the other planets. On Mars and Jupiter, for instance, the longitudes **are** measured positively to the west, and this is why the longitude of their central meridian, as seen from the Earth, is **increasing** with time.

Transformation from ecliptical into equatorial coordinates :

$$\tan \alpha = \frac{\sin \lambda \cos \varepsilon - \tan \beta \sin \varepsilon}{\cos \lambda} \tag{12.3}$$

$$\sin \delta = \sin \beta \cos \varepsilon + \cos \beta \sin \varepsilon \sin \lambda \tag{12.4}$$

Calculation of the local horizontal coordinates :

$$\tan A = \frac{\sin H}{\cos H \sin \phi - \tan \delta \cos \phi} \tag{12.5}$$

$$\sin h = \sin \phi \sin \delta + \cos \phi \cos \delta \cos H \tag{12.6}$$

If one wishes to reckon the azimuth from the North instead of the South, add 180° to the value of A given by formula (12.5).

Transformation from horizontal into equatorial coordinates :

$$\tan H = \frac{\sin A}{\cos A \sin \phi + \tan h \cos \phi}$$

$$\sin \delta = \sin \phi \sin h - \cos \phi \cos h \cos A$$

The current galactic system of coordinates has been defined by the International Astronomical Union in 1959. In the standard equatorial system of B1950.0, the galactic (Milky Way) North Pole has the coordinates

$$\alpha_{1950} = 12^h49^m = 192^\circ.25, \qquad \delta_{1950} = +27^\circ.4$$

and the origin of the galactic longitudes is the point (in western Sagittarius) of the galactic equator which is 33° distant from the ascending node (in western Aquila) of the galactic equator with the equator of B1950.0.

These values have been fixed *conventionally* and therefore must be considered as *exact* for the mentioned equinox of B1950.0.

Transformation from equatorial coordinates, referred to the standard equinox of B1950.0, into galactic coordinates:

$$\tan x = \frac{\sin(192°.25 - \alpha)}{\cos(192°.25 - \alpha)\sin 27°.4 - \tan\delta\cos 27°.4}$$

$$l = 303° - x \tag{12.7}$$

$$\sin b = \sin\delta\sin 27°.4 + \cos\delta\cos 27°.4\cos(192°.25 - \alpha) \tag{12.8}$$

Transformation from galactic coordinates into equatorial coordinates referred to the standard equinox of B1950.0 :

$$\tan y = \frac{\sin(l - 123°)}{\cos(l - 123°)\sin 27°.4 - \tan b\cos 27°.4}$$

$$\alpha = y + 12°.25$$

$$\sin\delta = \sin b\sin 27°.4 + \cos b\cos 27°.4\cos(l - 123°)$$

If the 2000.0 mean place of the star is given instead of the 1950.0 mean place, then, before using formulae (12.7) and (12.8), convert α_{2000} and δ_{2000} to α_{1950} and δ_{1950}. See Chapter 20.

The formulae (12.1), (12.3), etc., give $\tan\lambda$, $\tan\alpha$, etc., and then λ, α, etc., by the function arctangent. However, the exact quadrant in which the angle is situated is then unknown. To remove the ambiguity of 180°, apply the ATN2 function to the numerator and the denominator of the fraction (instead of performing the actual division), or use another trick — see 'The correct quadrant' in Chapter 1.

Example 12.a — Calculate the ecliptical coordinates of the star Pollux (β Gem), whose equatorial coordinates are

$$\alpha_{2000} = 7^h45^m18^s.946, \quad \delta_{2000} = +28°01'34''.26.$$

Using the values $\alpha = 116°.328\,942$, $\delta = +28°.026\,183$, and $\varepsilon = 23°.439\,2911$, formulae (12.1) and (12.2) give

$$\tan\lambda = \frac{+1.034\,039\,86}{-0.443\,523\,98} \quad \text{whence} \quad \lambda = 113°.215\,630\ ;$$

$$\beta = +6°.684\,170.$$

Because α and δ are referred to the standard equinox of 2000.0, λ and β too are referred to that equinox.

Exercise. — Using the values of λ and β found above, find α and δ again by means of formulae (12.3) and (12.4).

Example 12.b — Find the azimuth and the altitude of Venus on 1987 April 10 at $19^h21^m00^s$ UT at the U.S. Naval Observatory at Washington, D.C. (longitude = $+77°03'56''$ = $+5^h08^m15^s.7$, latitude = $+38°55'17''$).

The planet's apparent equatorial coordinates, interpolated from an ephemeris, are

$$\alpha = 23^h09^m16^s.641, \qquad \delta = -6°43'11''.61$$

These are the *apparent* right ascension and declination of the planet. We need the *apparent* sidereal time for the given instant.

We first calculate the *mean* sidereal time at Greenwich on 1987 April 10 at $19^h21^m00^s$ UT, and find $8^h34^m57^s.0896$ (see Example 11.b).

By means of the method described in Chapter 21, we find for the same instant :

nutation in longitude : $\Delta\psi = -3''.868$

true obliquity of the ecliptic : $\varepsilon = 23°26'36''.87$

The apparent sidereal time at Greenwich is

$$\theta_0 = 8^h34^m57^s.0896 + \left(\frac{-3.868}{15} \cos\varepsilon\right) \text{seconds} = 8^h34^m56^s.853$$

Hour angle of Venus at Washington :

$$\begin{aligned} H &= \theta_0 - L - \alpha \\ &= 8^h34^m56^s.853 - 5^h08^m15^s.7 - 23^h09^m16^s.641 \\ &= -19^h42^m35^s.488 = -19^h.709\,8578 = -295°.647\,867 \\ &= +64°.352\,133 \end{aligned}$$

Formulae (12.5) and (12.6) then give

$$\tan A = \frac{+0.901\,4712}{+0.363\,6015} \qquad \text{whence} \quad A = +68°.0337$$

$$h = +15°.1249$$

so the planet is 15 degrees above the horizon between the southwest and the west.

It should be noted that formula (12.6) does not take into account the effect of the atmospheric refraction, nor that of the planet's parallax, nor the dip of the horizon. For the atmospheric refraction, see Chapter 15. The correction for parallax is dealt with in Chapter 39.

As an exercise, find the galactic coordinates of Nova Serpentis 1978, whose equatorial coordinates are

$$\alpha_{1950} = 17^h48^m59^s.74, \qquad \delta_{1950} = -14°43'08''.2$$

Answer : $l = 12°.9593$, $b = +6°.0463$.

Ecliptic and Horizon

If ε is the obliquity of the ecliptic, ϕ the latitude of the observer, and θ the local sidereal time, then the longitudes of the two points of the ecliptic which are (180° apart) on the horizon, are given by

$$\tan\lambda = \frac{-\cos\theta}{\sin\varepsilon\ \tan\phi + \cos\varepsilon\ \sin\theta} \tag{12.9}$$

The angle I between the ecliptic and the horizon is given by

$$\cos I = \cos\varepsilon\ \sin\phi - \sin\varepsilon\ \cos\phi\ \sin\theta \tag{12.10}$$

In the course of one sidereal day, the angle I varies between two extreme values. For example, for latitude 48°00' N, with ε = 23°26', the extreme values of I are

$$90° - \phi + \varepsilon = 65°26' \quad \text{for} \quad \theta = 90°$$
$$90° - \phi - \varepsilon = 18°34' \quad \text{for} \quad \theta = 270°$$

It should be noted that I is *not* the angle which the daily path of the Sun makes with the horizon.

Example 12.c — For $\varepsilon = 23°.44$, $\phi = +51°$, $\theta = 5^h00^m = 75°$, we find, from formula (12.9),

$$\tan\lambda = -0.1879, \quad \text{whence} \quad \lambda = 169°21' \quad \text{and} \quad \lambda = 349°21'.$$

Formula (12.10) gives $I = 62°$.

References

1. *The Nautical Almanac and Astronomical Ephemeris for the year 1835*, p. 508 (London, 1833).
2. *The American Ephemeris and Nautical Almanac for the Year 1857*, p. 491 (Washington, 1854).
3. *The Astronomical Ephemeris for the Year 1960*, pp. 434 & fol. (London, 1958).
4. W. Chauvenet, *A Manual of Spherical and Practical Astronomy*, Vol. I, pp. 317 & fol. (Philadelphia, 1891).
5. A. Danjon, *Astronomie Générale*, p. 46 (Paris, 1959).
6. S. Newcomb, *A Compendium of Spherical Astronomy*, p. 119 (New York, 1906).

Chapter 13

The Parallactic Angle

Suppose that on a bright morning we are looking at the Sun through a piece of dark glass, and that we see a large sunspot near the western ('right') limb of the Sun (Fig. 1, A). At noon, the Sun being near the southern meridian, we note that the spot is lower (Fig. 1, B). And in the afternoon, we see that the spot has moved still farther along the Sun's limb (Fig. 1, C).

The spot did not actually move that much over the solar disk. It is the whole image of the Sun which rotated clockwise. This can be seen easier with the Moon (Figure 2).

This apparent rotation is easily understood when we consider the diurnal motion of the celestial sphere. Each celestial body describes a parallel circle, a diurnal arc (Figure 3). Only when the Sun (or the Moon) is exactly on the southern meridian, will the celestial north be up, in the direction of the zenith.

The constellations show a similar effect. For an observer in the northern hemisphere of the Earth, the constellation of Orion is inclined to the 'left' in the southeast, is upright in the south, and

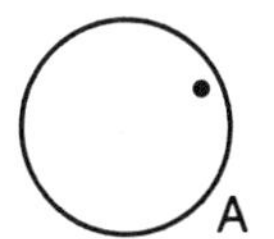

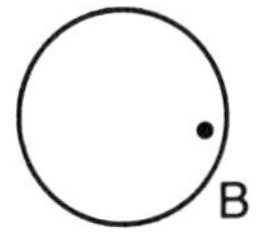

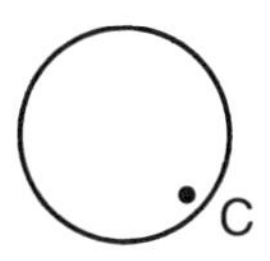

Fig. 1. *The apparent displacement of a sunspot in the course of the day: in the morning (A), near noon (B), and in the afternoon (C). In each of the three sketches, the circle represents the solar disk, and the zenith is at the top.*

Fig. 2 *The First-Quarter Moon for an observer in the northern hemisphere: (A) near the south, around the time of sunset; and (B) later that evening. The zenith is up.*

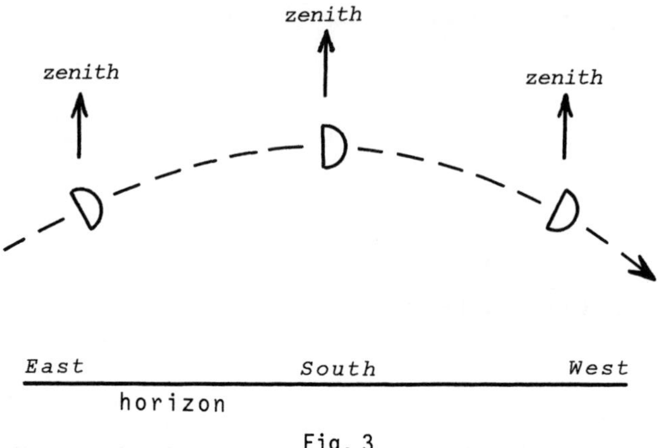

Fig. 3

is inclined to the 'right' in the southwest.

In Figure 4, the circle represents the disk of the Sun (or that of the Moon). The arc AB is a part of its diurnal arc on the celestial sphere. C is the center of the disk. The direction of the zenith and that of the celestial North are indicated. The latter direction is perpendicular to the arc AB. Z is the *zenith point* of the disk; it is the uppermost point of the disk at the sky as seen by the observer at the given instant. N is the *north point* of the disk; the direction CN points towards the northern celestial pole.

The angle ZCN is called the *parallactic angle* and is generally designed by q. This parallactic angle has absolutely nothing to do with the parallax! The name arises from the fact that the celestial body moves along a *parallel* circle. Compare with the 'parallactic' mounting of a telescope.

By convention, the angle q is negative before, and positive after the passage through the southern meridian. Exactly on the meridian, we have $q = 0°$.

The parallactic angle q can be calculated by means of the formula

$$\tan q = \frac{\sin H}{\tan \phi \cos \delta - \sin \delta \cos H} \tag{13.1}$$

where, as in the preceding Chapter, ϕ is the geographical latitude of the observer, δ the declination of the celestial body, and H its hour angle at the given instant.

Exactly in the zenith, the angle q is not defined. Indeed, in that case we have $H = 0°$ and $\delta = \phi$, so formula (13.1) yields $\tan q = 0/0$. This can be compared with somebody who is exactly at the North Pole of the Earth: his geographical longitude is not defined, because all meridians of the Earth converge to his place. For that special observer, all points of the horizon are in the southern direction!

When a celestial body passes exactly through the zenith, the parallactic angle q suddenly jumps from $-90°$ to $+90°$.

If the celestial body is on the horizon (hence rising or setting), formula (13.1) simplifies greatly, namely

$$\cos q = \frac{\sin \phi}{\cos \delta}$$

and in that case it is not necessary to know the value of the hour angle H.

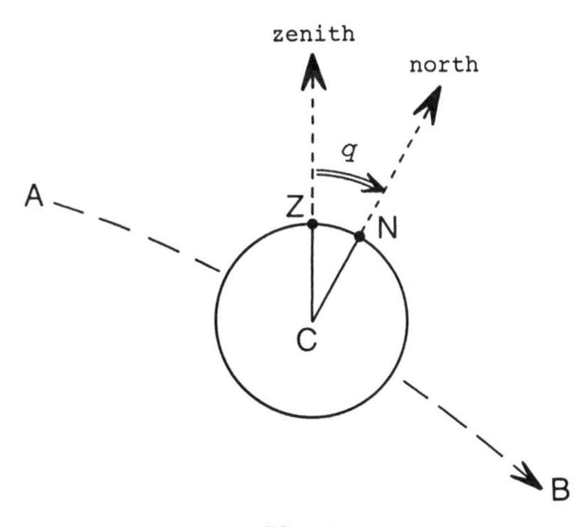

Fig. 4

Chapter 14

Rising, Transit and Setting

The hour angle corresponding to the time of rise or set of a celestial body is obtained by putting $h = 0$ in formula (12.6). This gives

$$\cos H_0 = -\tan \phi \ \tan \delta$$

However, the instant so obtained refers to the geometric rise or set of the center of the celestial body. By reason of the atmospheric refraction, the body is actually below the horizon at the instant of its apparent rise or set. The value of 0°34' is generally adopted for the effect of refraction at the horizon. For the Sun, the calculated times generally refer to the apparent rise or set of the upper limb of the disk; hence, 0°16' should be added for the semidiameter.

Actually, the amount of refraction changes with temperature, pressure, and the elevation of the observer (see Chapter 15). A change of temperature from winter to summer can shift the times of sunrise and sunset by about 20 seconds in mid-northern and mid-southern latitudes. Similarly, observing sunrise or sunset over a range of barometric pressures leads to a variation of a dozen seconds in the times. However, in this Chapter we shall use a mean value for the atmospheric refraction at the horizon, namely the value of 0°34' mentioned above.

We will use the following symbols :

L = geographic longitude of the observer in degrees, measured *positively west from Greenwich*, negatively to the east;

ϕ = geographic latitude of the observer, positive in the northern hemisphere, negative in the southern hemisphere;

ΔT = the difference TD - UT in *seconds* of time;

h_0 = the 'standard' altitude, *i.e.* the geometric altitude of the center of the body at the time of apparent rising or setting, namely

$h_0 = -0°34' = -0°.5667$ for stars and planets;

$h_0 = -0°50' = -0°.8333$ for the Sun.

For the Moon, the problem is more complicated because h_0 is not constant. Taking into account the variations of semidiameter and parallax, we have for the Moon

$$h_0 = 0.7275\,\pi - 0°34'$$

where π is the Moon's horizontal parallax. If no great accuracy is required, the mean value $h_0 = +0°.125$ can be used for the Moon.

Suppose we wish to calculate the times, in *Universal Time*, of rising, of transit (when the body crosses the local meridian at upper culmination) and of setting of a celestial body at the observer's place on a given date D. We take the following values from an almanac, or we calculate them ourselves with a computer program :

— the apparent sidereal time θ_0 at 0^h *Universal Time* on day D for the Greenwich meridian, converted into *degrees* ;

— the apparent right ascensions and declinations of the body

α_1 and δ_1 on day $D-1$ at 0^h Dynamical Time
α_2 and δ_2 on day D —
α_3 and δ_3 on day $D+1$ —

The right ascensions should be expressed in *degrees* too.

We first calculate *approximate* times as follows.

$$\cos H_0 = \frac{\sin h_0 - \sin\phi\,\sin\delta_2}{\cos\phi\,\cos\delta_2} \tag{14.1}$$

Attention! First test if the second member is between -1 and +1 before calculating H_0. See Note 2 at the end of this Chapter.

Express H_0 in degrees. H_0 should be taken between 0° and +180°. Then we have :

$$\left.\begin{aligned}
&\text{for the transit :} & m_0 &= \frac{\alpha_2 + L - \theta_0}{360} \\
&\text{for the rising :} & m_1 &= m_0 - \frac{H_0}{360} \\
&\text{for the setting :} & m_2 &= m_0 + \frac{H_0}{360}
\end{aligned}\right\} \tag{14.2}$$

These three values m are times, on day D, expressed as fractions of a day. Hence, they should be between 0 and +1. If one or more of them are outside of this range, add or subtract 1. For instance, +0.3744 should remain unchanged, but -0.1709 should be changed to +0.8291, and +1.1853 should be changed to +0.1853.

Now, for *each* of the three m-values *separately*, perform the following calculation.

Find the sidereal time at Greenwich, in *degrees*, from

$$\theta = \theta_0 + 360.985\,647\,m$$

where m is either m_0, m_1 or m_2.

For $n = m + \Delta T/86400$, interpolate α from α_1, α_2, α_3 and δ from δ_1, δ_2, δ_3, using the interpolation formula (3.3). For the calculation of the time of transit, δ is not needed.

Find the local hour angle of the body from $H = \theta - L - \alpha$, and then the body's altitude h by means of formula (12.6). This altitude is not needed for the calculation of the time of transit.

Then the correction to m will be found as follows :

— in the case of a transit,

$$\Delta m = -\frac{H}{360}$$

where H is expressed in degrees and *must* be between -180 and +180 degrees. (In most cases, H will be a small angle and be between -1° and +1°);

— in the case of a rising or a setting,

$$\Delta m = \frac{h - h_0}{360 \cos\delta \cos\phi \sin H}$$

where h and h_0 are expressed in degrees.

The corrections Δm are small quantities, in most cases being between -0.01 and +0.01.

The corrected value of m is then $m + \Delta m$. If necessary, a new calculation should be performed using the new value of m.

At the end of the calculation, each value of m should be converted into hours by multiplication by 24.

Example 14.a — Venus on 1988 March 20 at Boston,

longitude = +71°05' = +71°.0833,
latitude = +42°20' = +42°.3333.

From an accurate ephemeris, we take the following values :

1988 March 20, 0^h UT : $\theta_0 = 11^h50^m58^s.10 = 177°.74208$

1988 March 19, 0^h TD :	α_1 =	$2^h42^m43^s.25$	= 40°.68021
March 20, 0^h	α_2 =	2 46 55.51	= 41.73129
March 21, 0^h	α_3 =	2 51 07.69	= 42.78204

1988 March 19, 0^h TD :	δ_1 = +18°02′51″.4	= +18°.04761
March 20, 0^h	δ_2 = +18 26 27.3	= +18.44092
March 21, 0^h	δ_3 = +18 49 38.7	= +18.82742

We take $h_\circ = -0°.5667$, $\Delta T = +56^s$, and find by formula (14.1) $\cos H_\circ = -0.317\,8735$, $H_\circ = 108°.5344$, whence the approximate values :

transit : $m_\circ = -0.18035$, whence $m_\circ = +0.81965$
rising : $m_1 = m_\circ - 0.30148 = +0.51817$
setting : $m_2 = m_\circ + 0.30148 = +1.12113$, whence +0.12113

Calculation of more exact times :

		rising	transit	setting
	m	+0.51817	+0.81965	+0.12113
	θ	4°.79401	113°.62397	221°.46827
	n	+0.51882	+0.82030	+0.12178
inter-	α	42°.27648	42°.59324	41°.85927
polation	δ	+18°.64229		+18°.48835
	H	−108°.56577	−0°.05257	+108°.52570
	h	−0°.44393		−0°.52711
	Δm	−0.00051	+0.00015	+0.00017
corrected	m	+0.51766	+0.81980	+0.12130

A new calculation, using these new values of m, yields the new corrections −0.000 003, −0.000 004, and −0.000 004, respectively, which can be neglected. So we have, finally :

rising : $m_1 = +0.51766$, $24^h \times 0.51766 = 12^h25^m$ UT
transit : $m_\circ = +0.81980$, $24^h \times 0.81980 = 19^h41^m$ UT
setting : $m_2 = +0.12130$, $24^h \times 0.12130 = 2^h55^m$ UT

NOTES

1. In Example 14.*a* we found that at Boston the time of setting was 2^h55^m UT on March 20. However, converted to *local* standard time this corresponds to an instant on the evening of the previous day! If really the time of setting on March 20 is needed in local time, the calculation should be performed using the value $m_2 = +1.12113$ first found, instead of +0.12113.

2. If the body is circumpolar, the second member of formula (14.1) will be larger than 1 in absolute value, and there will be no angle $H_\circ$. In such a case, the body will remain the whole day either above or below the horizon.

3. If *approximate* times are sufficient, just use the *initial* values $m_\circ$, m_1 and m_2 given by (14.2).

Chapter 15

Atmospheric Refraction

Atmospheric refraction is the bending of light while passing through the Earth's atmosphere. As a ray of light penetrates the atmosphere, it encounters layers of air of increasing density, resulting in the continuous bending of the light. As a result, a star (or the Sun's limb, etc.) will appear higher in the sky than its true position. The atmospheric refraction, which is zero in the zenith, increases towards the horizon. At an altitude of 45°, the refraction is about one arcminute; at the horizon, it amounts to about 35'. Thus the Sun and the Moon are actually below the horizon when they appear to be rising. Moreover, the rapidly changing refraction at low altitudes gives the rising or setting Sun its familiar oval appearance.

Allowance must be made for atmospheric refraction when determining positions, and one distinguishes two cases :

— the apparent altitude h_0 of a celestial body has been *measured*, and one should find the refraction R to be *subtracted* from h_0 to obtain the true altitude h;

— the true 'airless' altitude h has already been *calculated* from celestial coordinates and formulae of spherical trigonometry, and we want to calculate the refraction R to be *added* to h in order to predict the apparent altitude h_0.

Almost all refraction formulae we have come across consider the first case only: they are designed for deriving true altitudes from observed ones. But here we will consider both cases.

For many purposes, 'average' meteorological conditions may be assumed. However, anomalous refraction near the horizon, exemplified by distortions of the setting Sun, should remind us that rigorous exactness at very low altitudes cannot be reached.

When the altitude of the celestial body is larger than 15°, one of the following two formulae may be used, as the case may be :

$$R = 58''.294 \tan(90° - h_0) - 0''.0668 \tan^3(90° - h_0) \qquad (15.1)$$

$$R = 58''.276 \tan(90° - h) - 0''.0824 \tan^3(90° - h) \qquad (15.2)$$

The first formula is given by Smart [1], while the second one has been derived by us from the first formula. For altitudes below 15°, these expressions will give inaccurate, or even completely meaningless results.

It appears that, at high altitudes, the refraction is proportional to the tangent of the zenithal distance.

A surprisingly simple formula for refraction, with good accuracy at all altitudes from 90° to 0°, was given by G.G. Bennett of the University of New South Wales [2]. If the refraction R is expressed in *minutes* of arc, Bennett's formula is

$$R = \frac{1}{\tan\left(h_0 + \dfrac{7.31}{h_0 + 4.4}\right)} \tag{15.3}$$

where h_0 is the *apparent* altitude in degrees. According to Bennett, this formula is accurate to 0.07 arcminute for all values of h_0. The largest error, 0.07 arcminute, occurs at 12° altitude.

It should be noted that for the zenith ($h_0 = 90°$) formula (15.3) yields $R = -0''.08$ instead of exactly zero. This can be rectified by adding +0.001 3515 to the second member of the formula.

Bennett also showed how his formula can be refined. Calculate R by means of formula (15.3); then a correction to R, expressed in minutes of arc, is

$$-0.06 \sin(14.7R + 13)$$

where the expression between parentheses is expressed in degrees. Calculated in this way, the maximum error is stated to be only 0.015 arcminute, or 0''.9, for the whole altitude range 90°–0°. [At the zenith, one finds $R = -0''.89$, so expression (15.3), without further correction, is better in this case.]

For the inverse problem, that of calculating the effect of refraction when the *true* altitude h is known, Sæmundsson, of the University of Iceland, proposes the following formula [3] :

$$R = \frac{1.02}{\tan\left(h + \dfrac{10.3}{h + 5.11}\right)} \tag{15.4}$$

This formula is consistent with Bennett's (15.3) to within 4''. Again, it does not give exactly $R = 0$ for $h = 90°$. This can be remedied by adding +0.001 9279 to the second member.

Formulae (15.1) to (15.4) assume that the observation is made at sea level, when the atmospheric pressure is 1010 millibars, and when the temperature is 10° Celsius. The effect of refraction increases when the pressure *increases* or when the temperature *decreases*.

If the pressure at the Earth's surface is P millibars, and the air temperature is T degrees Celsius, then the values of R given by the formulae (15.1) to (15.4) should be multiplied by

$$\frac{P}{1010} \times \frac{283}{273 + T}$$

However, this is only approximately correct. The problem is more complicated because the refraction depends on the wave-length of the light too! The expressions given in this Chapter are for yellow light, where the human eye has maximum sensitivity.

Example 15.a — Calculate the apparent flattening of the solar disk near the horizon, when the lower limb is at an apparent altitude of exactly 0°30'. Assume a true solar diameter of exactly 0°32', and mean conditions of air pressure and temperature.

For $h_0 = 0°.5$, formula (15.3) gives $R = 28'.754$, so the true altitude of the Sun's lower limb is

$$0°30' - 0°28'.754 = 0°01'.246$$

and hence the true altitude of the upper limb is

$$h = 0°01'.246 + 0°32' = 0°33'.246 = 0°.5541$$

For this value of h, formula (15.4) yields $R = 24'.618$, so the apparent altitude of the Sun's upper limb is $33'.246 + 24'.618 = 57'.864$, and the apparent vertical diameter of the solar disk is $57'.864 - 30' = 27'.864$.

Consequently, the ratio of the apparent vertical diameter to the horizontal diameter of the solar disk, under the conditions of this Problem, is $27.864/32 = 0.871$.

It should be noted that, while of course the azimuth is unchanged by refraction, the *horizontal* diameter of the solar disk is very slightly contracted by reason of the refraction. This is due to the fact that the extremities of this diameter are raised along vertical circles that meet at the zenith. Danjon [4] writes that the apparent contraction of the horizontal diameter of the Sun is practically constant and independent of the altitude, and that this contraction is approximately 0".6.

For heights of a few degrees the results of the formulae should be judged with care. Near the horizon unpredictable disturbances of the atmosphere become rather important. According to investigations by Schaefer and Liller [5], the refraction at the horizon fluctuates by 0°.3 around a mean value normally, and in some cases apparently much more. Remembering our Chapter about accuracy, it should be mentioned here that giving rising or setting times of a body more accurately than to the nearest minute makes no sense.

References

1. W.M. Smart, *Text-Book on Spherical Astronomy*; Cambridge (Engl.), University Press (1956); page 68.

2. G.G. Bennett, 'The Calculation of Astronomical Refraction in Marine Navigation', *Journal of the Institute for Navigation*, Vol. 35, pages 255-259 (1982).

3. Þorsteinn Sæmundsson, *Sky and Telescope*, Vol. 72, page 70 (July 1986).

4. A. Danjon, *Astronomie Générale* (Paris, 1959); page 156.

5. B.E. Schaefer, W. Liller, 'Refraction near the Horizon', *Publ. Astron. Society of the Pacific*, Vol. 102, pages 796-805 (July 1990).

Chapter 16

Angular Separation

The angular distance d between two celestial bodies, whose right ascensions and declinations are known, is given by the formula

$$\cos d = \sin \delta_1 \sin \delta_2 + \cos \delta_1 \cos \delta_2 \cos(\alpha_1 - \alpha_2) \qquad (16.1)$$

where α_1 and δ_1 are the right ascension and declination of one body, and α_2 and δ_2 those of the other body.

The same formula may be used when the ecliptical (celestial) longitudes λ and latitudes β of the two bodies are given, provided that α_1, α_2, δ_1 and δ_2 are replaced by λ_1, λ_2, β_1 and β_2, respectively.

Formula (16.1) may not be used when d is very near to 0° or to 180° because in those cases $|\cos d|$ is nearly equal to 1 and varies very slowly with d, so that d cannot be found accurately. For instance,

$$\cos 0°01'00'' = 0.999\,999\,958$$
$$\cos 0°00'30'' = 0.999\,999\,989$$
$$\cos 0°00'15'' = 0.999\,999\,997$$
$$\cos 0°00'00'' = 1.000\,000\,000$$

If the angular separation is small, say less than 0°10', then this separation should be calculated from

$$d = \sqrt{(\Delta\alpha \,.\, \cos\delta)^2 + (\Delta\delta)^2} \qquad (16.2)$$

where $\Delta\alpha$ is the difference between the right ascensions, $\Delta\delta$ the difference between the declinations, while δ is the average of the declinations of the two bodies. It should be noted that $\Delta\alpha$ and $\Delta\delta$ should be expressed in the same angular units.

If $\Delta\alpha$ is expressed in hours (and decimals), $\Delta\delta$ in degrees (and decimals), then d expressed in seconds of a degree (") is given by

$$d = 3600 \sqrt{(15\,\Delta\alpha \,.\, \cos\delta)^2 + (\Delta\delta)^2} \qquad (16.3)$$

If $\Delta\alpha$ is expressed in seconds of time (^{s}), and $\Delta\delta$ in seconds of a degree ($''$), then d expressed in $''$ is given by

$$d = \sqrt{(15\ \Delta\alpha \cdot \cos\delta)^2 + (\Delta\delta)^2} \tag{16.4}$$

Formulae (16.2), (16.3) and (16.4) may be used only when d is small.

However, see also the alternative formulae at the end of this Chapter.

Example 16.a — Calculate the angular distance between Arcturus (α Boo) and Spica (α Vir).

The J2000.0 coordinates of these stars are

α Boo : $\alpha_1 = 14^h15^m39^s.7 = 213°.9154$
$\delta_1 = +19°10'57'' = +19°.1825$

α Vir : $\alpha_2 = 13^h25^m11^s.6 = 201°.2983$
$\delta_2 = -11°09'41'' = -11°.1614$

Formula (16.1) gives $\cos d = +0.840\,633$, whence $d = 32°.7930 = 32°48'$.

Of course, this distance holds only for the epoch for which the stars' distances are given, namely 2000.0. It varies slowly with time, by reason of the proper motions of the stars.

Exercise. — Calculate the angular distance between Aldebaran and Antares. (Answer : 169°58')

One or both bodies may be moving objects. For example: a planet and a star, or two planets. In that case, a program may be written where first the quantities δ_1, δ_2 and $(\alpha_1 - \alpha_2)$ are interpolated, after which d is calculated by means of one of the formulae (16.1) or (16.2). Hint: from the interpolated quantities, calculate $\cos d$ by means of formula (16.1). Then, if $\cos d < 0.999\,995$, find d; but if $\cos d > 0.999\,995$, use formula (16.2).

Exercise. — Using the following coordinates, calculate the instant and the value of the least distance between Mercury and Saturn.

1978	Mercury		Saturn	
0 h TD	α_1	δ_1	α_2	δ_2
Sep 12	$10^h23^m17^s.65$	$+11°31'46''.3$	$10^h33^m01^s.23$	$+10°42'53''.5$
13	10 29 44.27	+11 02 05.9	10 33 29.64	+10 40 13.2
14	10 36 19.63	+10 29 51.7	10 33 57.97	+10 37 33.4
15	10 43 01.75	+ 9 55 16.7	10 34 26.22	+10 34 53.9
16	10 49 48.85	+ 9 18 34.7	10 34 54.39	+10 32 14.9

Answer : The least angular separation between the two planets was 0°03'44", on 1978 September 13 at $15^h06^m.5$ TD = 15^h06^m UT.

As we see, this was a rather close conjunction. We must insist on the fact that, in such a case, first the quantities δ_1, δ_2 and $(\alpha_1 - \alpha_2)$ should be interpolated, *not* the distances themselves. The distance is to be deduced from the *interpolated coordinates*.

Suppose that, nevertheless, we try to interpolate the distances themselves. By means of formula (16.1), we find the following distances, in degrees and decimals, for the five given times :

1978 Sep 12.0 TD	$d_1 = 2^\circ.5211$
13.0	$d_2 = 0.9917$
14.0	$d_3 = 0.5943$
15.0	$d_4 = 2.2145$
16.0	$d_5 = 3.8710$

It is evident that the least distance occurs between 13.0 and 14.0 September, and closer to 14.0 than to 13.0.

If we now use the *three* central values d_2, d_3, d_4 and we calculate the value of the minimum by means of formula (3.4), we obtain 0°.5017 = 0°30'06". Taking the *five* values d_1 to d_5, formula (3.9) yields a 'better' value for n_m, after which (3.8) is used to calculate the value of the function for that value of n; this gives 0°.4865 = 0°29'11".

Both results are completely wrong, however; as has been mentioned above, the exact value of the least distance is only 0°03'44". So, what happened?

The reason is that the conjunction was a close one. Until a short time before the least distance, Mercury was moving almost exactly straight towards Saturn, and the angular distance between the two planets was decreasing almost exactly linearly with time. Similarly, some short time after the least distance, Mercury was moving almost straight away from Saturn.

In the Figure on next page, the solid curve represents the true variation of the angular separation between the two planets. Except very close to the least distance, this curve consists of two almost exactly *straight* segments (one near B, the other from C to D), and in such a case the interpolation formulae are no longer valid !

Formulae (3.3), (3.4) and (3.5), for instance, suppose that the function, in the considered part of the curve, is a *parabola*. But the curve is no parabola, except very close to the minimum, inside the small rectangle.

If we make use of the three points B, C, D, corresponding to the three central distances d_2, d_3, d_4, then by the interpolation formula (3.3), we in fact draw a parabola through those three points; it is the dashed curve in the Figure. This parabola differs considerably from the true curve, and in particular its minimum is too high.

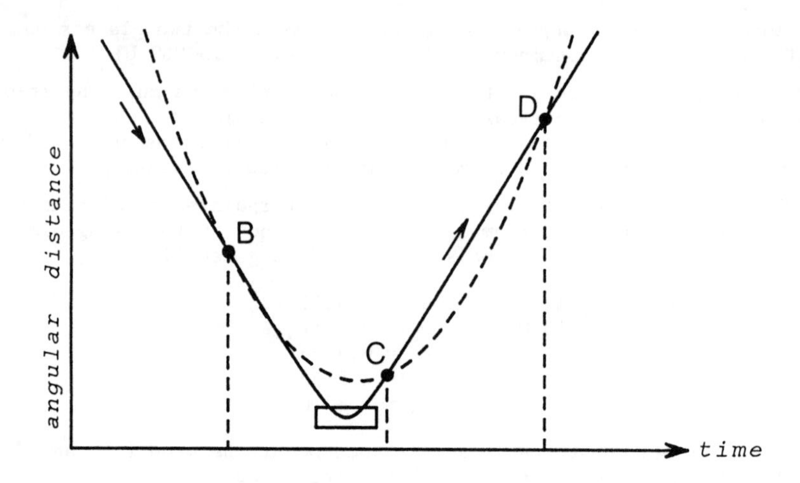

And it would be of no help to use the *five* values d_1 to d_5 instead of the three central ones, because the solid curve differs even considerably from a polynomial of the fourth degree!

Hence, performing an interpolation from the *distances* cannot give accurate results. As we have said, we must interpolate the original *coordinates* separately, and only then can the accurate distance for an intermediate instant be deduced. Using the interpolation formula (3.8), we so find the value of the distance for several values of the interpolating factor n :

n = -0.50	distance = 0°.21437
-0.45	0.14057
-0.40	0.07790
-0.35	0.07028
-0.30	0.12815

The least separation occurs for n between -0.40 and -0.35, so we calculate the angular distance for three more values, at smaller intervals :

n = -0.38	distance = 0°.06408
-0.37	0.06229
-0.36	0.06448

The tabular interval is now small enough so that formulae (3.4) and (3.5) may be used. We find that the least separation is 0°.06228 = 0°03'44", for n = -0.370 502, corresponding to September 13.629 498 = September 13 at $15^h06^m.5$ TD, as mentioned earlier.

It is possible, however, to find the angular separation without trying several values of the interpolating factor n, namely by appealing to rectangular coordinates. These coordinates u and v, in seconds of arc, can be calculated as follows [1].

Calculate the auxiliary quantity

$$K = \frac{206\,264.8062}{1 + \sin^2 \delta_1 \tan \Delta\alpha \tan \frac{\Delta\alpha}{2}}$$

where 206 264.8062 is the number of arcseconds in one radian. Then

$$u = -K\,(1 - \tan \delta_1 \sin \Delta\delta) \cos \delta_1 \tan \Delta\alpha$$

$$v = K\,(\sin \Delta\delta + \sin \delta_1 \cos \delta_1 \tan \Delta\alpha \tan \frac{\Delta\alpha}{2})$$

In the above expressions, α_1, δ_1 are the right ascension and declination of the first planet, and $\Delta\alpha = \alpha_2 - \alpha_1$, $\Delta\delta = \delta_2 - \delta_1$, where α_2, δ_2 are the right ascension and declination of the second planet.

Let us calculate the values of u and v for three equidistant times. For any intermediate time, then, their values can be interpolated by means of formula (3.3), while their variation (in arcseconds per unit of the tabular interval) is given by

$$u' = \frac{u_3 - u_1}{2} + n\,(u_1 + u_3 - 2u_2)$$

where n is the interpolating factor, and u_1, u_2, u_3 are the three calculated values of u, and with a similar expression for the variation v'.

Start from any value for the interpolating factor n; a good choice is $n = 0$. For this value of n, interpolate u and v by means of formula (3.3), and find the variations u' and v'. Then the correction to n is given by

$$\Delta n = -\frac{uu' + vv'}{u'^2 + v'^2}$$

So the new value of n is $n + \Delta n$. Repeat the calculation for the new value of n, until the correction Δn is a very small quantity, for instance less than 0.000 001 in absolute value.

For the final value of n, calculate u and v again. Then the least distance, in arcseconds, will be $\sqrt{u^2 + v^2}$.

Let us apply this method to the above-mentioned conjunction between Mercury and Saturn. The three chosen instants are 13.0, 14.0 and 15.0 September 1978. We find the following values for u and v, retaining one extra decimal to avoid rounding errors :

	u	v
Sept. 13.0	−3322″.44	−1307″.48
14.0	+2088.54	+ 463.66
15.0	+7605.36	+2401.71

For $n = 0$, we have

$$u = +2088.54 \qquad u' = +5463.90$$
$$v = +\ 463.66 \qquad v' = +1854.595$$

whence $\Delta n = -0.368\,582$, and the corrected value of n is $0 - 0.368\,582 = -0.368\,582$.

For this new value of n we find

$$u = +\ 81.83 \qquad u' = +5424.89$$
$$v = -208.57 \qquad v' = +1793.07$$

whence $\Delta n = -0.002\,142$, and the new corrected value of n is $-0.368\,582 - 0.002\,142 = -0.370\,724$.

A new iteration gives $\Delta n = -0.000\,003$, so the final value of n is $-0.370\,724 - 0.000\,003 = -0.370\,727$.

[This value differs from the value $n = -0.370\,502$ found before, because in the present calculation we used the planet's positions for only three instants instead of five. But the difference is only 0.000 225 day, or 19 seconds.]

For the value $n = -0.370\,727$, we find $u = +70''.20$, $v = -212''.42$, and consequently the least distance between the two planets is

$$\sqrt{u^2 + v^2} = 224'' = 3'44'',$$

as found before.

The same methods can be used if one of the bodies is a star. The latter's coordinates are then constant, but it is important to note that the α and δ of the star should be referred to the same equinox as that of the moving body.

If the moving body is a major planet, whose apparent right ascension and apparent declination referred to the equinox of the date are given, then for the star the apparent coordinates too must be used. If one takes the star's position from a catalogue, where it is referred to a standard equinox (for instance that of 2000.0), then the apparent α and δ are found by taking into account the proper motion of the star and the effects of precession, nutation and aberration, as explained in Chapter 22.

If the α and δ of the moving body are referred to a standard equinox (astrometric coordinates), then the α and δ of the star should be referred to this same standard equinox, the only corrections being those for the proper motion of the star.

Alternative formulae

Although formula (16.1) is truly exact, mathematically speaking, its accuracy is very poor for small values of the angle d, as has been seen at the beginning of this Chapter. For this reason, several alternative methods have been proposed.

One of them [2] consists in using the old *haversine* (hav) function, which can be a great aid in certain astronomical calculations involving small angles, as it can preserve significant digits. By definition, for any angle θ, we have

$$\text{hav } \theta = \frac{1 - \cos \theta}{2}$$

The cosine formula (16.1) for angular separation is precisely equivalent to

$$\text{hav } d = \text{hav } \Delta\delta + \cos \delta_1 \cos \delta_2 \text{ hav } \Delta\alpha \qquad (16.5)$$

where $\Delta\alpha = \alpha_1 - \alpha_2$, $\Delta\delta = \delta_1 - \delta_2$. To use this formula on a computer we can get the help of another identity, namely

$$\text{hav } \theta = \sin^2 \left(\frac{\theta}{2}\right)$$

By means of formula (16.5), angular separations can be calculated accurately for angles from nearly 180° all the way down to exactly 0 degree!

V.J. Slabinski [3] offers another approach that can be used :

$$\sin^2 d = (\cos \delta_1 \sin \Delta\alpha)^2 + (\sin \delta_2 \cos \delta_1 \cos \Delta\alpha - \cos \delta_2 \sin \delta_1)^2$$

However, this formula cannot distinguish between supplementary angles, for instance 144° and 36°, and has a poor accuracy when d is close to 90°.

References

1. A. Danjon, *Astronomie Générale*, page 36, formulae 3 bis (Paris, 1959).

2. *Sky and Telescope*, Vol. 68, page 159 (August 1984).

3. *Sky and Telescope*, Vol. 69, page 158 (February 1985).

[illegible] using the [illegible] values [illegible] [illegible] significant digits [illegible]

[illegible]

The [illegible] for [illegible]

[illegible]

[illegible]

However, [illegible]

References

1. [illegible] (19[illegible]).
2. [illegible], Vol. 56, [illegible]
3. [illegible], Vol. [illegible], page [illegible] (19[illegible]).

Chapter 17

Planetary Conjunctions

Given three or five ephemeris positions of two planets passing near each other, a program can be written which calculates the time of conjunction in *right ascension*, and the difference in declination between the two bodies at that time. The method consists of calculating the differences $\Delta\alpha$ of the corresponding right ascensions, and then calculating the instant when $\Delta\alpha = 0$ by means of formula (3.6) or (3.7) in the case of three positions, or (3.10) or (3.11) in the case of five points. When that instant is found, direct interpolation of the differences $\Delta\delta$ of the declinations, by means of formula (3.3) or (3.8), yields the required difference in declination at the time of conjunction.

Conjunctions in celestial *longitude* can be calculated in the same way, using of course the planets' geocentric longitudes and latitudes instead of their right ascensions and declinations.

It should be noted that neither the instant of the conjunction in right ascension, nor that of the conjunction in longitude, coincides with that of the least angular separation between the two bodies.

Example 17.a — Calculate the circumstances of the Mercury-Venus conjunction of 1991 August 7.

The following positions, for 0^h TD of the date, are taken from an accurate ephemeris :

Mercury	Aug. 5	$\alpha = 10^h24^m30\overset{s}{.}125$	$\delta = +6°26'32\overset{''}{.}05$
	6	10 25 00.342	+6 10 57.72
	7	10 25 12.515	+5 57 33.08
	8	10 25 06.235	+5 46 27.07
	9	10 24 41.185	+5 37 48.45

Venus	Aug. 5	$\alpha' = 10^h27^m27^s.175$	$\delta' = +4°04'41''.83$
	6	10 26 32.410	+3 55 54.66
	7	10 25 29.042	+3 48 03.51
	8	10 24 17.191	+3 41 10.25
	9	10 22 57.024	+3 35 16.61

We first calculate the differences of the right ascensions and those of the declinations, both in degrees and decimals :

Aug. 5	$\Delta\alpha$ = −0.737 708	$\Delta\delta$ = +2.363 950
6	−0.383 617	+2.250 850
7	−0.068 863	+2.158 214
8	+0.204 350	+2.088 006
9	+0.434 004	+2.042 178

Applying formula (3.10) to the values of $\Delta\alpha$, we find that $\Delta\alpha$ is zero for the value $n = +0.23797$ of the interpolation factor. Hence, the conjunction in right ascension takes place on

$$\begin{aligned} \text{1991 August 7.23797} &= \text{1991 August 7 at } 5^h42^m.7 \text{ TD} \\ &= \text{1991 August 7 at } 5^h42^m \text{ UT.} \end{aligned}$$

With the value of n just found, and applying formula (3.8) to the values of $\Delta\delta$, we find $\Delta\delta = +2°.13940$ or $+2°08'$. Hence, at the time of the conjunction in right ascension, Mercury is 2°08' north of Venus.

If the second body is a star, its coordinates may be considered as being constant during the time interval considered. We then have

$$\alpha_1' = \alpha_2' = \alpha_3' = \alpha_4' = \alpha_5'$$
$$\delta_1' = \delta_2' = \delta_3' = \delta_4' = \delta_5'$$

The program can be written in such a manner that, if the second object is a star, its coordinates must be entered only once.

The important remark given on page 110 does apply here too : *the coordinates of the star and those of the moving body must be referred to the same equinox.*

As an exercise, calculate the conjunction in right ascension between the minor planet 4 Vesta and the star θ Leonis in May 1992. The minor planet's right ascension and declination, referred to the standard equinox of B1950.0, are as follows (from an ephemeris calculated by Edwin Goffin) :

0 h TD	α_{1950}	δ_{1950}
1992 May 8	$11^h06^m30^s.379$	$+16°13'37''.98$
13	11 08 22.410	+15 44 26.59
18	11 10 52.398	+15 11 26.24
23	11 13 57.547	+14 34 58.49
28	11 17 35.311	+13 55 21.19

The star's coordinates for the epoch and equinox of 1950.0 are $\alpha' = 11^h11^m37^s.089$ and $\delta' = +15°42'11''.49$, and the centennial proper motions (that is, the proper motions per 100 years) are $-0^s.420$ in right ascension and $-7''.87$ in declination.

Consequently, from the proper motions during the 42.38 years (0.4238 century) since 1950.0, we find that the star's position referred to the equinox of 1950.0, but for the epoch 1992.38, is

$$\alpha' = 11^h11^m36^s.911, \qquad \delta' = +15°42'08''.15$$

Now, calculate the conjunction.

Answer : Vesta passes 0°40' south of θ Leo on 1992 May 19 at 7^h TD.

Chapter 18

Bodies in Straight Line

In this Chapter and in the next one, we shall deal with two problems which have no importance 'scientifically', but which may be of value to persons interested in nice celestial events or to authors of popular articles.

Let (α_1, δ_1), (α_2, δ_2), (α_3, δ_3) be the equatorial coordinates of the three heavenly bodies. These bodies are in 'straight line' — that is, they lie on the same great circle of the celestial sphere — if

$$\tan \delta_1 \sin(\alpha_2 - \alpha_3) + \tan \delta_2 \sin(\alpha_3 - \alpha_1) + \tan \delta_3 \sin(\alpha_1 - \alpha_2) = 0 \qquad (18.1)$$

This formula is valid for ecliptical coordinates too, the right ascensions α being replaced by the longitudes λ, and the declinations δ by the latitudes β.

Do not forget that the right ascensions α are generally expressed in hours, minutes and seconds. They should be converted to hours and decimals, and then into degrees by multiplication by 15.

If one or two of the bodies are stars, then once again the important remark given on page 110 does apply: *the coordinates of the star(s) must be referred to the same equinox as that of the planets.*

Example 18.a — Find the instant when Mars is seen in straight line with Pollux and Castor in 1994.

From an ephemeris of Mars and a star atlas, it is found that the planet is in straight line with the two stars about 1994 October 1. For this date, the apparent equatorial coordinates of the stars are:

Castor (α Gem) : $\alpha_1 = 7^h34^m16^s.40 = 113°.56833$
$\delta_1 = +31°53'51''.2 = +31°.89756$

Pollux (β Gem) : $\alpha_2 = 7^h45^m00^s.10 = 116°.25042$
$\delta_2 = +28°02'12''.5 = +28°.03681$

For our problem, these values of α_1, δ_1, α_2 and δ_2 may be considered as constants for several days.

The apparent coordinates of Mars (α_3, δ_3) are variable. Here are their values, taken from an accurate ephemeris :

TD	α_3	δ_3
1994 Sep. 29.0	$7^h55^m55^s.36$ = 118°.98067	+21°41'03".0 = +21°.68417
30.0	7 58 22.55 = 119.59396	+21 35 23.4 = +21.58983
Oct. 1.0	8 00 48.99 = 120.20413	+21 29 38.2 = +21.49394
2.0	8 03 14.66 = 120.81108	+21 23 47.5 = +21.39653
3.0	8 05 39.54 = 121.41475	+21 17 51.4 = +21.29761

Using these values, the first member of formula (18.1) takes the following values :

Sep. 29.0	+0.001 9767
30.0	+0.001 0851
Oct. 1.0	+0.000 1976
2.0	−0.000 6855
3.0	−0.001 5641

Using formula (3.10), we find that the value is zero for

1994 October 1.2233 = 1994 October 1, at 5^h TD (UT)

In the preceding Example, we made use of the geocentric positions of Mars. For this reason the result is, strictly speaking, valid only for a geocentric observer, and for an observer for whom Mars is at the zenith. But for the present problem, it is not worthwhile to take into account the parallax of the planet, which is very small. This is no longer true in the case of the Moon, whose parallax can reach 1°. In this case, the *topocentric* position of the Moon should be used (see Chapter 39).

Chapter 19

Smallest Circle containing three Celestial Bodies

Let A, B, C be three celestial bodies situated not too far from each other on the celestial sphere, say closer than about 6 degrees. We wish to calculate the angular diameter of the smallest circle containing these three bodies. Two cases can occur :

type I : the smallest circle has as diameter the longest side of the triangle ABC, and one point is inside of the circle ;

type II : the smallest circle is the circle passing through the three points A, B, C.

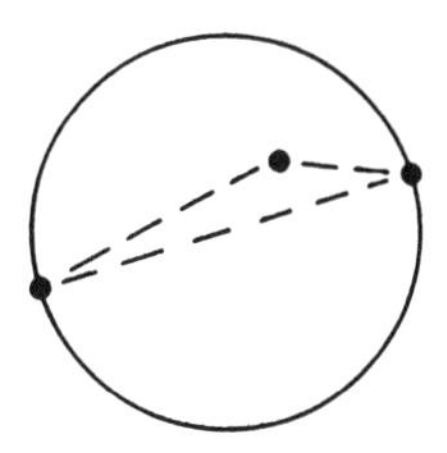

Type I

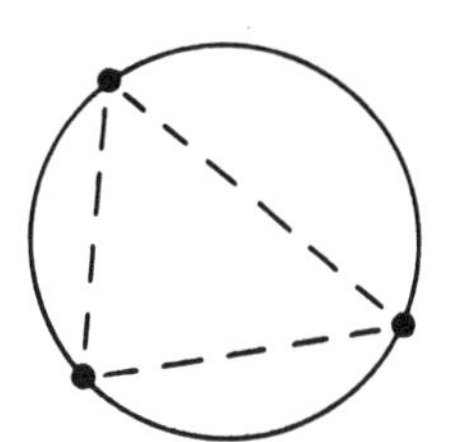

Type II

The diameter Δ of the smallest circle can be found as follows. Calculate the lengths (in degrees) of the sides of the triangle ABC by means of the method given in Chapter 16.

Let a be the length of the *longest* side of the triangle, and b and c the lengths of the two other sides.

If $a > \sqrt{b^2 + c^2}$, then the grouping is of type I, and $\Delta = a$;

if $a < \sqrt{b^2 + c^2}$, then the grouping is of type II, and

$$\Delta = \frac{2abc}{\sqrt{(a+b+c)\,(a+b-c)\,(b+c-a)\,(a+c-b)}} \tag{19.1}$$

Example 19.a — Calculate the diameter of the smallest circle containing Mercury, Jupiter and Saturn on 1981 September 11 at 0^h Dynamical Time. The positions of these planets at that instant are :

	α	δ
Mercury	$\alpha = 12^h41^m08^s.63$	$\delta = -5°37'54''.2$
Jupiter	12 52 05.21	-4 22 26.2
Saturn	12 39 28.11	-1 50 03.7

The three angular separations are found by means of (16.1) :

Mercury – Jupiter	3°.00152
Mercury – Saturn	3.82028
Jupiter – Saturn	4.04599 = a

Because $4.04599 < \sqrt{(3.00152)^2 + (3.82028)^2} = 4.85836$, we calculate Δ by means of formula (19.1). The result is

$$\Delta = 4°.26364 = 4°16'$$

This is an example of Type II.

As an exercise, perform the calculation for the planets Venus, Mars and Jupiter on 1991 June 20 at 0^h TD, using the following positions :

	α	δ
Venus	$\alpha = 9^h05^m41^s.44$	$\delta = +18°30'30''.0$
Mars	9 09 29.00	+17 43 56.7
Jupiter	8 59 47.14	+17 49 36.8

Show that this case is of type I, and that $\Delta = 2°19'$.

A program can be written in which first the right ascensions and the declinations of the planets are interpolated, after which a, b, c, and finally Δ are calculated. With such a program, it is possible to calculate (by trial) the minimum value of Δ of a grouping of three planets. Indeed, Δ varies with time, and the method described in this Chapter provides the value of Δ for only one given instant.

It is important to note that, while the *positions* of the planets can be interpolated by means of the usual formulae, the values of the circle's diameter Δ cannot. The reason is that the variation of Δ generally cannot be represented by a polynomial; see for instance the graph in Example 19.c.

Example 19.b — In September 1981, there was a grouping of the planets Mercury, Jupiter and Saturn. The positions of these planets were as follows; instead of right ascensions and declinations, we will use ecliptical coordinates (longitudes and latitudes) here.

1981	Mercury		Jupiter		Saturn	
0 h TD	*long.*	*latit.*	*long.*	*latit.*	*long.*	*latit.*
Sep. 7	186°045	-0°560	192°866	+1°117	189°324	+2°226
8	187.482	-0.696	193.069	+1.116	189.439	+2.225
9	188.897	-0.833	193.272	+1.114	189.555	+2.224
10	190.290	-0.971	193.476	+1.113	189.671	+2.223
11	191.661	-1.109	193.681	+1.112	189.788	+2.222
12	193.008	-1.246	193.886	+1.110	189.906	+2.221
13	194.332	-1.384	194.092	+1.109	190.023	+2.220
14	195.631	-1.521	194.299	+1.108	190.142	+2.219

We will not give details here, and leave it as an exercise to the reader. Let us just mention that from September 7.00 to 8.81 the grouping was of type I, the diameter Δ of the smallest circle decreasing almost linearly from 7°01' to 5°00'. From September 8.81 to 12.19, the grouping was of type II, and Δ reached a minimum of 4°14' on September 10.53. From September 12.19 on, the grouping was of type I again, Δ increasing almost linearly with time.

Example 19.c — Let us now consider the following fictitious case. On March 12.0, the ecliptical coordinates (in degrees) of three planets are as follows.

	longitude	*latitude*	*daily motion in longitude*
planet P1	214.23	+0.29	+0.11
planet P2	211.79	+0.48	+0.20
planet P3	208.41	+0.75	+1.08

We suppose that the latitudes are constant and that the longitudes increase at the constant rates mentioned in the last column.

Again, we leave the actual calculation as an exercise to the reader. Let us just illustrate the variation of the diameter Δ of the smallest circle (see the Figure on next page). Note the discontinuities at points A and B. Except during two short periods (March 15.87 to 15.91 near A, and March 17.93 to 18.05 near B), where the grouping is of type II, we have type I. The least value of Δ, namely 1°55', occurs at B, on March 17.94.

If one of the bodies is a star, once again the important remark made on page 110 does apply: the coordinates of the star should be referred to the same equinox as that for the planets.

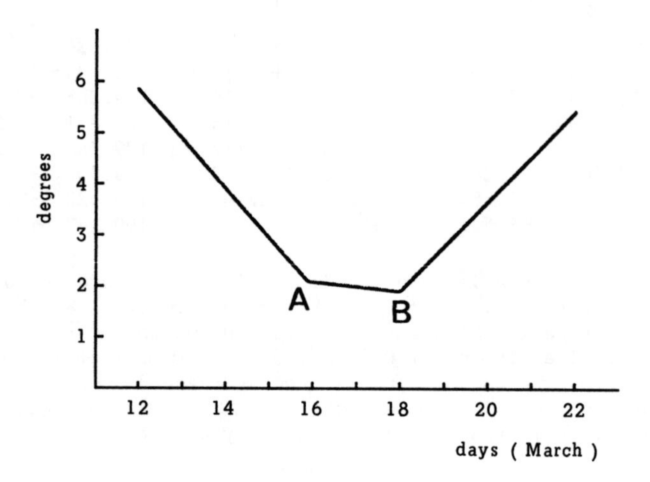

Chapter 20

Precession

The direction of the rotational axis of the Earth is not really fixed in space. Over time it undergoes a slow drift, or *precession*, much like that of a spinning top. This effect stems from the gravitational attraction of the Sun and the Moon on the Earth's equatorial bulge.

Due to the precession, the northern celestial pole (presently situated near the star α Ursae Minoris, or Polaris) slowly turns around the pole of the ecliptic, with a period of about 26 000 years; as a consequence, the vernal equinox, the intersection of equator and ecliptic, regresses by about 50" per year along the ecliptic.

Moreover, the plane of the ecliptic itself is not fixed in space. Due to the gravitational attraction of the planets on the Earth, it slowly rotates around a 'line of nodes', the speed of this rotation being presently 47" per century.

The plane of the ecliptic and that of the equator, and the vernal equinox, are the fundamental planes and the origin of two important coordinate systems on the celestial sphere: the ecliptical coordinates (longitude λ and latitude β) and the equatorial coordinates (right ascension α and declination δ). So, due to the precession, the coordinates of the 'fixed' stars are continuously changing. Star catalogues, therefore, list the right ascensions and declinations of stars for a given epoch, such as 1900.0, or 1950.0, or 2000.0.

In this Chapter, we consider the problem of converting the right ascension α and the declination δ of a star, given for an epoch and an equinox, to the corresponding values for another epoch and equinox. Only the *mean* places of a star and the effect of the precession alone are considered here. The problem of finding the apparent place of a star will be considered in Chapter 22.

Low accuracy

If no great accuracy is required, if the two epochs are not widely separated, and if the star is not too close to one of the celestial poles, the following formulae may be used for the *annual* precessions in right ascension and declination :

$$\Delta\alpha = m + n \sin\alpha \tan\delta \qquad \Delta\delta = n \cos\alpha \tag{20.1}$$

where m and n are two quantities which vary slowly with time. They are given by

$$m = 3^s.07496 + 0^s.00186\,T$$
$$n = 1^s.33621 - 0^s.00057\,T$$
$$n = 20''.0431 - 0''.0085\,T$$

T being the time measured in centuries from 2000.0 (the beginning of the year 2000). Here are the values of m and n for some epochs :

Epoch	m	n	n
1700.0	$3^s.069$	$1^s.338$	$20''.07$
1800.0	3.071	1.337	20.06
1900.0	3.073	1.337	20.05
2000.0	3.075	1.336	20.04
2100.0	3.077	1.336	20.03
2200.0	3.079	1.335	20.03

For the calculation of $\Delta\alpha$, the value of n expressed in seconds of time (^{s}) must be used. Remember that 1^s corresponds to $15''$.

In the case of a star, the effect of the proper motion should be added to the values given by formulae (20.1).

Example 20.a — The coordinates of Regulus (α Leonis) for the epoch and equinox of 2000.0 are

$$\alpha_0 = 10^h08^m22^s.3 \qquad \delta_0 = +11°58'02''$$

and the annual proper motions are

$-0^s.0169$ in right ascension,
$+0''.006$ in declination.

Reduce these coordinates to the epoch and the equinox of 1978.0.

Here we have

$\alpha = 152°.093$ $\qquad m = 3^s.075$
$\delta = +11°.967$ $\qquad n = 1^s.336 = 20''.04$

From the formulae (20.1) we deduce

$$\Delta\alpha = +3^s.208, \qquad \Delta\delta = -17''.71$$

to which we must add the annual proper motion, giving an annual variation of $+3^s.191$ in right ascension, and $-17''.70$ in declination.

Variations during −22 years (from 2000.0 to 1978.0) :

$$\text{in } \alpha : \quad +3^s.191 \times (-22) = -70^s.2 = -1^m10^s.2$$
$$\text{in } \delta : \quad -17''.70 \times (-22) = +389'' = +6'29''$$

Required right ascension : $\alpha = \alpha_0 - 1^m10^s.2 = 10^h07^m12^s.1$

Required declination : $\delta = \delta_0 + 6'29'' = +12°04'31''$

Besselian and Julian Year

The International Astronomical Union has decided that from 1984 onwards the astronomical ephemerides should use the following system.

The new standard epoch is 2000 January 1 at 12^h TD, corresponding to JDE 2451 545.0. This epoch is designated J2000.0. For purposes of calculating positions of stars, the beginning of a 'year' differs from the standard epoch J2000.0 by an integral multiple of the Julian year, or 365.25 days. For example, the epoch J1986.0 is 14 × 365.25 days before J2000.0, and hence the corresponding JDE is 2451 545.0 − 14 × 365.25 = 2446 431.50.

The letter J, in notations such as J2000.0 or J1986.0, indicates that the unit of time (for star catalogues) is the Julian year. Previously, star position catalogues used for a standard epoch the beginning of a Besselian year. The beginning of the Besselian solar year is the instant when the mean longitude of the Sun, affected by the aberration ($-20''.5$) and measured from the mean equinox of the date, is exactly 280°. This instant is always near the beginning of the Gregorian civil year. The length of the Besselian year, equal to that of the tropical year, was 365.242 1988 days in A.D. 1900, according to Newcomb.

To distinguish an old epoch, based on the Besselian year, from the new system, the letter B is used. For example,

B1900.0 = JDE 2415 020.3135 = 1900 January 0.8135
B1950.0 = JDE 2433 282.4235 = 1950 January 0.9235

but

J2000.0 = JDE 2451 545.00 exactly
J2050.0 = JDE 2469 807.50 exactly

and so on. The notation .0 after a year number (as in 1986.0 or 2000.0) signifies that the start of the year is meant.

Rigorous method

Let T be the time interval, in Julian centuries, between J2000.0 and the starting epoch, and let t be the interval, in the same units, between the starting epoch and the final epoch.

In other words, if $(\mathrm{JD})_0$ and (JD) are the Julian Days corresponding to the initial and the final epoch, respectively, we have

$$T = \frac{(\mathrm{JD})_0 - 2451\,545.0}{36525} \qquad t = \frac{(\mathrm{JD}) - (\mathrm{JD})_0}{36525}$$

Then we have the following numerical expressions for the quantities ζ, z and θ which are needed for the accurate reduction of positions from one equinox to another [1] :

$$\begin{aligned}
\zeta &= (2306''.2181 + 1''.39656T - 0''.000\,139T^2)t \\
&\quad + (0''.30188 - 0''.000\,344T)t^2 + 0''.017\,998t^3 \\
z &= (2306''.2181 + 1''.39656T - 0''.000\,139T^2)t \\
&\quad + (1''.09468 + 0''.000\,066T)t^2 + 0''.018\,203t^3 \\
\theta &= (2004''.3109 - 0''.85330T - 0''.000\,217T^2)t \\
&\quad - (0''.42665 + 0''.000\,217T)t^2 - 0''.041\,833t^3
\end{aligned} \qquad (20.2)$$

If the starting epoch is J2000.0 itself, we have $T = 0$ and the expressions (20.2) reduce to

$$\begin{aligned}
\zeta &= 2306''.2181t + 0''.30188t^2 + 0''.017\,998t^3 \\
z &= 2306''.2181t + 1''.09468t^2 + 0''.018\,203t^3 \\
\theta &= 2004''.3109t - 0''.42665t^2 - 0''.041\,833t^3
\end{aligned} \qquad (20.3)$$

Then, the rigorous formulae for the reduction of the given equatorial coordinates α_0 and δ_0 of the starting epoch to the coordinates α and δ of the final epoch are :

$$\left.\begin{aligned}
A &= \cos\delta_0 \sin(\alpha_0 + \zeta) \\
B &= \cos\theta \cos\delta_0 \cos(\alpha_0 + \zeta) - \sin\theta \sin\delta_0 \\
C &= \sin\theta \cos\delta_0 \cos(\alpha_0 + \zeta) + \cos\theta \sin\delta_0 \\
\tan(\alpha - z) &= \frac{A}{B} \qquad \sin\delta = C
\end{aligned}\right\} \qquad (20.4)$$

The angle $\alpha - z$ can be obtained in the correct quadrant by applying the 'second' arctangent function ATN2 to the quantities A and B, or by another procedure — see 'The correct quadrant' in Chapter 1.

If the star is close to the celestial pole, one should use the formula $\cos\delta = \sqrt{A^2 + B^2}$ instead of $\sin\delta = C$.

Before making the reduction from α_0, δ_0 to α, δ, the effect of the star's proper motion should be calculated.

Example 20.b — The star θ Persei has the following mean coordinates for the epoch and equinox of J2000.0 :

$$\alpha_0 = 2^h44^m11^s.986 \qquad \delta_0 = +49°13'42''.48$$

and its annual proper motions referred to that same equinox are

$+0^s.03425$ in right ascension,
$-0''.0895$ in declination.

Reduce the coordinates to the epoch and mean equinox of 2028 November 13.19 TD.

The initial epoch is J2000.0 or JD 2451 545.0, and the final one is JD 2462 088.69. Hence, t = +0.288 670 500 Julian centuries, or 28.867 0500 Julian years.

We first calculate the effect of the proper motion. The variations over 28.86705 years are

$+0^s.03425 \times 28.86705 = +0^s.989$ in right ascension,
$-0''.0895 \times 28.86705 = -2''.58$ in declination.

Thus the star's coordinates, for the mean equinox of J2000.0, but for the epoch 2028 November 13.19, are

$$\alpha_0 = 2^h44^m11^s.986 + 0^s.989 = 2^h44^m12^s.975 = +41°.054\,063$$
$$\delta_0 = +49°13'42''.48 - 2''.58 = +49°13'39''.90 = +49°.227\,750$$

Since the initial equinox is that of J2000.0, we can use the formulae (20.3). With the value t = +0.288 670 500, we obtain

$$\zeta = +665''.7627 = +0°.184\,9341$$
$$z = +665''.8288 = +0°.184\,9524$$
$$\theta = +578''.5489 = +0°.160\,7080$$

$$A = +0.430\,494\,05$$
$$B = +0.488\,948\,49$$
$$C = +0.758\,685\,86$$

$$\alpha - z = +41°.362\,262$$
$$\alpha = +41°.547\,214 = 2^h46^m11^s.331$$
$$\delta = +49°.348\,483 = +49°20'54''.54$$

Exercise. — The equatorial coordinates of α Ursae Minoris, for the epoch and mean equinox of J2000.0, are

$$\alpha = 2^h31^m48^s.704, \qquad \delta = +89°15'50''.72$$

and the star's annual proper motions for the same equinox are

$+0^s.19877$ in right ascension,
$-0''.0152$ in declination.

Find the coordinates of the star for the epochs and mean equinoxes of B1900.0, J2050.0, and J2100.0.

Answer :

B1900.0	$\alpha = 1^h22^m33^s.90$	$\delta = +88°46'26''.18$
J2050.0	3 48 16.43	+89 27 15.38
J2100.0	5 53 29.17	+89 32 22.18

It should be noted that the formulae (20.2) and (20.3) are valid only for a limited period of time. If we use them for the year 32700, for instance, we find for that epoch that α UMi will be at declination −87°, a completely wrong result !

Using ecliptical coordinates

If, instead of the star's equatorial coordinates (right ascension and declination), we use its ecliptical coordinates (longitude, latitude), the following rigorous method can be used [2].

T and t having the same meaning as before, calculate

$$\begin{aligned}
\eta &= (47''.0029 - 0''.06603T + 0''.000\,598T^2)t \\
&\quad + (-0''.03302 + 0''.000\,598T)t^2 + 0''.000\,060t^3 \\
\Pi &= 174°.876\,384 + 3289''.4789T + 0''.60622T^2 \\
&\quad - (869''.8089 + 0''.50491T)t + 0''.03536t^2 \\
p &= (5029''.0966 + 2''.22226T - 0''.000\,042T^2)t \\
&\quad + (1''.11113 - 0''.000\,042T)t^2 - 0''.000\,006t^3
\end{aligned} \tag{20.5}$$

The quantity η is the angle between the ecliptic at the starting epoch and the ecliptic at the final epoch.

If the starting epoch is J2000.0, we have $T = 0$ and the above expressions reduce to

$$\begin{aligned}
\eta &= 47''.0029t - 0''.03302t^2 + 0''.000\,060t^3 \\
\Pi &= 174°.876\,384 - 869''.8089t + 0''.03536t^2 \\
p &= 5029''.0966t + 1''.11113t^2 - 0''.000\,006t^3
\end{aligned} \tag{20.6}$$

Then, the rigorous formulae for the reduction of the given ecliptical coordinates λ_0 and β_0 of the starting epoch to the coordinates λ and β of the final epoch are :

$$A' = \cos\eta \cos\beta_0 \sin(\Pi - \lambda_0) - \sin\eta \sin\beta_0$$
$$B' = \cos\beta_0 \cos(\Pi - \lambda_0)$$
$$C' = \cos\eta \sin\beta_0 + \sin\eta \cos\beta_0 \sin(\Pi - \lambda_0)$$

$$\tan(p + \Pi - \lambda) = \frac{A'}{B'} \qquad \sin\beta = C'$$

The old precessional elements

As we have said earlier, for star catalogues and for the purpose of calculating star positions, the standard epoch is now J2000.0 and the unit of time is now the *Julian* year (365.25 days) or the Julian century (36525 days). Previously, the beginning of the Besselian year was taken as reference time and the unit of time was the *tropical* year or the tropical century.

However, these are not the only differences between the old system (the FK4) and the new one (the FK5). ['FK' means *Fundamental Katalog*.]

Firstly, there is a small error (the 'equinox correction') in the zero point of the right ascensions of the FK4.

Secondly, as we shall see in Chapter 22, the aberrational displacements of a star in longitude ($\Delta\lambda$) and in latitude ($\Delta\beta$) resulting from the motion of the Earth in its elliptic orbit are given by

$$\Delta\lambda = -\kappa \frac{\cos(\odot - \lambda)}{\cos\beta} + e\kappa \frac{\cos(\pi - \lambda)}{\cos\beta}$$
$$\Delta\beta = -\kappa \sin(\odot - \lambda) \sin\beta + e\kappa \sin(\pi - \lambda) \sin\beta$$

where $\odot$ is the longitude of the Sun, π the longitude of the perihelion of the Earth's orbit, e the eccentricity of this orbit, and κ the constant of aberration.

Now, the second terms in the right-hand sides of these expressions are almost constant for a given star, because e, $\pi - \lambda$ and β vary very slowly with time. For this reason, it has been astronomical practice to leave this part of the aberration (the so-called E-terms) in the mean positions of the stars.

Presently, the terms depending on the ellipticity of the Earth's orbit are no longer included in the mean places of stars; they are, instead, calculated in the reduction from mean to apparent places (see Chapter 22).

A procedure for performing the conversion of mean positions and proper motions of stars referred to the mean equinox and equator B1950.0 and based on Newcomb's expressions for the precession (the

FK4 system) to the new IAU system at J2000.0 (the FK5) can be found, for instance, in the *Astronomical Almanac* for 1984 [3].

The precessional formulae (20.2) and (20.3) may be used only for the stars referred to the FK5 system. If only FK4 positions and proper motions are available, then one should proceed as follows to calculate apparent star positions in the FK5 system:

1. use must be made of Newcomb's precessional formulae (see below);
2. in the reduction from mean to apparent place, the *E*-terms of the aberration should be dropped;
3. to the final right ascension of the star, add the equinox correction

$$\Delta\alpha = 0^s.0775 + 0^s.0850\,T$$

where T is the time in Julian centuries from J2000.0.

Newcomb's precessional expressions are the following ones.

Let $(JD)_0$ and (JD) be the Julian Days corresponding to the initial and the final epoch, respectively. Then

$$T = \frac{(JD)_0 - 2415\,020.3135}{36524.2199} \qquad t = \frac{(JD) - (JD)_0}{36524.2199}$$

$$\zeta = (2304''.250 + 1''.396\,T)t + 0''.302t^2 + 0''.018t^3$$
$$z = \zeta + 0''.791t^2 + 0''.001t^3$$
$$\theta = (2004''.682 - 0''.853T)t - 0''.426t^2 - 0''.042t^3$$

If the starting epoch is B1950.0, we have $T = 0.5$, and the above expressions become

$$\zeta = 2304''.948t + 0''.302t^2 + 0''.018t^3$$
$$z = 2304''.948t + 1''.093t^2 + 0''.019t^3$$
$$\theta = 2004''.255t - 0''.426t^2 - 0''.042t^3$$

References

1. *Astronomical Almanac* for the year 1984 (Washington, D.C.; 1983), page S19.
2. *Connaissance des Temps* pour 1984 (Paris, 1983), pages XXX and XL.
3. *Astronomical Almanac* for the year 1984 (Washington, D.C.; 1983), pages S34 – S35.

Chapter 21

Nutation and the Obliquity of the Ecliptic

The nutation, discovered by the British astronomer James Bradley (1693 - 1762), is a periodic oscillation of the rotational axis of the Earth around its 'mean' position. Due to the nutation, the instantaneous pole of rotation of the Earth oscillates around a mean pole which advances by the precession around the pole of the ecliptic.

The nutation is due principally to the action of the Moon, and can be described by a sum of periodic terms. The most important term has a period of 6798.4 days (18.6 years), but some other terms have a very short period (less than 10 days).

Nutation is conveniently partitioned into a component parallel to, and one perpendicular to the ecliptic. The component along the ecliptic is denoted by $\Delta\psi$ and is called the *nutation in longitude*; it affects the celestial longitude of all heavenly bodies. The component perpendicular to the ecliptic is denoted by $\Delta\varepsilon$ and is called the *nutation in obliquity*, since it affects the obliquity of the equator to the ecliptic.

The quantities $\Delta\psi$ and $\Delta\varepsilon$ are needed for the calculation of the apparent place of a heavenly body and for that of the apparent sidereal time. For any given instant, $\Delta\psi$ and $\Delta\varepsilon$ can be calculated as follows.

Find the time T, measured in Julian centuries from the Epoch J2000.0 (JDE 2451 545.0),

$$T = \frac{\text{JDE} - 2451\,545}{36525} \qquad (21.1)$$

where JDE is the Julian Ephemeris Day; it differs from the Julian Day (JD) by the small quantity ΔT (see Chapter 7). Then calculate the following angles expressed in degrees and decimals. These expressions are those which are provided by the International Astronomical Union [1]; they differ slightly from those used in Chapront's lunar theory (Chapter 45).

Mean elongation of the Moon from the Sun :

$$D = 297.85036 + 445\,267.111\,480T - 0.001\,9142T^2 + T^3/189\,474$$

Mean anomaly of the Sun (Earth) :

$$M = 357.52772 + 35\,999.050\,340T - 0.000\,1603T^2 - T^3/300\,000$$

Mean anomaly of the Moon :

$$M' = 134.96298 + 477\,198.867\,398T + 0.008\,6972T^2 + T^3/56\,250$$

Moon's argument of latitude :

$$F = 93.27191 + 483\,202.017\,538T - 0.003\,6825T^2 + T^3/327\,270$$

Longitude of the ascending node of the Moon's mean orbit on the ecliptic, measured from the mean equinox of the date :

$$\Omega = 125.04452 - 1934.136\,261T + 0.002\,0708T^2 + T^3/450\,000$$

The nutations in longitude ($\Delta\psi$) and in obliquity ($\Delta\varepsilon$) are then obtained by making the sum of the terms given in Table 21.A, where the coefficients are given in units of $0''.0001$. These terms are those of the '1980 IAU Theory of Nutation' [2] where, however, we have neglected the terms with a coefficient smaller than $0''.0003$. The argument of each sine (for $\Delta\psi$) and cosine (for $\Delta\varepsilon$) is a linear combination of the five fundamental arguments D, M, M', F and Ω. For instance, the argument on the second line is $-2D + 2F + 2\Omega$.

Of course, if no great accuracy is needed, only the periodic terms with the largest coefficients can be used.

If an accuracy of $0''.5$ in $\Delta\psi$ and of $0''.1$ in $\Delta\varepsilon$ are sufficient, then we may drop the terms in T^2 and in T^3 in the above expression for Ω, and then use the following simplified expressions :

$$\Delta\psi = -17''.20 \sin\Omega - 1''.32 \sin 2L - 0''.23 \sin 2L' + 0''.21 \sin 2\Omega$$

$$\Delta\varepsilon = +9''.20 \cos\Omega + 0''.57 \cos 2L + 0''.10 \cos 2L' - 0''.09 \cos 2\Omega$$

where L and L' are the mean longitudes of the Sun and the Moon, respectively :

$$L = 280^\circ.4665 + 36\,000.7698T$$

$$L' = 218^\circ.3165 + 481\,267^\circ.8813T$$

TABLE 21.A

Periodic terms for the nutation in longitude ($\Delta\psi$) and in obliquity ($\Delta\varepsilon$). The unit is 0".0001.

Argument multiple of					$\Delta\psi$ Coefficient of the sine of the argument		$\Delta\varepsilon$ Coefficient of the cosine of the argument	
D	*M*	*M'*	*F*	Ω				
0	0	0	0	1	-171996	-174.2 T	+92025	+8.9 T
-2	0	0	2	2	-13187	-1.6 T	+5736	-3.1 T
0	0	0	2	2	-2274	-0.2 T	+977	-0.5 T
0	0	0	0	2	+2062	+0.2 T	-895	+0.5 T
0	1	0	0	0	+1426	-3.4 T	+54	-0.1 T
0	0	1	0	0	+712	+0.1 T	-7	
-2	1	0	2	2	-517	+1.2 T	+224	-0.6 T
0	0	0	2	1	-386	-0.4 T	+200	
0	0	1	2	2	-301		+129	-0.1 T
-2	-1	0	2	2	+217	-0.5 T	-95	+0.3 T
-2	0	1	0	0	-158			
-2	0	0	2	1	+129	+0.1 T	-70	
0	0	-1	2	2	+123		-53	
2	0	0	0	0	+63			
0	0	1	0	1	+63	+0.1 T	-33	
2	0	-1	2	2	-59		+26	
0	0	-1	0	1	-58	-0.1 T	+32	
0	0	1	2	1	-51		+27	
-2	0	2	0	0	+48			
0	0	-2	2	1	+46		-24	
2	0	0	2	2	-38		+16	
0	0	2	2	2	-31		+13	
0	0	2	0	0	+29			
-2	0	1	2	2	+29		-12	
0	0	0	2	0	+26			
-2	0	0	2	0	-22			
0	0	-1	2	1	+21		-10	
0	2	0	0	0	+17	-0.1 T		
2	0	-1	0	1	+16		-8	
-2	2	0	2	2	-16	+0.1 T	+7	
0	1	0	0	1	-15		+9	

TABLE 21.A (cont.)

Argument					Δψ	Δε
D	*M*	*M'*	*F*	Ω	*sine*	*cosine*
-2	0	1	0	1	-13	+7
0	-1	0	0	1	-12	+6
0	0	2	-2	0	+11	
2	0	-1	2	1	-10	+5
2	0	1	2	2	-8	+3
0	1	0	2	2	+7	-3
-2	1	1	0	0	-7	
0	-1	0	2	2	-7	+3
2	0	0	2	1	-7	+3
2	0	1	0	0	+6	
-2	0	2	2	2	+6	-3
-2	0	1	2	1	+6	-3
2	0	-2	0	1	-6	+3
2	0	0	0	1	-6	+3
0	-1	1	0	0	+5	
-2	-1	0	2	1	-5	+3
-2	0	0	0	1	-5	+3
0	0	2	2	1	-5	+3
-2	0	2	0	1	+4	
-2	1	0	2	1	+4	
0	0	1	-2	0	+4	
-1	0	1	0	0	-4	
-2	1	0	0	0	-4	
1	0	0	0	0	-4	
0	0	1	2	0	+3	
0	0	-2	2	2	-3	
-1	-1	1	0	0	-3	
0	1	1	0	0	-3	
0	-1	1	2	2	-3	
2	-1	-1	2	2	-3	
0	0	3	2	2	-3	
2	-1	0	2	2	-3	

The obliquity of the ecliptic

The obliquity of the ecliptic, or inclination of the Earth's axis of rotation, is the angle between the equator and the ecliptic. One distinguishes the *mean* and the *true* obliquity, being the angles which the ecliptic makes with the mean and with the true (instantaneous) equator, respectively.

The mean obliquity of the ecliptic is given by the following formula, adopted by the International Astronomical Union [1] :

$$\varepsilon_0 = 23°26'21''.448 - 46''.8150\,T - 0''.00059\,T^2 + 0''.001\,813\,T^3 \qquad (21.2)$$

where, again, T is the time measured in Julian centuries from the epoch J2000.0.

The accuracy of formula (21.2) is not satisfactory over a long period of time: the error in ε_0 reaches 1″ over a period of 2000 years, and about 10″ over a period of 4000 years. The following improved expression is due to Laskar [3]. Here, U is the time measured in units of 10 000 Julian years from J2000.0, or $U = T/100$.

$$\begin{aligned}\varepsilon_0 = 23°26'21''.448 &- 4680''.93\,U \\ &- 1.55\,U^2 \\ &+ 1999.25\,U^3 \\ &- 51.38\,U^4 \\ &- 249.67\,U^5 \\ &- 39.05\,U^6 \\ &+ 7.12\,U^7 \\ &+ 27.87\,U^8 \\ &+ 5.79\,U^9 \\ &+ 2.45\,U^{10}\end{aligned} \qquad (21.3)$$

The accuracy of this expression is estimated at 0″.01 after 1000 years (that is, between A.D. 1000 and 3000), and a few seconds of arc after 10 000 years.

It is important to note that formula (21.3) is valid only over a period of 10 000 years on each side of J2000.0, that is, for $|U| < 1$. For $U = +2.834$, for example, the formula would yield $\varepsilon_0 = 90°$, a completely erroneous result !

The Figure on the next page shows the variation of ε_0 from 10 000 years before to 10 000 years after A.D. 2000. According to Laskar's formula, the inclination of the Earth's axis of rotation was a maximum (24°14'07″) about the year −7530. And near the year +12 030 a minimum (22°36'41″) will be reached. By a mere chance we are presently approximately half-way between these extreme values, near the middle of the curve in the Figure. Here the curve is almost linear; this is the reason why in (21.3) the coefficient of U^2 is very small.

The *true* obliquity of the ecliptic is $\varepsilon = \varepsilon_0 + \Delta\varepsilon$, where $\Delta\varepsilon$ is the nutation in obliquity.

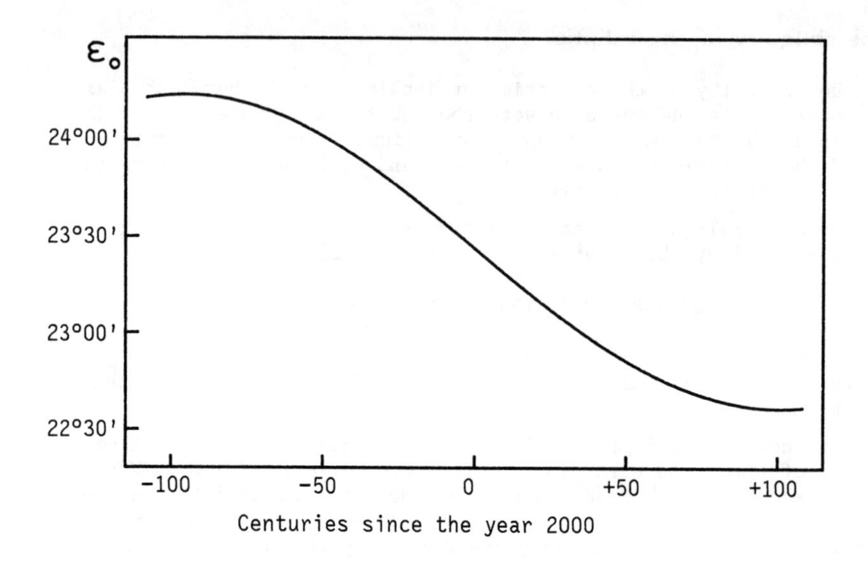

Example 21.a — Calculate $\Delta\psi$, $\Delta\varepsilon$ and the true obliquity of the ecliptic for 1987 April 10 at 0^h TD.

This date corresponds to JDE 2446 895.5, and we find

T = −0.127 296 372 348
D = −56 383°.0377 = 136°.9623
M = −4225°.0208 = 94°.9792
M' = −60 610°.7216 = 229°.2784
F = −61 416°.5921 = 143°.4079
Ω = 371°.2531 = 11°.2531

$\Delta\psi$ = −3".788
$\Delta\varepsilon$ = +9".443

ε_0 = 23°26'27".407
ε = 23°26'36".850

References

1. *Astronomical Almanac* for the year 1984 (Washington, D.C.; 1983), page S26.
2. *Ibid.*, page S23.
3. J. Laskar, *Astronomy and Astrophysics*, Vol. 157, page 68 (1986).

Chapter 22

Apparent Place of a Star

The *mean place* of a star at any time is its apparent position on the celestial sphere, as it would be seen by an observer at rest on the Sun (or, more exactly, at the barycenter of the solar system), and referred to the ecliptic and mean equinox of the date (or to the mean equator and mean equinox of the date).

The *apparent* place of a star at any time is its position on the celestial sphere as it is actually seen from the center of the moving Earth, and referred to the instantaneous equator, ecliptic and equinox. It should be noted that :

- the *mean equinox* is the intersection of the ecliptic of the date with the mean equator of the date;
- the *true equinox* is the intersection of the ecliptic with the true (instantaneous) equator (that is, the equator affected by the nutation);
- there is no 'mean' ecliptic, because the ecliptic has a regular motion.

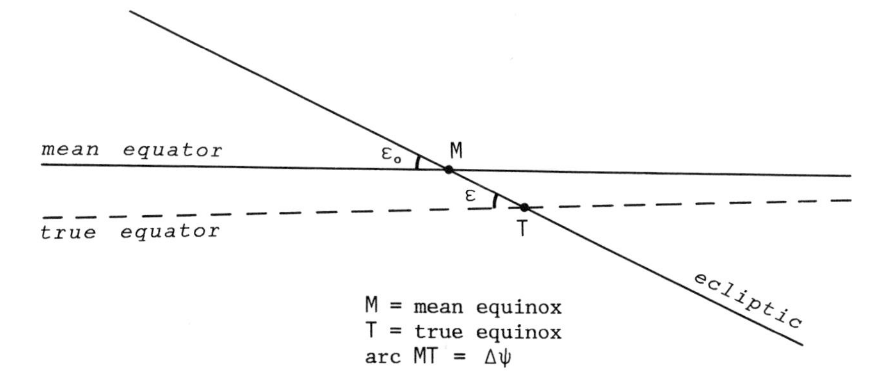

The problem of the reduction of the place of a star from the mean place at one time (for instance, of a standard epoch and equinox) to the apparent place of another time, involves the following corrections:

(A) The *proper motion* of the star between the two epochs. We may assume that by its proper motion each star moves on a great circle with an invariable angular speed. Except when the proper motion is an important fraction of the polar distance of the star, not only the proper motion itself, but also its components in right ascension and declination *with respect to a fixed equinox* may be considered as constants during several centuries. Therefore, we start by finding the effect of the proper motion when the axes of reference remain fixed, as in Example 20.b;

(B) The effect of *precession*. This has been explained in Chapter 20;

(C) The effect of *nutation* (see below);

(D) The effect of *annual aberration* (see below);

(E) The effect of the *annual parallax*. Of course, stellar parallaxes are of fundamental importance in astronomy. As George Lovi writes [1]:

> "Parallax is the only true geometrical link between us and our nearer neighbors in that vast interstellar void. It has enabled astronomers to create and calibrate procedures to take us much farther out."

However, for the person wishing to calculate accurate star positions, the stellar parallax is a nuisance. Fortunately, stellar parallaxes never exceed 0″.8 and they may be neglected in most cases. According to R. Burnham [2], only 13 stars brighter than magnitude 9.0 are nearer than 13 light-years (4 parsecs) and have a parallax exceeding 0″.25. These stars are α Centauri, Lalande 21185 (in Ursa Major), Sirius, ε Eridani, 61 Cygni, Procyon, ε Indi, Σ 2398 (in Draco), Groombridge 34 (in Andromeda), τ Ceti, Lacaille 9352 (in Piscis Austrinus), Cordoba 29191 (in Microscopium), and the Star of Kapteyn (in Pictor). None of these stars is near the ecliptic, and so none is involved in occultations by the Moon or in close conjunctions with planets.

For this reason, in what follows we shall neglect the effect of the annual parallax in the calculation of the apparent position of a star.

The effect of nutation

The simplest and most direct method of applying the effect of nutation to mean positions is to add $\Delta\psi$ to the ecliptical longitude of the objects. The ecliptic and therefore the latitude of a body is unchanged by nutation.

This procedure can profitably be used in the calculation of apparent positions of *planets*, where ecliptical coordinates are calculated first. Stellar positions, however, are generally given in the equatorial system, so we prefer to calculate the correction in right ascension and in declination directly.

First-order corrections to a star's right ascension α and declination δ due to the nutation are

$$\begin{aligned} \Delta\alpha_1 &= (\cos\varepsilon + \sin\varepsilon \sin\alpha \tan\delta)\,\Delta\psi - (\cos\alpha \tan\delta)\,\Delta\varepsilon \\ \Delta\delta_1 &= (\sin\varepsilon \cos\alpha)\,\Delta\psi + (\sin\alpha)\,\Delta\varepsilon \end{aligned} \quad (22.1)$$

These expressions are invalid if the star is close to one of the celestial poles. If this is the case, it is better to work in ecliptical coordinates and just add $\Delta\psi$ to the longitude, as mentioned above.

The quantities $\Delta\psi$ and $\Delta\varepsilon$ can be calculated by means of the method described in Chapter 21, while ε is the obliquity of the ecliptic given by formula (21.2).

The effect of aberration

Let λ and β be the star's celestial longitude and latitude, κ the constant of aberration (20''.49552), $\odot$ the true (geometric) longitude of the Sun, e the eccentricity of the Earth's orbit, and π the longitude of the perihelion of this orbit.

$\odot$ can be calculated by the method described in Chapter 24, while

$$e = 0.016\,708\,617 - 0.000\,042\,037\,T - 0.000\,000\,1236\,T^2$$

$$\pi = 102^\circ.93735 + 1^\circ.71953\,T + 0^\circ.00046\,T^2$$

where T is the time in Julian centuries from the epoch J2000.0, as obtained by formula (21.1).

Then the changes in longitude and in latitude of the star due to the annual aberration are

$$\left.\begin{aligned} \Delta\lambda &= \frac{-\kappa\cos(\odot-\lambda) + e\kappa\cos(\pi-\lambda)}{\cos\beta} \\ \Delta\beta &= -\kappa\sin\beta\,(\sin(\odot-\lambda) - e\sin(\pi-\lambda)) \end{aligned}\right\} \quad (22.2)$$

In equatorial coordinates, the changes in the right ascension α and in the declination δ of the star due to the annual aberration are

$$\left.\begin{aligned}
\Delta\alpha_2 &= -\kappa\,\frac{\cos\alpha\ \cos\odot\ \cos\varepsilon + \sin\alpha\ \sin\odot}{\cos\delta} \\
&\quad + e\kappa\,\frac{\cos\alpha\ \cos\pi\ \cos\varepsilon + \sin\alpha\ \sin\pi}{\cos\delta} \\[2ex]
\Delta\delta_2 &= -\kappa\,\big[\cos\odot\ \cos\varepsilon\ (\tan\varepsilon\ \cos\delta - \sin\alpha\ \sin\delta) \\
&\qquad\qquad + \cos\alpha\ \sin\delta\ \sin\odot\,\big] \\
&\quad + e\kappa\,\big[\cos\pi\ \cos\varepsilon\ (\tan\varepsilon\ \cos\delta - \sin\alpha\ \sin\delta) \\
&\qquad\qquad + \cos\alpha\ \sin\delta\ \sin\pi\,\big]
\end{aligned}\right\}\quad (22.3)$$

The total corrections to α and δ, due to the nutation and the aberration, are therefore $\Delta\alpha_1 + \Delta\alpha_2$ and $\Delta\delta_1 + \Delta\delta_2$, respectively. Calculated from the above formulae, both are expressed in seconds of a degree (if $\Delta\psi$, $\Delta\varepsilon$ and κ are expressed in the same units).

Important remark. — Formulae (22.2) and (22.3) are the complete expressions for the components of the aberration. They include the so-called *E*-terms, and should be used for the star positions given in the FK5 [3] and in all catalogues based on it.

If, however, FK4 positions are used, those parts of formulae (22.2) and (22.3) that contain the eccentricity e of the orbit of the Earth should be dropped, as explained in Chapter 20.

Example 22.a — Calculate the apparent place of θ Persei for 2028 November 13.19 TD.

The mean position of this star for that instant, including the effect of proper motion, was found in Example 20.b, namely

$$\alpha = 2^h46^m11^s.331 = 41^\circ.5472 \qquad \delta = +49^\circ20'54''.54 = +49^\circ.3485$$

The nutations in longitude and in obliquity, for the same instant, can be found by means of the method given in Chapter 21. We obtain

$$\Delta\psi = +14''.861 \qquad \Delta\varepsilon = +2''.705$$

Formula (21.2) gives $\varepsilon = 23^\circ.436$, while the Sun's true longitude, calculated by means of the method ('low accuracy') given in Chapter 24, is $\odot = 231^\circ.328$. (An accuracy of 0.01 degree is sufficient in this case.) We further find

$$T = +0.288\,6705 \qquad e = 0.016\,696\,47 \qquad \pi = 103^\circ.434$$

Putting the values of α, δ, ε, $\Delta\psi$, $\Delta\varepsilon$, $\odot$, e and π in formulae (22.1) and (22.3), one finds

$\Delta\alpha_1 = +15''.843 \qquad \Delta\delta_1 = +6''.218$
$\Delta\alpha_2 = +30''.047 \qquad \Delta\delta_2 = +6''.696$

and the total corrections in right ascension and in declination are

$$\Delta\alpha = +15''.843 + 30''.047 = +45''.890 = +3^s.059$$
$$\Delta\delta = +6''.218 + 6''.696 = +12''.91$$

Hence, the required apparent coordinates of the star are

$$\alpha = 2^h46^m11^s.331 + 3^s.059 = 2^h46^m14^s.390$$
$$\delta = +49°20'54''.54 + 12''.91 = +49°21'07''.45$$

The Ron-Vondrák expression for aberration

Expressions (22.2) and (22.3) contain the effect of the eccentricity of the Earth's orbit and will provide quite accurate results. Nevertheless, these results are not rigorously exact because the said formulae are based on an unperturbed motion of the Earth in its elliptical orbit. Actually, the Earth's motion is somewhat perturbed by the attraction of the Moon and that of the planets. And the Sun itself is slowly moving around the center of mass of the solar system, mainly due to the actions of the giants Jupiter and Saturn.

If a very accurate result is required, stellar aberration must, in fact, be computed from the total velocity of the Earth referred to this barycenter. One method for performing this calculation has been presented by Ron and Vondrák [4].

If $T = (\mathrm{JD} - 2451\,545)/36525$ is, as before, the time in Julian centuries elapsed since J2000.0, then calculate, for the given instant, the following angles expressed in *radians* :

$$L2 = 3.1761467 + 1021.328\,5546\,T$$
$$L3 = 1.7534703 + 628.307\,5849\,T$$
$$L4 = 6.2034809 + 334.061\,2431\,T$$
$$L5 = 0.5995465 + 52.969\,0965\,T$$
$$L6 = 0.8740168 + 21.329\,9095\,T$$
$$L7 = 5.4812939 + 7.478\,1599\,T$$
$$L8 = 5.3118863 + 3.813\,3036\,T$$
$$L' = 3.8103444 + 8399.684\,7337\,T$$
$$D = 5.1984667 + 7771.377\,1486\,T$$
$$M' = 2.3555559 + 8328.691\,4289\,T$$
$$F = 1.6279052 + 8433.466\,1601\,T$$

The quantities $L2$ up to $L8$ are the mean longitudes of the planets Venus to Neptune referred to the mean equinox of J2000.0 (the effects of Mercury and Pluto are negligible), while L' is the mean longitude of the Moon.

TABLE 22.A

Velocity components of the Earth with respect to the center of mass of the solar system

No.	Argument	X' sin	X' cos	Y' sin	Y' cos	Z' sin	Z' cos
1	$L3$	-1719914 -2T	-25	25 -13T	1578089 +156T	10 +32T	684185 -358T
2	$2L3$	6434 +141T	28007 -107T	25697 -95T	-5904 -130T	11141 -48T	-2559 -55T
3	$L5$	715	0	6	-657	-15	-282
4	L'	715	0	0	-656	0	-285
5	$3L3$	486 -5T	-236 -4T	-216 -4T	-446 +5T	-94	-193
6	$L6$	159	0	2	-147	-6	-61
7	F	0	0	0	26	0	-59
8	$L' + M'$	39	0	0	-36	0	-16
9	$2L5$	33	-10	-9	-30	-5	-13
10	$2L3 - L5$	31	1	1	-28	0	-12
11	$3L3 - 8L4 + 3L5$	8	-28	25	8	11	3
12	$5L3 - 8L4 + 3L5$	8	-28	-25	-8	-11	-3
13	$2L2 - L3$	21	0	0	-19	0	-8
14	$L2$	-19	0	0	17	0	8
15	$L7$	17	0	0	-16	0	-7
16	$L3 - 2L5$	16	0	0	15	1	7
17	$L8$	16	0	1	-15	-3	-6
18	$L3 + L5$	11	-1	-1	-10	-1	-5
19	$2L2 - 2L3$	0	-11	-10	0	-4	0
20	$L3 - L5$	-11	-2	-2	9	-1	4
21	$4L3$	-7	-8	-8	6	-3	3
22	$3L3 - 2L5$	-10	0	0	9	0	4
23	$L2 - 2L3$	-9	0	0	-9	0	-4
24	$2L2 - 3L3$	-9	0	0	-8	0	-4

TABLE 22.A (cont.)

25	$2L6$	0	-9	-8	0	-3	0
26	$2L2 - 4L3$	0	-9	8	0	3	0
27	$3L3 - 2L4$	8	0	0	-8	0	-3
28	$L' + 2D - M'$	8	0	0	-7	0	-3
29	$8L2 - 12L3$	-4	-7	-6	4	-3	2
30	$8L2 - 14L3$	-4	-7	6	-4	3	-2
31	$2L4$	-6	-5	-4	5	-2	2
32	$3L2 - 4L3$	-1	-1	-2	-7	1	-4
33	$2L3 - 2L5$	4	-6	-5	-4	-2	-2
34	$3L2 - 3L3$	0	-7	-6	0	-3	0
35	$2L3 - 2L4$	5	-5	-4	-5	-2	-2
36	$L' - 2D$	5	0	0	-5	0	-2

Then the components X', Y', Z' of the velocity of the Earth with respect to the barycenter of the solar system, in the equatorial J2000.0 reference frame, are equal to the sums of the terms given in Table 22.A. Here, the argument of each sine and cosine is a linear combination of some of the angles $L2$, $L3$, etc. For instance, the terms on line 12 of the table have as argument the angle

$$A = 5L3 - 8L4 + 3L5$$

and the contributions to the velocity components are :

to X' : $+ 8 \sin A - 28 \cos A$

to Y' : $-25 \sin A - 8 \cos A$

to Z' : $-11 \sin A - 3 \cos A$

The values of X', Y', Z' thus obtained are expressed in units of 10^{-8} astronomical unit per day. Let c be the velocity of light in the same units, namely

$$c = 17\,314\,463\,350.$$

Then the changes in the star's right ascension and declination due to the annual aberration are, in *radians*, given by formulae (22.4).

$$\left.\begin{aligned}\Delta\alpha &= \frac{Y'\cos\alpha - X'\sin\alpha}{c\cos\delta}\\ \Delta\delta &= -\frac{(X'\cos\alpha + Y'\sin\alpha)\sin\delta - Z'\cos\delta}{c}\end{aligned}\right\}\qquad(22.4)$$

It is important to note that the Earth's velocity components, as calculated by means of Table 22.A, are given in a rectangular coordinate system based on the *fixed* equator and equinox of FK5 for the epoch J2000.0, *not* with respect to the mean equinox of the date. Consequently, if the Ron-Vondrák method for the calculation of the aberration is preferred instead of the formulae (22.3), then the corrections (22.4) should be performed *before* the calculation of the effects of precession and nutation. In other words, the sequence of the calculations will be: FK5 position (J2000.0), proper motion, aberration (Table 22.A and expressions 22.4), precession (expressions 20.3 and 20.4), nutation (Chapter 21 and expressions 22.1).

Example 22.b — Let us again calculate the apparent place of θ Persei for 2028 November 13.19 TD, but now using the Ron-Vondrák algorithm.

As in Example 20.b, we find that the star's coordinates for the epoch 2028 November 13.19, but referred to the mean equinox of J2000.0, are (allowing for proper motion)

$$\alpha = 2^h44^m12^s.9747 = +41^\circ.054\,0613$$
$$\delta = +49^\circ13'39''.896 = +49^\circ.227\,7489$$

We keep extra decimals here, in order to avoid rounding errors. We further find

T = +0.288 670 500
$L2$ = 298.003 5712 rad.
$L3$ = 183.127 3350
$L4$ = 102.637 1070
$L5$ = 15.890 1621
$L6$ = 7.031 3324
$L7$ = 7.640 0181
$L8$ = 6.412 6746

L' = 2428.551 5363 rad.
D = 2248.565 7939
M' = 2406.603 0750
F = 2436.120 7984

X' = −1363 700
Y' = + 990 286
Z' = + 429 285

Formulae (22.4) then give

$$\Delta\alpha = +0.000\,145\,252 \text{ radian} = +0^\circ.008\,3223$$
$$\Delta\delta = +0.000\,032\,723 \text{ radian} = +0^\circ.001\,8749$$

so that the new values for α and δ, corrected for aberration, but still in the J2000.0 reference frame, are

$$\alpha = 41^\circ.054\,0613 + 0^\circ.008\,3223 = +41^\circ.062\,3836$$
$$\delta = 49^\circ.227\,7489 + 0^\circ.001\,8749 = +49^\circ.229\,6238$$

The effect of precession is obtained by means of formulae (20.4). The values of ζ, z and θ, for the same instant, were found in Example 20.b. We now find

$$A = +0.430\,549\,036$$
$$B = +0.488\,867\,290$$
$$C = +0.758\,706\,993$$

$$\text{new } \alpha = +41°.555\,5635$$
$$\text{new } \delta = +49°.350\,3415$$

Finally, the corrections for the nutation are given by (22.1). As in Example 22.a, we have $\Delta\psi = +14''.861$, $\Delta\varepsilon = +2''.705$, and $\varepsilon = 23°.436$. We find

$$\Delta\alpha_1 = +15''.844 = +0°.004\,4011$$
$$\Delta\delta_1 = +6''.217 = +0°.001\,7270$$

Hence, the required apparent right ascension and declination are

$$\alpha = 41°.555\,5635 + 0°.004\,4011 = 41°.559\,9646 = 2^h46^m14^s.392$$

$$\delta = 49°.350\,3415 + 0°.001\,7270 = +49°.352\,0685 = +49°21'07''.45$$

Compare these results with those of Example 22.a.

References

1. *Sky and Telescope*, Vol. 77, page 288 (March 1989).
2. Robert Burnham, *Burnham's Celestial Handbook*, Vol. III, page 2126 (Dover Publications, New York; 1978).
3. Fifth Fundamental Catalogue (FK5), *Veröffentlichungen Astronomisches Rechen-Institut Heidelberg*, No. 32 (Karlsruhe, 1988).
4. C. Ron, J. Vondrák, "Expansion of Annual Aberration into Trigonometric Series", *Bull. Astron. Inst. Czechosl.*, Vol. 37, pages 96 - 103 (1986).

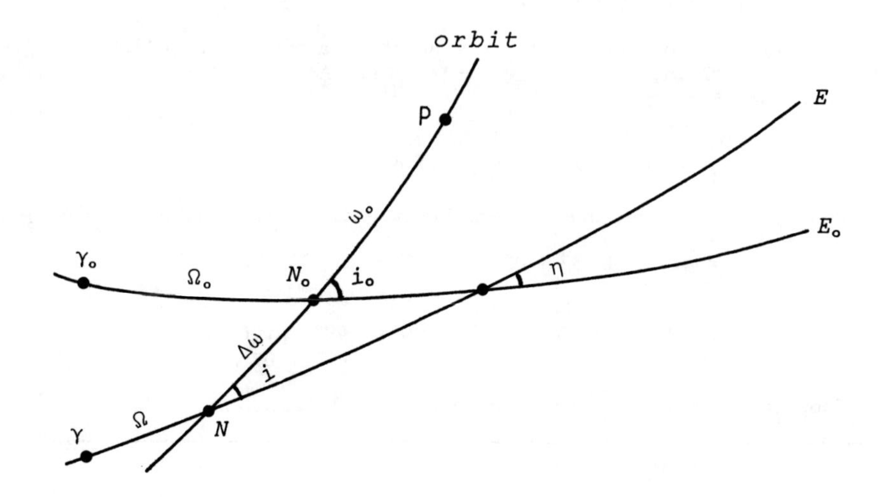
orbit
P
E
ω₀
E₀
γ₀
Ω₀
N₀
i₀
η
Δω
i
Ω
N
γ

Chapter 23

Reduction of Ecliptical Elements from one Equinox to another one

For some problems, it may be necessary to reduce orbital elements of a planet, a minor planet or a comet from one equinox to another one. Of course, the semimajor axis a and the eccentricity e do not change when the orbit is referred to another equinox, and hence only the three elements

i = inclination,
ω = argument of perihelion,
Ω = longitude of ascending node

should be taken into consideration here. Let i_0, ω_0, Ω_0 be the known values of these elements at the initial epoch, and i, ω, Ω their (unknown) values at the final epoch.

In the Figure on page 146, E_0 and γ_0 are the ecliptic and the vernal equinox at the initial epoch, and E and γ the ecliptic and equinox at the final epoch. The angle between the two ecliptics is denoted by η, and the orbit's perihelion by P.

As in Chapter 20, let T be the time interval, in Julian centuries, between J2000.0 and the initial epoch, and t the interval, in the same units, between the initial epoch and the final epoch.

Then calculate the angles η, Π and p by means of formulae (20.5) or, if the initial epoch is J2000.0, by means of (20.6).

Find $\psi = \Pi + p$. Then the quantities i and $\Omega - \psi$, and hence Ω, can be calculated from

$$\cos i = \cos i_0 \cos \eta + \sin i_0 \sin \eta \cos(\Omega_0 - \Pi) \qquad (23.1)$$

$$\begin{aligned} \sin i \sin(\Omega - \psi) &= \sin i_0 \sin(\Omega_0 - \Pi) \\ \sin i \cos(\Omega - \psi) &= -\sin \eta \cos i_0 + \cos \eta \sin i_0 \cos(\Omega_0 - \Pi) \end{aligned} \qquad (23.2)$$

Formula (23.1) should not be used when the inclination is small.

Then $\omega = \omega_0 + \Delta\omega$, where $\Delta\omega$ is found from

$$\begin{aligned} \sin i \ \sin\Delta\omega &= -\sin\eta \ \sin(\Omega_0 - \Pi) \\ \sin i \ \cos\Delta\omega &= \sin i_0 \ \cos\eta - \cos i_0 \ \sin\eta \ \cos(\Omega_0 - \Pi) \end{aligned} \tag{23.3}$$

If $i_0 = 0$, then Ω_0 is not determined, and we have $i = \eta$ and $\Omega = \psi + 180°$.

It is important to note that the method described here reduces the orbital elements i, ω and Ω from one *equinox* to another one, but the new orbital elements remain valid for the same *epoch* as the initial elements. It is, in fact, the same orbit. The calculation of the orbital elements for another *epoch* is a completely different problem (celestial mechanics!) which we cannot discuss here.

Example 23.a — In their *Catalogue Général des Orbites de Comètes de l'an -466 à 1952* [Observatoire de Paris, Section d'Astrophysique de Meudon (1952)], F. Baldet and G. De Obaldia give the following orbital elements for comet Klinkenberg (1744), referred to the mean equinox of B1744.0 :

$$i_0 = 47°.1220$$
$$\omega_0 = 151°.4486$$
$$\Omega_0 = 45°.7481$$

Reduce these elements to the standard equinox of B1950.0.

The final epoch is B1950.0, or (JD) = 2433 282.4235 (see Chapter 20), and the initial epoch is 206 *tropical* years earlier (because both epochs correspond to the beginning of a Besselian year), whence

$$(\mathrm{JD})_0 = 2433\,282.4235 - (206 \times 365.242\,1988) = 2358\,042.5305.$$

We then find

$$T = -2.559\,958\,097$$
$$t = +2.059\,956\,002$$
$$\eta = +97''.0341 = +0°.026\,954$$
$$\Pi = 174°.876\,384 - 10205''.9108 = 172°.041\,409$$
$$p = +10\,352''.7137 = +2°.875\,754$$
$$\psi = 174°.917\,163$$

Then formulae (23.2) give

$$\sin i \ \sin(\Omega - \psi) = -0.5906\,3831 = A$$
$$\sin i \ \cos(\Omega - \psi) = -0.4340\,8084 = B$$

from which we deduce $\sin i = \sqrt{A^2 + B^2} = 0.7329\,9372, \quad i = 47°.1380$

$$\Omega - \psi = \mathrm{ATN2}\,(A, B) = -126°.313\,473$$
$$\Omega = 48°.6037$$

Formulae (23.3) give $\sin i \ \sin\Delta\omega = +0.0003\,7917$

$$\sin i \ \cos\Delta\omega = +0.7329\,9362$$

whence $\Delta\omega = +0^\circ.0296$, and $\omega = 151^\circ.4782$.

In his *Catalogue of Cometary Orbits* (sixth edition; 1989), Marsden gives the values $i = 47^\circ.1378$, $\omega = 151^\circ.4783$, $\Omega = 48^\circ.6030$. The discrepancy of 0°.0007 between the values of Ω results from the fact that the new IAU precession formulae yield for the general precession in longitude a value which is a little larger (+1″.1 per century) than that adopted by Newcomb. The effect over 206 years (from 1744 to 1950) amounts to 0.0006 degree.

If the initial equinox is that of B1950.0, and the final equinox that of J2000.0, then the formulae simplify to the following ones.

$$\left.\begin{aligned}
&S = 0.000\,113\,9788 \qquad C = 0.999\,999\,9935\\
&W = \Omega_0 - 174^\circ.298\,782\\
&A = \sin i_0 \sin W\\
&B = C \sin i_0 \cos W - S \cos i_0\\
&\sin i = \sqrt{A^2 + B^2} \qquad \tan x = \frac{A}{B}\\
&\Omega = 174^\circ.997\,194 + x\\
&\text{and finally} \quad \omega = \omega_0 + \Delta\omega, \quad \text{with}\\
&\tan \Delta\omega = \frac{-S \sin W}{C \sin i_0 - S \cos i_0 \cos W}
\end{aligned}\right\} \qquad (23.4)$$

Care must be taken for the correct quadrant of the angles x and $\Delta\omega$. For safety, they should be calculated by means of the ATN2 function, if the latter is available in the computer language, for instance $x = \text{ATN2}(A, B)$. Except when the orbital inclination is *very* small, the new value of Ω should be approximately 0°.7 larger than the initial value Ω_0, and $\Delta\omega$ must lie near 0°, not near 180°.

Example 23.b — S. Nakano calculated the following orbital elements for the 1990 return of periodic comet Encke (*Minor Planet Circular* 12577) :

Epoch = 1990 Nov 5.0 TD = JDE 2448 200.5

T = 1990 Oct. 28.54502 TD

q = 0.330 8858

a = 2.209 1404

e = 0.850 2196

$$\left.\begin{aligned} i &= 11^\circ.93911\\ \Omega &= 334^\circ.04096\\ \omega &= 186^\circ.24444 \end{aligned}\right\} 1950.0$$

We wish to reduce i, Ω and ω to the equinox J2000.0, and we find successively

$$W = +159^\circ.742\,178$$
$$A = +0.071\,628\,4465$$
$$B = -0.194\,187\,3149$$

$$\sin i = 0.206\,9767 \qquad \Omega = 334^\circ.75006$$
$$i = 11^\circ.94524 \qquad \Delta\omega = -0^\circ.01092$$
$$x = +159^\circ.752\,866 \qquad \omega = 186^\circ.23352$$

The other orbital elements (T, q, a, e) remain unchanged, and the Epoch is still 1990 November 5.0.

However, formulae (23.4) assume that the elements i_o, ω_o and Ω_o are given in the FK5 system. To convert elements from B1950.0/FK4 to J2000.0/FK5, one may use the following algorithm due to Yeomans (Note from D. K. Yeomans, Chairman IAU System Transition Committee, to Richard West, President of IAU Commission 20; 1990 August 10).

Let

$$L' = 4.500\,016\,88 \text{ degrees}$$
$$L = 5.198\,562\,09 \text{ degrees}$$
$$J = 0.006\,519\,66 \text{ degrees}$$
$$W = L + \Omega_o$$

Then we have

$$\sin(\omega - \omega_o) \sin i = \sin J \sin W$$
$$\cos(\omega - \omega_o) \sin i = \sin i_o \cos J + \cos i_o \sin J \cos W$$
$$\cos i = \cos i_o \cos J - \sin i_o \sin J \cos W$$
$$\sin(L' + \Omega) \sin i = \sin i_o \sin W$$
$$\cos(L' + \Omega) \sin i = \cos i_o \sin J + \sin i_o \cos J \cos W$$

from which i, Ω and ω can be deduced.

Example 23.c — Same starting values i_o, Ω_o and ω_o as in Example 23.b.

We obtain

$$\left.\begin{array}{l} i = 11^\circ.94521 \\ \Omega = 334.75043 \\ \omega = 186.23327 \end{array}\right\} \text{FK5, J2000.0}$$

Chapter 24

Solar Coordinates

Low accuracy

When an accuracy of 0.01 degree is sufficient, the position of the Sun may be calculated by assuming a purely elliptical motion of the Earth; that is, the perturbations by the Moon and the planets may be neglected. The calculation can be performed as follows.

Let JD be the Julian (Ephemeris) Day, which can be calculated by means of the method described in Chapter 7. Then the time T, measured in Julian centuries of 36525 ephemeris days from the epoch J2000.0 (2000 January 1.5 TD), is given by

$$T = \frac{\text{JD} - 2451\,545.0}{36525} \tag{24.1}$$

This quantity should be calculated with a sufficient number of decimals. For instance, five decimals are not sufficient (unless the Sun's longitude is required with an accuracy not better than one degree): remember that T is expressed in centuries, so that an error of 0.00001 in T corresponds to an error of 0.37 day in the time.

Then the geometric mean longitude of the Sun, referred to the mean equinox of the date, is given by

$$L_0 = 280°.46645 + 36\,000°.769\,83\,T + 0°.000\,3032\,T^2 \tag{24.2}$$

The mean anomaly of the Sun is

$$M = 357°.52910 + 35\,999°.050\,30\,T - 0°.000\,1559\,T^2 - 0°.000\,000\,48\,T^3 \tag{24.3}$$

The eccentricity of the Earth's orbit is

$$e = 0.016\,708\,617 - 0.000\,042\,037\,T - 0.000\,000\,1236\,T^2 \tag{24.4}$$

Then find the Sun's equation of center C as follows:

$$C = +(1°.914\,600 - 0°.004\,817\,T - 0°.000\,014\,T^2)\ \sin M$$
$$+ (0°.019\,993 - 0°.000\,101\,T)\ \sin 2M$$
$$+ 0°.000\,290\ \sin 3M$$

Then the Sun's true longitude is

$$\odot = L_o + C$$

and its true anomaly is $v = M + C$.

The Sun's radius vector, or the distance from the Earth to the Sun, expressed in astronomical units, is given by

$$R = \frac{1.000\,001\,018\ (1 - e^2)}{1 + e \cos v} \qquad (24.5)$$

The numerator of the fraction is a quantity which varies slowly with time. It is equal to

0.999 7190	in the year 1800
0.999 7204	1900
0.999 7218	2000
0.999 7232	2100

The Sun's longitude $\odot$, obtained by the method described above, is the true *geometric* longitude referred to the *mean* equinox of the date. This longitude is the quantity required for instance in the calculation of geocentric planetary positions.

If the *apparent* longitude λ of the Sun, referred to the *true* equinox of the date, is required, it is necessary to correct $\odot$ for the nutation and the aberration. Unless high accuracy is required, this can be performed as follows.

$$\Omega = 125°.04 - 1934°.136\,T$$
$$\lambda = \odot - 0°.00569 - 0°.00478\ \sin \Omega$$

In some instances, for example in meteor work, it is necessary to have the Sun's longitude referred to the standard equinox of J2000.0. Between the years 1900 and 2100, this can be performed with sufficient accuracy from

$$\odot_{2000} = \odot - 0°.01397\ (\text{year} - 2000)$$

If the Sun's longitude, referred to the standard equinox of J2000.0, should be obtained with a higher accuracy than 0.01 degree, then the method given in Chapter 25 can be used.

The Sun's latitude, referred to the ecliptic of the date, never exceeds 1".2. Unless high accuracy is required, this latitude may be put equal to zero. In that case, the Sun's right ascension α and declination δ can be calculated from (24.6) and (24.7), where ε, the obliquity of the ecliptic, is given by (21.2).

$$\tan \alpha = \frac{\cos \varepsilon \ \sin \odot}{\cos \odot} \qquad (24.6)$$

$$\sin \delta = \sin \varepsilon \ \sin \odot \qquad (24.7)$$

If the *apparent* position of the Sun is required, then in formulae (24.6) and (24.7) one should use λ instead of $\odot$, and ε should be corrected by the quantity

$$+0^\circ.00256 \cos \Omega \qquad (24.8)$$

Formula (24.6) may of course be transformed to

$$\tan \alpha = \cos \varepsilon \ \tan \odot$$

and then it must be remembered that α must be in the same quadrant as $\odot$. However, if the ATN2 function is available in the computer language, it is better to leave formula (24.6) unchanged, and to apply the ATN2 function to the numerator and the denominator of the fraction: α = ATN2 ($\cos \varepsilon \ \sin \odot$, $\cos \odot$).

Example 24.a — Calculate the Sun's position on 1992 October 13 at 0^h TD = JDE 2448 908.5.

We find successively :

T = −0.072 183 436
L_0 = −2318°.192 81 = 201°.807 19
M = −2241°.006 04 = 278°.993 96
e = 0.016 711 651
C = −1°.897 32
$\odot$ = 199°.909 87 = 199°54'36"
R = 0.997 66
Ω = 264°.65
λ = 199°.908 94 = 199°54'32"
ε_0 = 23°26'24".83 = 23°.440 23 [by (21.2)]
ε = 23°.439 99

α_{app} = −161°.619 18 = +198°.380 82 = $13^h.225\,388$ = $13^h 13^m 31^s.4$
δ_{app} = −7°.785 07 = −7°47'06"

The correct values, calculated by means of the complete VSOP87 theory (see Chapter 31), are :

geometric long., mean equinox of date :	$\odot$ = 199°54'26".18
apparent longitude :	λ = 199°54'21".56
apparent latitude :	β = +0".72
radius vector :	R = 0.997 608 53
apparent right ascension :	$13^h 13^m 30^s.749$
apparent declination :	−7°47'01".74

Higher accuracy

In their book [1], Bretagnon and Simon give a method for the calculation of the longitude of the Sun with an accuracy that is sufficient for many applications. Their method yields an accuracy of 0.0006 degree (2″.2) between the years 0 and +2800, and of 0.0009 degree (3″.2) between -4000 and +8000, yet only 49 periodic terms are used.

A very high accuracy, better than 0.01 arcsecond, is obtained when use is made of the complete VSOP87 theory (see Chapter 31), but for the Earth this theory contains 2425 periodic terms on the magnetic tape provided by the Bureau des Longitudes, namely 1080 terms for the Earth's longitude, 348 for the latitude, and 997 for the radius vector. Evidently, this big amount of numerical data cannot be reproduced in this book. Instead, we give in Appendix II the most important terms from the VSOP87, allowing the calculation of the position of the Sun with an error not exceeding 1″ between the years -2000 and +6000. The procedure is as follows.

Using from Appendix II the data for the *Earth*, calculate the latter's heliocentric longitude L, latitude B and radius vector R for the given instant, as explained in Chapter 31. Don't forget that the time τ is measured from JDE 2451 545.0 in Julian *millennia* (365250 days), not in centuries, and that the final values obtained for L and B are in radians.

To obtain the *geocentric* longitude $\odot$ and latitude β of the Sun, add 180° (or π radians) to L, and change the sign of B :

$$\odot = L + 180^\circ, \qquad \beta = -B$$

Conversion to the FK5 system. — The Sun's longitude $\odot$ and latitude β obtained thus far are referred to the mean *dynamical* ecliptic and equinox of the date defined by the VSOP planetary theory of P. Bretagnon. This reference frame differs very slightly from the standard FK5 system mentioned in Chapter 20. The conversion of $\odot$ and β to the FK5 system can be performed as follows, where T is the time in centuries from 2000.0, or $T = 10\,\tau$.

Calculate

$$\lambda' = \odot - 1^\circ.397\,T - 0^\circ.00031\,T^2$$

Then the corrections to $\odot$ and β are

$$\begin{aligned} \Delta\odot &= -0''.09033 \\ \Delta\beta &= +0''.03916\,(\cos\lambda' - \sin\lambda') \end{aligned} \qquad (24.9)$$

These corrections are needed only for very accurate calculations. They may be dropped if use is made of the abridged version of the VSOP87 given in Appendix II.

Apparent Place of the Sun. — The Sun's longitude $\odot$ obtained thus far is the true ('geometric') longitude of the Sun referred to the mean equinox of the date. To obtain the apparent longitude λ, the effects of nutation and aberration should be taken into account.

For the nutation, simply add to Θ the nutation in longitude $\Delta\psi$ (see Chapter 21).

To take the aberration into account, apply to the Sun's geometric longitude the correction

$$- \frac{20''.4898}{R} \qquad (24.10)$$

where R is the Earth's radius vector in astronomical units. The numerator of the fraction is equal to the constant of aberration (κ = 20".49552) multiplied by $a(1 - e^2)$, the same as the numerator in formula (24.5). Therefore, the numerator of (24.10) actually varies very slowly with time, being equal to 20".4893 in the year 0, and to 20".4904 in the year +4000.

But, more important, formula (24.10) will not give a rigorously exact result, because it assumes an unperturbed motion of the Earth in its elliptical orbit. By reason of perturbations, mainly due to the Moon, the result can be up to 0".01 in error.

When a very high accuracy is needed — this is not the case when the data of Appendix II are used for the calculation — the correction to the Sun's longitude due to the aberration can be obtained as follows. Find the variation $\Delta\lambda$ of the Sun's longitude, in arcseconds per day, as explained below. The correction for aberration is then

$$- 0.005\,775\,518 \; R \; \Delta\lambda \qquad (24.11)$$

where R is, as before, the Sun's radius vector in astronomical units. The numerical constant is the light-time for unit distance, in days (= 8.3 minutes).

After the Sun's longitude has been corrected for nutation and aberration, we have obtained the Sun's *apparent longitude* λ.

The apparent longitude λ and latitude β of the Sun can then be transformed into the apparent right ascension α and declination δ by means of formulae (12.3) and (12.4), where ε is the true obliquity of the ecliptic, that is, affected by the nutation in obliquity $\Delta\varepsilon$.

The variation $\Delta\lambda$ of the geocentric longitude of the Sun, in arcseconds per day, in the fixed reference frame J2000.0, can be obtained by means of the formula given on the next page, where τ is the time in millennia from J2000.0 (as in Chapter 31), and the arguments of the sines are in *degrees* and decimals.

In that expression given here, only the most important periodic terms have been retained. Consequently, the result will not be rigorous, but $\Delta\lambda$ will not be more than 0".1 in error. If the resulting value of $\Delta\lambda$ is used to calculate the Sun's aberration by means of (24.11), the error will be less than 0".001.

If, for some other application, the value of $\Delta\lambda$ is needed with respect to the mean equinox of date instead of to a fixed reference frame, the constant term 3548.193 should be replaced by 3548.330.

Daily variation, in arcseconds, of the geocentric longitude of the Sun in a fixed reference frame

The time τ is measured from J2000.0 (JDE 2451 545.0) in Julian millenia.

The arguments of the sines are in degrees.

$$
\begin{aligned}
\Delta\lambda = 3548.193 & \\
& + 118.568 \sin(87.5287 + 359\,993.7286\,\tau) \\
& + 2.476 \sin(85.0561 + 719\,987.4571\,\tau) \\
& + 1.376 \sin(27.8502 + 4452\,671.1152\,\tau) \\
& + 0.119 \sin(73.1375 + 450\,368.8564\,\tau) \\
& + 0.114 \sin(337.2264 + 329\,644.6718\,\tau) \\
& + 0.086 \sin(222.5400 + 659\,289.3436\,\tau) \\
& + 0.078 \sin(162.8136 + 9224\,659.7915\,\tau) \\
& + 0.054 \sin(82.5823 + 1079\,981.1857\,\tau) \\
& + 0.052 \sin(171.5189 + 225\,184.4282\,\tau) \\
& + 0.034 \sin(30.3214 + 4092\,677.3866\,\tau) \\
& + 0.033 \sin(119.8105 + 337\,181.4711\,\tau) \\
& + 0.023 \sin(247.5418 + 299\,295.6151\,\tau) \\
& + 0.023 \sin(325.1526 + 315\,559.5560\,\tau) \\
& + 0.021 \sin(155.1241 + 675\,553.2846\,\tau) \\
& + 7.311\,\tau \sin(333.4515 + 359\,993.7286\,\tau) \\
& + 0.305\,\tau \sin(330.9814 + 719\,987.4571\,\tau) \\
& + 0.010\,\tau \sin(328.5170 + 1079\,981.1857\,\tau) \\
& + 0.309\,\tau^2 \sin(241.4518 + 359\,993.7286\,\tau) \\
& + 0.021\,\tau^2 \sin(205.0482 + 719\,987.4571\,\tau) \\
& + 0.004\,\tau^2 \sin(297.8610 + 4452\,671.1152\,\tau) \\
& + 0.010\,\tau^3 \sin(154.7066 + 359\,993.7286\,\tau)
\end{aligned}
$$

The periodic terms where τ has the coefficient 359 993.7, 719 987, or 1079 981, are due to the eccentricity of the Earth's orbit. The terms with 4452 671, 9224 660, or 4092 677 are due to the action of the Moon; those with 450 369, 225 184, 315 560, or 675 553 are due to Venus; those with 329 645, 659 289, or 299296 are due to Jupiter; finally, the term with 337 181 is due to the action of Mars.

Example 24.b — Let us again, as in Example 24.a, calculate the position of the Sun for 1992 October 13.0 TD = JDE 2448 908.5.

Using from Appendix II the data for the Earth, we find by the method explained in Chapter 31,

$$L = -43.634\,847\,96 \text{ radians} = -2500.092\,628 \text{ degrees}$$
$$+19.907\,372 \text{ degrees}$$

$$B = -0.000\,003\,12 \text{ radian} = -0°.000\,179 = -0''.644$$

$$R = 0.997\,607\,75$$

Whence

$$\odot = L + 180° = 199°.907\,372$$
$$\beta = +0''.644$$

Converting to the FK5 system, we find

$$\lambda' = 200°.01 \qquad \Delta\odot = -0''.09033 = -0°.000\,025 \qquad \Delta\beta = -0''.023$$

whence

$$\odot = 199°.907\,347 = 199°54'26''.449 \qquad \beta = +0''.62$$

The nutation is calculated by means of the method described in Chapter 21. We find

$$\Delta\psi = +15''.908 \qquad \Delta\varepsilon = -0''.308 \qquad \text{true } \varepsilon = 23°.440\,1443$$

and by (24.10) the correction for aberration is $-20''.539$.

Hence, the Sun's apparent longitude is

$$\lambda = \odot + 15''.908 - 20''.539 = 199°54'21''.818$$

Then, by (12.3) and (12.4),

$$\alpha = 198°.378\,178 = 13^h13^m30^s.763$$
$$\delta = -7°.783\,871 = -7°47'01''.94$$

Resuming, the final results are

$$\odot = 199°54'26''.45 \qquad R = 0.997\,607\,75$$
$$\lambda = 199°54'21''.82 \qquad \alpha = 13^h13^m30^s.763$$
$$\beta = +0''.62 \qquad \delta = -7°47'01''.94$$

Compare these results with the correct values mentioned at the end of Example 24.a. Our results are now much better than those obtained with the low-accuracy method.

---

1. P. Bretagnon and J.-L. Simon, *Planetary Programs and Tables from -4000 to +2800* (Willmann-Bell, Richmond, 1986).

Chapter 25

Rectangular Coordinates of the Sun

The rectangular geocentric equatorial coordinates X, Y, Z of the Sun are needed for the calculation of an ephemeris of a minor planet (see Chapter 32) or a comet. The origin of these coordinates is the center of the Earth. The X-axis is directed towards the vernal equinox (longitude 0°); the Y-axis lies in the plane of the equator too and is directed towards longitude 90°, while the Z-axis is directed towards the north celestial pole.

The values of X, Y, Z are given for each day at 0^h TD in the great astronomical almanacs; they are expressed in astronomical units. Generally they are not referred to the mean equator and mean equinox of the date, but to a standard equinox, for instance that of J2000.0.

Reference to the mean equinox of the date

Calculate the *geometric* coordinates of the Sun by means of the method ('higher accuracy') described in Chapter 24, that is, *with* the corrections (24.9) for reduction to the FK5 system, but *without* the corrections for nutation and aberration.

If $\odot$ and β are the geometric longitude and latitude of the Sun, and R its radius vector in astronomical units, then the required rectangular coordinates of the Sun, referred to the mean equator and equinox of the date, are given by

$$\begin{aligned} X &= R \cos\beta \cos\odot \\ Y &= R\,(\cos\beta \sin\odot \cos\varepsilon - \sin\beta \sin\varepsilon) \\ Z &= R\,(\cos\beta \sin\odot \sin\varepsilon + \sin\beta \cos\varepsilon) \end{aligned} \qquad (25.1)$$

where ε is the *mean* obliquity of the ecliptic given by (21.2).

Since the Sun's latitude, referred to the ecliptic *of the date*, never exceeds 1.2 arcsecond, one may safely put $\cos\beta = 1$ in the formulae (25.1).

Example 25.a — For 1992 October 13.0 TD = JDE 2448 908.5, we have found in Example 24.b :

$$\odot = 199°.907\,347 \qquad \beta = +0''.62$$

$$R = 0.997\,607\,75$$

For the given instant, formula (21.2) gives

$$\varepsilon = 23°26'24''.827 = 23°.440\,2297$$

whence, by (25.1),

$$X = -0.937\,9952$$
$$Y = -0.311\,6544$$
$$Z = -0.135\,1215$$

Reference to the standard equinox J2000.0

As explained in Chapter 31, calculate for the given instant the Earth's heliocentric longitude L and latitude B referred to the equinox of J2000.0, and the radius vector. For this purpose, use from Appendix II the data for the Earth, *with the following exceptions* :

- in section L1, replace the first value of the coefficient 'A', namely 628 331 966 747, by 628 307 584 999 ;
- sections L2, L3 and L4 should be replaced by those given in Table 25.A ;
- drop section L5 ;
- for the calculation of the latitude B, use section B0 from Appendix II, but sections B1 to B4 from Table 25.A.

Obtain the geocentric longitude $\odot$ of the Sun by adding 180° (or π radians) to L, and the Sun's latitude β by changing the sign of B. That is,

$$\odot = L + 180°, \qquad \beta = -B$$

[At this stage, if *only* the Sun's geometric longitude referred to the standard equinox of J2000.0 is required, subtract 0''.09033 from $\odot$ in order to convert the longitude from the VSOP dynamical equinox to the FK5 equinox, as in (24.9). — Otherwise, do *not* perform this correction and proceed as follows.]

Calculate

$$\begin{aligned} X &= R \cos\beta \cos\odot \\ Y &= R \cos\beta \sin\odot \\ Z &= R \sin\beta \end{aligned} \qquad (25.2)$$

TABLE 25.A

EARTH J2000.0 (some terms only)

	No.	A	B	C
L2	1	8722	1.0725	6283.0758
	2	991	3.1416	0
	3	295	0.437	12566.152
	4	27	0.05	3.52
	5	16	5.19	26.30
	6	16	3.69	155.42
	7	9	0.30	18849.23
	8	9	2.06	77713.77
	9	7	0.83	775.52
	10	5	4.66	1577.34
	11	4	1.03	7.11
	12	4	3.44	5573.14
	13	3	5.14	796.30
	14	3	6.05	5507.55
	15	3	1.19	242.73
	16	3	6.12	529.69
	17	3	0.30	398.15
	18	3	2.28	553.57
	19	2	4.38	5223.69
	20	2	3.75	0.98
L3	1	289	5.842	6283.076
	2	21	6.05	12566.15
	3	3	5.20	155.42
	4	3	3.14	0
	5	1	4.72	3.52
	6	1	5.97	242.73
	7	1	5.54	18849.23
L4	1	8	4.14	6283.08
	2	1	3.28	12566.15
B1	1	227778	3.413766	6283.075850
	2	3806	3.3706	12566.1517
	3	3620	0	0
	4	72	3.33	18849.23
	5	8	3.89	5507.55
	6	8	1.79	5223.69
	7	6	5.20	2352.87
B2	1	9721	5.1519	6283.07585
	2	233	3.1416	0
	3	134	0.644	12566.152
	4	7	1.07	18849.23
B3	1	276	0.595	6283.076
	2	17	3.14	0
	3	4	0.12	12566.15
B4	1	6	2.27	6283.08
	2	1	0	0

(Of course, these expressions are equivalent to $X = -R \cos B \cos L$, $Y = -R \cos B \sin L$, and $Z = -R \sin B$, respectively.)

The rectangular coordinates X, Y, Z, calculated by means of (25.2), are still defined in the ecliptical dynamical reference frame (VSOP) of J2000.0. They can be transformed into the equatorial FK5 J2000.0 reference frame as follows :

$$
\begin{aligned}
X_0 &= X + 0.000\,000\,440\,360\,Y - 0.000\,000\,190\,919\,Z \\
Y_0 &= -0.000\,000\,479\,966\,X + 0.917\,482\,137\,087\,Y - 0.397\,776\,982\,902\,Z \\
Z_0 &= 0.397\,776\,982\,902\,Y + 0.917\,482\,137\,087\,Z
\end{aligned}
\tag{25.3}
$$

Reference to the mean equinox of B1950.0

Proceed as above for J2000.0, except that expressions (25.3) should be replaced by the following ones.

$$
\begin{aligned}
X_0 &= 0.999\,925\,702\,634\,X + 0.012\,189\,716\,217\,Y + 0.000\,011\,134\,016\,Z \\
Y_0 &= -0.011\,179\,418\,036\,X + 0.917\,413\,998\,946\,Y - 0.397\,777\,041\,885\,Z \\
Z_0 &= -0.004\,859\,003\,787\,X + 0.397\,747\,363\,646\,Y + 0.917\,482\,111\,428\,Z
\end{aligned}
$$

It should be noted that the rectangular coordinates obtained in this manner are referred to the mean equator and equinox of the epoch B1950.0 in the FK5 system, not in the FK4 system which is affected by the 'equinox error' as mentioned in Chapter 20.

Reference to any other mean equinox

First, calculate the Sun's rectangular equatorial coordinates X_0, Y_0, Z_0 referred to the standard equinox of J2000.0 as explained above, that is, by means of the formulae (25.2) and (25.3).

Then, if JD is the Julian Day corresponding to the epoch of the given equinox, calculate

$$t = \frac{\text{JD} - 2451\,545.0}{36525}$$

and then the angles ζ, z and θ from (20.3).

Then the required rectangular coordinates of the Sun are given by

$$
\begin{aligned}
X' &= X_x X_0 + Y_x Y_0 + Z_x Z_0 \\
Y' &= X_y X_0 + Y_y Y_0 + Z_y Z_0 \\
Z' &= X_z X_0 + Y_z Y_0 + Z_z Z_0
\end{aligned}
$$

where

$$X_x = \cos\zeta \cos z \cos\theta - \sin\zeta \sin z$$

$$X_y = \sin\zeta \cos z + \cos\zeta \sin z \cos\theta$$

$$X_z = \cos\zeta \sin\theta$$

$$Y_x = -\cos\zeta \sin z - \sin\zeta \cos z \cos\theta$$

$$Y_y = \cos\zeta \cos z - \sin\zeta \sin z \cos\theta$$

$$Y_z = -\sin\zeta \sin\theta$$

$$Z_x = -\cos z \sin\theta$$

$$Z_y = -\sin z \sin\theta$$

$$Z_z = \cos\theta$$

It should be noted that the coordinates X', Y', Z' are referred to the mean equinox of an epoch which differs from the date for which the values are calculated.

Example 25.b — For 1992 October 13.0 TD = JDE 2448 908.5, calculate the equatorial rectangular coordinates of the Sun referred to

(a) the standard equinox of J2000.0;
(b) that of B1950.0;
(c) the mean equinox of J2044.0.

We find successively

$\tau = -0.007\,218\,343\,6003$

$L = -43.633\,088\,03$ radians $= -2499.991\,791$ degrees
$= +20.008\,209$ degrees

$B = +0.000\,003\,86$ radian $= +0^\circ.000\,221 = +0''.796$

$R = 0.997\,607\,75$ (as in Example 24.b, of course)

$$\left.\begin{array}{l} X = -0.937\,395\,75 \\ Y = -0.341\,336\,25 \\ Z = -0.000\,003\,85 \end{array}\right\} \text{ecliptic, dynamical equinox, J2000.0}$$

$$\left.\begin{array}{l} X_0 = -0.937\,395\,90 \\ Y_0 = -0.313\,167\,93 \\ Z_0 = -0.135\,779\,24 \end{array}\right\} \text{equatorial FK5 frame, J2000.0}$$

The correct values, obtained by means of an accurate calculation using the complete VSOP87 theory, are −0.937 397 07, −0.313 167 25 and −0.135 778 42, respectively.

$$\left.\begin{aligned} X_\circ &= -0.941\,487 \\ Y_\circ &= -0.302\,666 \\ Z_\circ &= -0.131\,214 \end{aligned}\right\} \begin{array}{l}\text{equatorial,}\\ \text{FK5 system,}\\ \text{B1950.0 frame}\end{array}$$

JD = 2467 616.0 (since the epoch J2044.0 is 44 × 365.25 days later than J2000.0)

t = +0.440 000

ζ = +1014".7959 = +0°.281 8878
z = +1014".9494 = +0°.281 9304
θ = + 881".8106 = +0°.244 9474

X_x = +0.999 9424 Y_x = −0.009 8403 Z_x = −0.004 2751

X_y = +0.009 8403 Y_y = +0.999 9516 Z_y = −0.000 0210

X_z = +0.004 2751 Y_z = −0.000 0210 Z_z = +0.999 9909

$$\left.\begin{aligned} X' &= -0.933\,680 \\ Y' &= -0.322\,374 \\ Z' &= -0.139\,779 \end{aligned}\right\} \begin{array}{l}\text{equatorial,}\\ \text{FK5 system,}\\ \text{J2044.0 frame}\end{array}$$

Chapter 26

Equinoxes and Solstices

The times of the equinoxes and solstices are the instants when the apparent geocentric longitude of the Sun (that is, calculated by including the effects of aberration and nutation) is a multiple of 90 degrees. (Because the latitude of the Sun is not exactly zero, the declination of the Sun is not exactly zero at the instant of an equinox.)

Approximate times can be obtained as follows. First, find the instant of the 'mean' equinox or solstice, using the relevant expression in Table 26.A or in Table 26.B. Note that Table 26.A should be used for the years -1000 to +1000 only, and Table 26.B for the years +1000 to +3000. In fact, Table 26.A may be used for several centuries before the year -1000, and Table 26.B for several centuries after +3000; the errors will still be quite small. In the formula for Y, given at the top of each table, 'year' is an *integer*; other values for 'year' would give meaningless results!

Then find

$$T = \frac{\mathrm{JDE}_0 - 2451\,545.0}{36525}$$

$$W = 35999^\circ.373\,T - 2^\circ.47$$

$$\Delta\lambda = 1 + 0.0334 \cos W + 0.0007 \cos 2W$$

Calculate the sum S of the 24 periodic terms given in Table 26.C. Each of these terms is of the form $A \cos(B + CT)$, and the argument of each cosine is given in *degrees*. In other words,

$$S = 485 \cos(324^\circ.96 + 1934^\circ.136\,T) + 203 \cos(337^\circ.23 + 32964^\circ.467\,T) + \ldots.$$

The required time, expressed as a Julian Ephemeris Day (hence, in Dynamical Time), is then

TABLE 26.A For the years −1000 to +1000 $Y = \frac{\text{year}}{1000}$

March equinox (beginning of astronomical spring) :

$$JDE_0 = 1721\,139.29189 + 365\,242.13740\,Y + 0.06134\,Y^2 + 0.00111\,Y^3 - 0.00071\,Y^4$$

June solstice (beginning of astronomical summer) :

$$JDE_0 = 1721\,233.25401 + 365\,241.72562\,Y - 0.05323\,Y^2 + 0.00907\,Y^3 + 0.00025\,Y^4$$

September equinox (beginning of astronomical autumn) :

$$JDE_0 = 1721\,325.70455 + 365\,242.49558\,Y - 0.11677\,Y^2 - 0.00297\,Y^3 + 0.00074\,Y^4$$

December solstice (beginning of astronomical winter) :

$$JDE_0 = 1721\,414.39987 + 365\,242.88257\,Y - 0.00769\,Y^2 - 0.00933\,Y^3 - 0.00006\,Y^4$$

TABLE 26.B For the years +1000 to +3000 $Y = \frac{\text{year} - 2000}{1000}$

March equinox (beginning of astronomical spring) :

$$JDE_0 = 2451\,623.80984 + 365\,242.37404\,Y + 0.05169\,Y^2 - 0.00411\,Y^3 - 0.00057\,Y^4$$

June solstice (beginning of astronomical summer) :

$$JDE_0 = 2451\,716.56767 + 365\,241.62603\,Y + 0.00325\,Y^2 + 0.00888\,Y^3 - 0.00030\,Y^4$$

September equinox (beginning of astronomical autumn) :

$$JDE_0 = 2451\,810.21715 + 365\,242.01767\,Y - 0.11575\,Y^2 + 0.00337\,Y^3 + 0.00078\,Y^4$$

December solstice (beginning of astronomical winter) :

$$JDE_0 = 2451\,900.05952 + 365\,242.74049\,Y - 0.06223\,Y^2 - 0.00823\,Y^3 + 0.00032\,Y^4$$

TABLE 26.C

$S = \Sigma A \cos(B + CT)$ B and C in degrees!

A	B	C	A	B	C
485	324.96	1934.136	45	247.54	29929.562
203	337.23	32964.467	44	325.15	31555.956
199	342.08	20.186	29	60.93	4443.417
182	27.85	445267.112	18	155.12	67555.328
156	73.14	45036.886	17	288.79	4562.452
136	171.52	22518.443	16	198.04	62894.029
77	222.54	65928.934	14	199.76	31436.921
74	296.72	3034.906	12	95.39	14577.848
70	243.58	9037.513	12	287.11	31931.756
58	119.81	33718.147	12	320.81	34777.259
52	297.17	150.678	9	227.73	1222.114
50	21.02	2281.226	8	15.45	16859.074

$$\text{JDE} = \text{JDE}_0 + \frac{0.00\,001\,S}{\Delta\lambda} \text{ days}$$

This final JDE can be converted into the ordinary calendar date by means of the method described in Chapter 7. The result will be expressed in Dynamical Time.

For the years 1951 - 2050, the accuracy of this method is seen from Table 26.D.

TABLE 26.D

	Number of errors < 20 sec.	*Number of errors < 40 sec.*	*Largest error (seconds)*
March equinox	76	97	51
June solstice	80	100	39
September equinox	78	99	44
December solstice	68	99	41

Example 26.a — Find the time of the June solstice of A.D. 1962.

We find successively

$$Y = -0.038$$
$$JDE_0 = 2437\,837.38589$$
$$T = -0.375\,294\,021$$
$$\Delta\lambda = 0.9681$$
$$S = +635$$
$$JDE = 2437\,837.38589 + \frac{0.00635}{0.9681} = 2437\,837.39245$$

which corresponds to 1962 June 21 at $21^h25^m08^s$ TD.

The correct instant, as calculated with the complete VSOP87 theory, is $21^h24^m42^s$ TD.

Of course, higher accuracy can be obtained by actually calculating the value of the apparent longitude of the Sun for two or three instants, and then finding by interpolation the time when that longitude is exactly 0°, or 90°, or 180°, or 270°.

One should keep in mind that the motion of the Sun along the ecliptic is only 3548 arcseconds per day approximately. Hence, an error of 1" in the calculated longitude of the Sun results in an error of approximately 24 seconds in the times of the equinoxes or solstices.

Alternatively, one may start from any approximate time. The value obtained from Table 26.A or 26.B is more than sufficient. For that instant, calculate the Sun's apparent longitude λ as explained in Chapter 24, including the corrections for reduction to the FK5 system, for aberration and for nutation. Then the correction to the assumed time, in *days*, is given by

$$+58 \sin(k.90° - \lambda) \qquad (26.1)$$

where

k = 0 for the March equinox,
1 for the June solstice,
2 for the September equinox,
3 for the December solstice.

The calculation is then repeated until the new correction is very small or, equivalently, until the new value for the Sun's apparent longitude is exactly $k.90°$.

Example 26.b — Let us again calculate the instant of the June solstice in the year 1962.

In Example 26.a, we found that the 'mean' solstice took place at $JDE_0 = 2437\,837.38589$ (from Table 26.A). Let us start from this approximate time, and calculate the Sun's apparent longitude for this instant, using the 'higher accuracy' procedure (Chapter 24). We find

$L = -234.048\,595\,59$ radians $= 270^\circ.003\,272$
$R = 1.016\,3018$

Nutation in longitude :	$\Delta\psi = -12''.965$	(Chapter 22)
FK5 correction :	$-\ 0''.09033$	(formula (24.9))
aberration :	$-20''.161$	(formula (24.10))

Apparent longitude of the Sun :

$$\lambda = 270^\circ.003\,272 - 180^\circ - 12''.965 - 0''.09033 - 20''.161$$
$$\lambda = 89^\circ.994\,045$$

Formula (26.1) then gives the correction to the assumed value of JDE_o :

$$\text{correction} = +58 \sin(90^\circ - \lambda) = +0.00603$$

and hence the corrected time is

$$JDE = 2437\,837.38589 + 0.00603 = 2437\,837.39192$$

Repeating the calculation for this new instant, we find

$$\lambda = 89^\circ.999\,797,$$

resulting in the correction +0.00021 day. This gives the improved instant JDE = 2437 837.39213.

A final calculation, performed for this new instant, yields

$$\lambda = 89^\circ.999\,998$$

and a correction smaller than 0.000 005 day.

Hence, the final instant is JDE = 2437 837.39213, which corresponds to 1962 June 21 at $21^h24^m40^s$ TD.

[This differs by only two seconds from the correct time which is mentioned at the end of Example 26.a.]

In 1962, the difference TD - UT was 34 seconds (see Table 9.A), so our result may be rounded to 21^h24^m UT.

Table 26.E gives the times of the equinoxes and solstices for the years 1991 to 2000, to the nearest second of time.

Table 26.F gives the durations of the four astronomical seasons for some epochs. About the year -4080, the Earth was in perihelion at the beginning of the autumn; then the summer had the same duration as the autumn, and the winter the same duration as the spring. In A.D. 1246, the Earth was in perihelion at the time of the winter solstice; then the spring had the same duration as the summer, and the autumn the same duration as the winter. Since the year +1246, the winter is the shortest season; it will reach its minimum value by about A.D. 3500, and remain the shortest season till about A.D. 6427, when the Earth will be in perihelion at the March equinox.

TABLE 26.E

Equinoxes and Solstices, 1991-2000, calculated by means of the complete VSOP87 theory. Instants in Dynamical Time.

Year	*March equinox*				*June solstice*				*Sept. equinox*				*Dec. solstice*			
	d	*h*	*m*	*s*	*d*	*h*	*m*	*s*	*d*	*h*	*m*	*s*	*d*	*h*	*m*	*s*
1991	21	3	02	54	21	21	19	46	23	12	49	04	22	8	54	38
1992	20	8	49	02	21	3	15	08	22	18	43	46	21	14	44	14
1993	20	14	41	38	21	9	00	44	23	0	23	29	21	20	26	49
1994	20	20	29	01	21	14	48	33	23	6	20	14	22	2	23	44
1995	21	2	15	27	21	20	35	24	23	12	14	01	22	8	17	50
1996	20	8	04	07	21	2	24	46	22	18	01	08	21	14	06	56
1997	20	13	55	42	21	8	20	59	22	23	56	49	21	20	08	05
1998	20	19	55	35	21	14	03	38	23	5	38	15	22	1	57	31
1999	21	1	46	53	21	19	50	11	23	11	32	34	22	7	44	52
2000	20	7	36	19	21	1	48	46	22	17	28	40	21	13	38	30

TABLE 26.F

Duration of the astronomical seasons, in days

Year	*Spring*	*Summer*	*Autumn*	*Winter*
-4000	93.54	89.18	89.08	93.43
-3500	93.82	89.53	88.82	93.07
-3000	94.04	89.92	88.62	92.67
-2500	94.19	90.33	88.48	92.24
-2000	94.28	90.76	88.40	91.81
-1500	94.30	91.20	88.38	91.37
-1000	94.25	91.63	88.42	90.94
- 500	94.14	92.05	88.53	90.52
0	93.96	92.45	88.70	90.14
+ 500	93.73	92.82	88.92	89.78
1000	93.44	93.15	89.18	89.47
1500	93.12	93.42	89.50	89.20
2000	92.76	93.65	89.84	88.99
2500	92.37	93.81	90.22	88.84
3000	91.97	93.92	90.61	88.74
3500	91.57	93.96	91.01	88.71
4000	91.17	93.93	91.40	88.73
4500	90.79	93.84	91.79	88.82
5000	90.44	93.70	92.15	88.96
5500	90.11	93.50	92.49	89.14
6000	89.82	93.25	92.79	89.38
6500	89.58	92.97	93.04	89.65

Chapter 27

Equation of Time

Due to the eccentricity of its orbit, and to a much less degree due to the perturbations by the Moon and the planets, the Earth's heliocentric longitude does not vary uniformly. It follows that the Sun appears to describe the ecliptic at a non-uniform rate. Due to this, and also to the fact that the Sun is moving in the ecliptic and not along the celestial equator, its right ascension does not increase uniformly.

Consider a first fictitious Sun travelling along the *ecliptic* with a constant speed and coinciding with the true Sun at the perigee and apogee (when the Earth is in perihelion and aphelion, respectively). Then consider a second fictitious Sun travelling along the *celestial equator* at a constant speed and coinciding with the first fictitious Sun at the equinoxes. This second fictitious Sun is the *mean Sun*, and by definition its right ascension increases at a uniform rate. [That is, there are no periodic terms, but its expression contains small secular terms in τ^2, τ^3, ...].

When the mean Sun crosses the observer's meridian, it is mean noon there. True noon is the instant when the true Sun crosses the meridian. The *equation of time* is the difference between apparent and mean time; or, in other words, it is the difference between the hour angles of the true Sun and the mean Sun.

Defined in this manner, the equation of time E, at a given instant, is given by

$$E = L_0 - 0°.0057183 - \alpha + \Delta\psi \cdot \cos \varepsilon \qquad (27.1)$$

In this formula, L_0 is the Sun's mean longitude. According to the VSOP87 theory (see Chapter 31) we have, in degrees,

$$\begin{aligned} L_0 = 280.4664567 &+ 360007.6982779\,\tau \\ &+ 0.03032028\,\tau^2 + \tau^3/49931 \\ &- \tau^4/15299 - \tau^5/1988000 \end{aligned} \qquad (27.2)$$

where τ is the time measured in Julian millennia (365 250 ephemeris days) from J2000.0 = JDE 2451 545.0. L_0 should be reduced to less than 360° by adding or subtracting a convenient multiple of 360°.

In the French almanacs and in older textbooks, the equation of time is defined with opposite sign, hence being equal to mean time minus apparent time.

In formula (27.1), the constant 0°.005 7183 is the sum of the mean value of the aberration in longitude (−20".49552) and the correction for reduction to the FK5 system (−0".09033); α is the apparent right ascension of the Sun, calculated by taking into account the aberration and the nutation. The quantity $\Delta\psi \cdot \cos\varepsilon$, where $\Delta\psi$ is the nutation in longitude and ε the obliquity of the ecliptic, is needed to refer the apparent right ascension of the Sun to the *mean* equinox of the date, as is the mean longitude L_0.

In formula (27.1), the quantities L_0, α and $\Delta\psi$ should be expressed in degrees. Then the equation of time E will be expressed in degrees too; it can be converted to minutes of time by multiplication by 4.

The equation of time E can be positive or negative. If $E > 0$, the true Sun crosses the observer's meridian before the mean Sun.

The equation of time is always less than 20 minutes in absolute value. If $|E|$ appears to be too large, add 24 hours to or subtract it from your result.

Example 27.a — Find the equation of time on 1992 October 13 at 0^h Dynamical Time.

This date corresponds to JDE = 2448 908.5, from which we deduce

$$\tau = \frac{\text{JDE} - 2451\,545.0}{365\,250} = -0.007\,218\,343\,600$$

$$L_0 = -2318°.192\,807 = +201°.807\,193$$

For the same instant we have, from Example 24.b,

$$\alpha = 198°.378\,178$$
$$\Delta\psi = +15''.908 = +0°.004\,419$$
$$\varepsilon = 23°.440\,1443$$

whence, by formula (27.1),

$$E = +3°.427\,351 = +13.70940 \text{ minutes} = +13^m42^s.6$$

Alternatively, the equation of time can be obtained, with somewhat less accuracy, by means of the following formula given by Smart [1] :

$$E = y \sin 2L_0 - 2e \sin M + 4ey \sin M \cos 2L_0 - \frac{1}{2} y^2 \sin 4L_0 - \frac{5}{4} e^2 \sin 2M \qquad (27.3)$$

where

$y = \tan^2 \frac{\varepsilon}{2}$, ε being the obliquity of the ecliptic,

L_0 = Sun's mean longitude,

e = eccentricity of the Earth's orbit,

M = Sun's mean anomaly.

The values of ε, L_0, e and M can be found by means of formulae (21.2), (27.2) or (24.2), (24.4), and (24.3), respectively.

The value of E given by formula (27.3) is expressed in radians. The result may be converted into degrees, and then into hours and decimals by division by 15.

Example 27.b — Find, once again, the value of the equation of time on 1992 October 13.0 TD = JDE 2448 908.5.

We find successively

T = −0.072 183 436 $\qquad$ e = 0.016 711 651

ε = 23°.44023 $\qquad$ M = 278°.99396

L_0 = 201°.80720 $\qquad$ y = 0.043 0381

Formula (27.3) then gives E = +0.059 825 557 radian
= +3.427 752 degrees
= +13 minutes 42.7 seconds

The curve representing the variation of the equation of time during the year is well-known and can be found in many astronomy books. Presently, this curve has a deep minimum near February 11, a high maximum near November 3, and a secondary maximum and minimum about May 14 and July 26, respectively.

However, the curve of the equation of time is gradually changing in the course of the centuries, because the obliquity of the ecliptic, the eccentricity of the Earth's orbit, and the longitude of the perihelion of this orbit are all slowly changing. The figure on the next page shows the curve of the equation of time at intervals of 1000 years, from 3000 B.C. to A.D. 4000. On the vertical scale, the tics are given at intervals of five minutes of time; the horizontal line represents the value E = zero. The tics on this horizontal line divide the year in four periods of three months each, beginning from January 1 at left. We see, for instance, that the minimum of February will be less deep in the future.

Between A.D. 1600 and 2100, the extreme values of the equation of

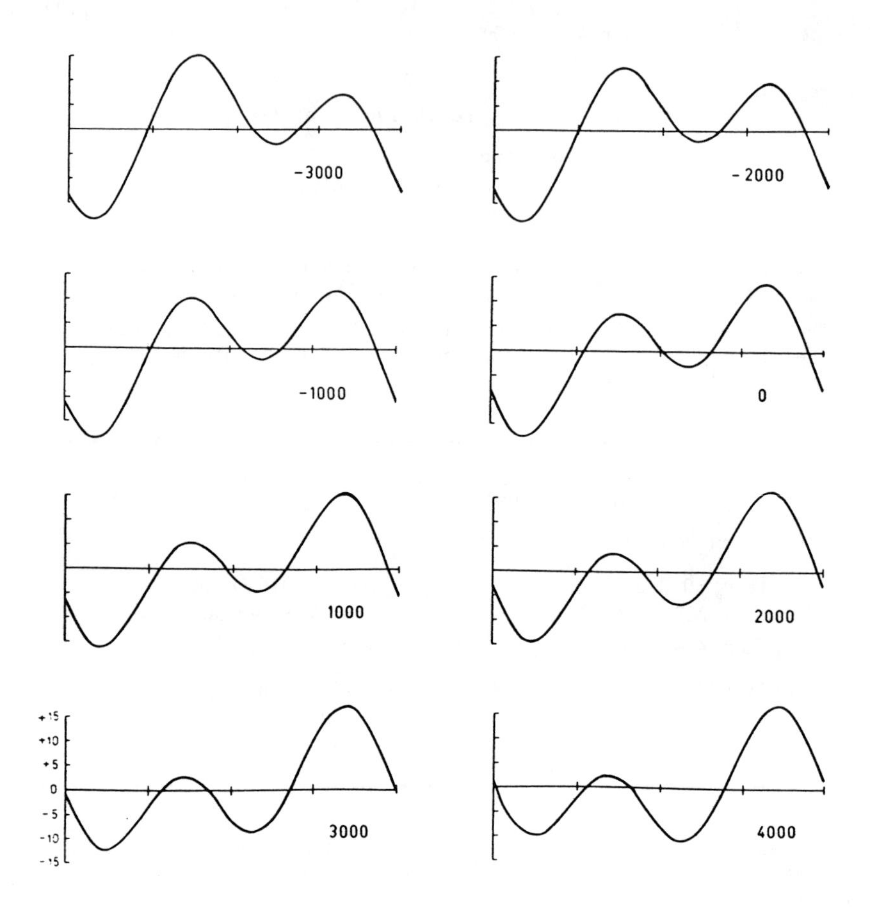

The curve of the equation of time, from -3000 to +4000

time vary as shown in Table 27.A. These are 'mean' values: the calculation is based on a non-perturbed elliptical motion of the Earth, and the nutation has not been taken into account.

In A.D. 1246, when the Sun's perigee coincided with the winter solstice, the curve representing the annual variation of the equation of time was exactly symmetrical with respect to the zero-line: the minimum of February was exactly as deep as the height of the November maximum; and the smaller May maximum was exactly as high as the value of the July minimum — see the last line of the Table.

TABLE 27.A

The extreme values of the equation of time in modern times

Year	*Minimum of February*	*Maximum of May*	*Minimum of July*	*Maximum of November*
	m s	*m s*	*m s*	*m s*
1600	−15 01	+4 19	−5 40	+16 03
1700	−14 50	+4 09	−5 53	+16 09
1800	−14 38	+3 59	−6 05	+16 15
1900	−14 27	+3 50	−6 18	+16 20
2000	−14 15	+3 41	−6 31	+16 25
2100	−14 03	+3 32	−6 44	+16 30
1246	−15 39	+4 58	−4 58	+15 39

Reference

1. W.M. Smart, *Text-Book on Spherical Astronomy*; Cambridge (Engl.), University Press (1956); page 149.

Chapter 28

Ephemeris for Physical Observations of the Sun

The formulae given in this Chapter are based on the elements determined by Carrington (1863), which have been in use for many years. For a given instant, the required quantities are :

P = the position angle of the northern extremity of the axis of rotation, measured eastwards from the North Point of the solar disk ;

B_0 = the heliographic latitude of the center of the solar disk ;

L_0 = the heliographic longitude of the same point.

Although position angles are generally counted from 0° to 360° (this is the case for the Moon, the planets, double stars, etc.), in the case of the Sun it is customary to keep P, in absolute value, less than 90°, and to assign to it a plus or a minus sign: P is positive when the northern extremity of the rotation axis of the Sun is tilted towards the East, negative if towards the West. Celestial and solar north can differ by up to 26 degrees. P reaches a minimum of −26°.3 about April 7, a maximum of +26°.3 about October 11, and is zero near January 5 and July 7.

B_0 represents the tilt of the Sun's north pole toward (+) or away (−) from Earth; it is zero about June 6 and December 7, and reaches maximum value about March 6 (−7°.25) and September 8 (+7°.25).

L_0 *decreases* by about 13.2 degrees per day. The mean synodic period is 27.2752 days. The beginning of each 'synodic rotation' is the instant at which L_0 passes through 0°. Rotation No. 1 commenced on 1853 November 9.

Let JD be the Julian Ephemeris Day, which can be calculated by means of the method described in Chapter 7. If the given instant is in Universal Time, add to JD the value ΔT = TD − UT expressed in days (see Chapter 9). If ΔT is expressed in seconds of time, the correction to JD will be $+\Delta T/86400$.

Then calculate the following quantities :

$$\theta = (\text{JD} - 2398\,220) \times \frac{360°}{25.38}$$

$$I = 7°.25 = 7°15'$$

$$K = 73°.6667 + 1°.395\,8333\ \frac{\text{JD} - 2396\,758}{36525}$$

where I is the inclination of the solar equator on the ecliptic, and K is the longitude of the ascending node of the solar equator on the ecliptic. In the formula for θ, 25.38 is the Sun's sidereal period of rotation in days. This value has been fixed *conventionally* by Carrington. It defines the zero meridian of the heliographic longitudes and therefore must be treated as *exact*.

Calculate the *apparent* longitude λ of the Sun (including the effect of aberration, but *not* that of nutation) by the method described in Chapter 24, and the obliquity of the ecliptic ε (including the effect of nutation) as explained in Chapter 21. Let λ' be λ corrected for the nutation in longitude.

Then calculate the angles x and y by means of

$$\tan x = -\cos\lambda' \tan\varepsilon$$

$$\tan y = -\cos(\lambda - K) \tan I$$

where both x and y should be taken between $-90°$ and $+90°$. Then the required quantities P, B_0 and L_0 are found as follows :

$$P = x + y$$

$$\sin B_0 = \sin(\lambda - K) \sin I$$

$$\tan\eta = \frac{-\sin(\lambda - K)\cos I}{-\cos(\lambda - K)} = \tan(\lambda - K)\cos I$$

η being in the same quadrant as $\lambda - K \pm 180°$,

$L_0 = \eta - \theta$, to be reduced to the interval 0 – 360 degrees.

Example 28.a — Calculate P, B_0 and L_0 for 1992 October 13 at 0^h Universal Time = JD 2448 908.5.

We will use the value $\Delta T = +59$ seconds $= +0.000\,68$ day. Consequently the corrected JD, or Julian Ephemeris Day, is 2448 908.50068 and we find successively

$$\theta = 718\,985°.8252 = 65°.8252$$

$$I = 7°.25$$

$$K = 75°.6597$$

From Chapters 24 and 21 :

$$L \text{ (Earth)} = -43.634\,836\,22 \text{ radians} = +19^\circ.908\,045$$
$$R = 0.997\,608$$
$$\Delta\psi = +15''.908 = +0^\circ.004\,419$$
$$\varepsilon = 23^\circ.440\,144$$

$$\text{correction for aberration} = -\frac{20''.4898}{R} = -0^\circ.005\,705$$

whence

$$\lambda = L + 180^\circ - 0^\circ.005\,705 = 199^\circ.902\,340$$
$$\lambda' = \lambda + \Delta\psi = 199^\circ.906\,759$$

$$\tan x = +0.407\,664 \qquad x = +22^\circ.1790$$
$$\tan y = +0.071\,584 \qquad y = +\ 4^\circ.0945$$

$$P = 26^\circ.27$$

$$\sin B_0 = +0.104\,324 \qquad B_0 = +5^\circ.99$$

$$\tan \eta = \frac{-0.820\,053}{+0.562\,699} \qquad \eta = -55^\circ.5431$$

$$L_0 = -121^\circ.3683 = 238^\circ.63$$

As mentioned above, a solar 'synodic rotation' begins when L_0 is equal to 0°. An approximate time for the beginning of Carrington's synodic rotation No. C is given by

$$\text{Julian Ephemeris Day} = 2398\,140.2270 + 27.275\,2316\,C \tag{28.1}$$

where, of course, C is an integer. The instant so obtained will not be more than 0.16 day in error.

However, the time obtained from the formula above can be corrected as follows. Calculate the angle M, in degrees, from

$$M = 281.96 + 26.882\,476\,C$$

Then the correction in days is

$$\begin{aligned} &+0.1454 \sin M \\ &-0.0085 \sin 2M \\ &-0.0141 \cos 2M \end{aligned} \tag{28.2}$$

Between the years 1850 and 2100, the resulting time will be less than 0.002 day in error.

Of course, a correct value for the time of the beginning of a synodic rotation can be obtained by calculating L_0 for two instants near the time given by the formula above, and then by performing an inverse interpolation to find when L_0 is zero.

Example 28.b — Find the instant of the beginning of solar rotation No. 1699.

For $C = 1699$, formula (28.1) gives JDE = 2444 480.8455.

We further find $M = 45\,955°.287 = 235°.287$, and the correction as given by (28.2) is −0.1225.

To convert from Dynamical Time to Universal Time, there is a further correction of −0.0006 day (in 1980, the value of ΔT = TD − UT was 51 seconds).

Hence, the final instant is

$$\text{JD} = 2444\,480.8455 - 0.1225 - 0.0006 = 2444\,480.7224$$

which corresponds to 1980 August 29.22.

The *Astronomical Ephemeris* for 1980, page 359, gives the same value.

It is customary to give the times of the commencement of the Sun's synodic rotations to the nearest 0.01 day, hence in days and *decimals*, not in hours and minutes.

Chapter 29

Equation of Kepler

There are several methods for calculating the position of a body (planet, minor planet, or periodic comet) on its elliptical orbit around the Sun at a given instant :

- by numerical integration, a subject which is outside the scope of this book;
- obtaining the body's heliocentric ccordinates (longitude, latitude, radius vector) by calculating the sum of periodic terms, as will be explained in Chapter 31;
- from the orbital elements of the body, as explained in Chapter 32.

In the latter case, we need to find the true anomaly of the object. This can be achieved either by solving Kepler's equation or, when the orbital eccentricity is not too large, by using series expressions (see 'The Equation of the Center' in Chapter 32).

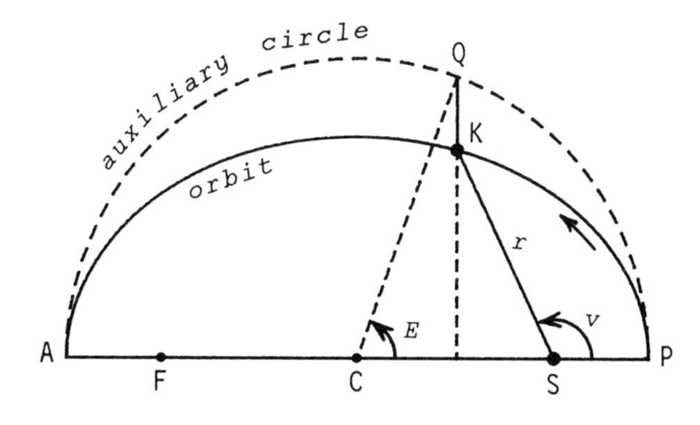

Figure 1

In Figure 1 we represent one half of an elliptical orbit (PKA). The Sun is situated in the focus S; the other, empty focus of the ellipse is F. The straight line AP is the *major axis* of the orbit. The center C of the ellipse is exactly half-way between the perihelion P and the aphelion A, as well as half-way between the foci F and S.

Suppose that, at a given instant, the moving body is at K. The distance SK is the *radius vector* of the body at that instant; this distance r is expressed in astronomical units. The *true anomaly* (v) at the same instant is the angle between the directions SP and SK; it is the angle over which the object moved, as seen from the Sun, since the previous passage through the perihelion P.

The semimajor axis, CP in Figure 1, is generally designated by a and is expressed in astronomical units. By definition, the eccentricity e of the orbit is equal to the ratio of the distances CS and CP, or $e = CS/CP$. For an ellipse, e is between 0 and 1. The perihelion and aphelion distances are designated by q and Q, respectively. In the perihelion, $v = 0°$ and $r = q$, while in the aphelion we have $v = 180°$ and $r = Q$. It follows that

distance CS = ae
distance SP = $q = a(1-e)$ = perihelion distance
distance SA = $Q = a(1+e)$ = aphelion distance
distance PA = $2a = q + Q$

Let us now consider (Figure 2) a fictitious planet or comet K′ describing around the Sun a circular orbit, hence with a constant velocity, with the *same period* as the real planet or comet K. Moreover, let us suppose that this fictitious body is at P′, on the line SP, at the instant when the real body is at the perihelion P. Some time later, when the true body is at K, the fictitious body is at K′. As we have seen, the angle v = angle PSK is the true anomaly of the body (at the given instant). The angle PSK′ at the same instant is called the *mean anomaly* and is generally designated by M.

In other words, the mean anomaly is the angular distance from perihelion which the planet would have if it moved around the Sun with a constant angular velocity.

By definition, the angle M increases linearly (uniformly) with time. The value of M at a given time is easily found, for $M = 0°$ when the planet is at perihelion, and it increases by exactly 360° in the course of one complete revolution of the planet.

The problem consists in finding the true anomaly v when the mean anomaly M and the orbital eccentricity e are known. Unless use is made of series expressions such as those given in Chapter 32, one has to solve Kepler's equation.

In this connection, it is necessary to introduce an auxiliary angle E, the *eccentric anomaly*, whose geometric definition is given in Figure 1. The exterior, dashed circle has diameter AP. We draw KQ perpendicularly to AP. The angle PCQ is the eccentric anomaly.

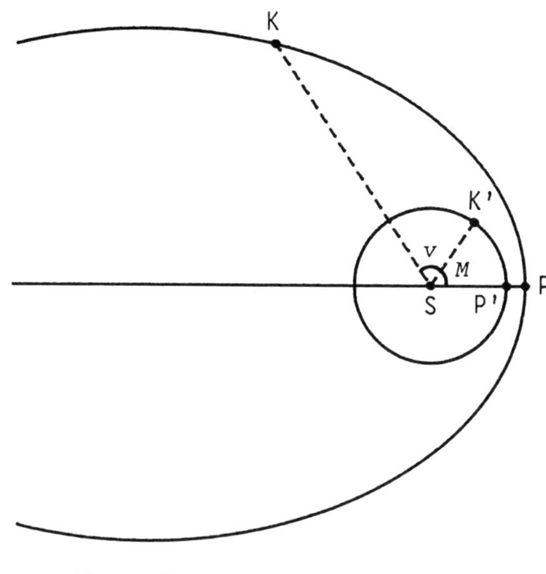

Figure 2

When the planet is at perihelion, the angles v, E and M are all zero. Near the perihelion, the true planet moves at a greater speed than the mean, fictitious planet. Therefore, between perihelion and aphelion, when the planet moves away from the Sun, we have $v > M$ and, because E is always between v and M, we then have

$$0° < M < E < v < 180°.$$

In the aphelion, $v = E = M = 180°$, and after aphelion passage, on its way back to perihelion, the true planet remains behind the mean planet.

When E is known, v can be obtained from

$$\tan \frac{v}{2} = \sqrt{\frac{1+e}{1-e}} \tan \frac{E}{2} \qquad (29.1)$$

while the radius vector can be calculated from one of the following expressions :

$$r = a\,(1 - e \cos E) \qquad (29.2)$$

$$r = \frac{a\,(1 - e^2)}{1 + e \cos v} \qquad (29.3)$$

$$r = \frac{q\,(1 + e)}{1 + e \cos v} \qquad (29.4)$$

But let us now consider the problem of finding the eccentric anomaly E.

The equation of Kepler is

$$E = M + e \sin E \qquad (29.5)$$

This equation must be solved for E. It is, however, a transcendental equation which cannot be solved directly. We will describe three iteration methods for finding E, and finally give a formula which yields an approximate result.

First Method

In formula (29.5) the angles M and E should be expressed in *radians*. Hence the calculation should be performed in 'radian mode', which is the case for many programming languages. If the calculation is made in 'degree mode', then in (29.5) one should multiply e by $180/\pi$, the factor for converting radians into degrees. Let e_o be the thus 'modified' eccentricity. Kepler's equation is then

$$E = M + e_o \sin E \qquad (29.6)$$

and now we can calculate with ordinary degrees.

To solve equation (29.6), give an approximate value to E in the right side of the formula. Then the formula will give a better approximation for E. This is repeated until the required accuracy is obtained; this process can be performed automatically in a computer program. For the first approximation, we may use $E = M$.

We thus have

$$\begin{aligned} E_0 &= M \\ E_1 &= M + e \sin E_0 \\ E_2 &= M + e \sin E_1 \\ E_3 &= M + e \sin E_2 \\ &\text{etc.} \end{aligned}$$

E_1, E_2, E_3, etc. are successive and better approximations for the eccentric anomaly E.

Example 29.a — Solve the equation of Kepler for $e = 0.100$ and $M = 5°$, to an accuracy of 0.000 001 degree.

We find

$$e_o = 0.100 \times 180/\pi = 5°.729\,577\,95,$$

and the equation of Kepler becomes

$$E = 5 + 5.729\,577\,95 \sin E$$

where all quantities are in degrees. Starting with $E = M = 5°$, we obtain successively

5.499 366
5.549 093
5.554 042
5.554 535
5.554 584
5.554 589
5.554 589

Hence, the required value is $E = 5°.554\,589$.

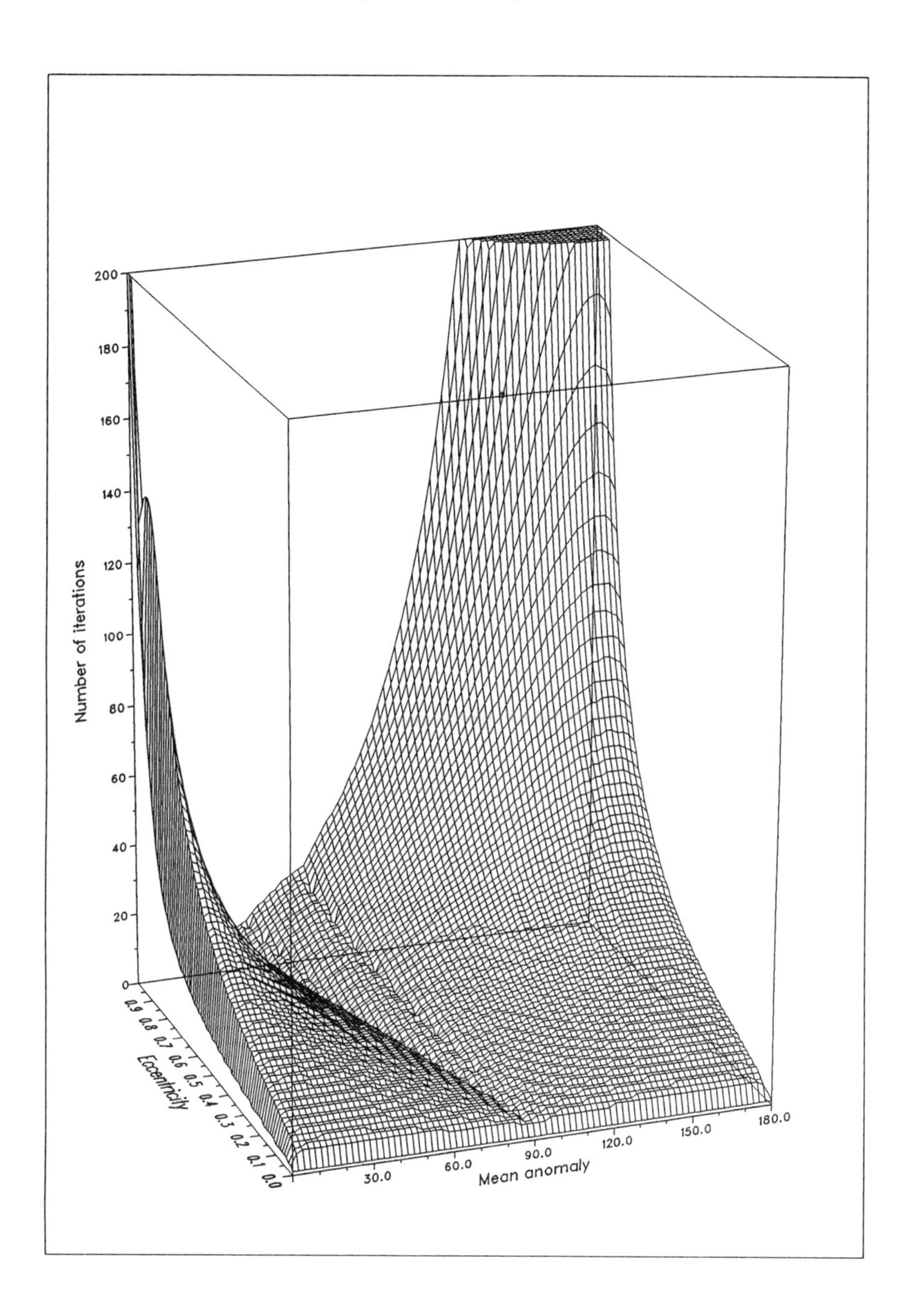
Number of iterations
200
180
160
140
120
100
80
60
40
20
0
Eccentricity
0.9 0.8 0.7 0.6 0.5 0.4 0.3 0.2 0.1 0.0
Mean anomaly
30.0 60.0 90.0 120.0 150.0 180.0

Figure 3

This method is very simple and does always converge. There will be no problems when e is small. However, the number of required iterations is generally increasing with e. For example, for e = 0.990 and $M = 2°$, the successive values of the iteration procedure are as follows :

2.000 000	15.168 909	24.924 579	29.813 009
3.979 598	16.842 404	25.904 408	30.200 940
5.936 635	18.434 883	26.780 556	30.533 515
7.866 758	19.937 269	27.557 863	30.817 592
9.763 644	21.341 978	28.242 483	.
11.619 294	22.643 349	28.841 471	:
13.424 417	23.837 929	29.362 399	:

After the 50th iteration, the result (32°.345 452) still differs from the correct result (32.361 007) by more than 0.01 degree.

Figure 3, due to the Belgian calculator Edwin Goffin, is a three-dimensional representation of the number of iterations needed to obtain an accuracy of 10^{-9} degree, as a function of the orbital eccentricity and the mean anomaly. We see that the number of required iterations becomes large when the eccentricity approaches 1 *and* when the mean anomaly is close to 0° or to 180°. [Note that 10^{-9} degree (4 millionth of an arcsecond) is an absurdly high accuracy; it has been retained here merely as a mathematical exercise.]

At the bottom of the drawing we note a horizontal straight 'valley'. This valley extends from the point $e = 0$, $M = 90°$ to the point $e = 1$, $M = 32°42'$. (This latter value is equal to $\pi/2 - 1$ radians.) This means that, for any eccentricity e, there is a value M_0 of the mean anomaly for which the number of iterations (to solve Kepler's equation by the method described above) is a minimum. This 'particular' mean anomaly is given by

$$M_0 = \frac{\pi}{2} - e \quad \text{radians}$$

and corresponds to the solution $E = \pi/2$ radians $= 90°$ exactly.

The number of required iterations increases as M differs more from M_0, on both sides of the 'valley'. For instance, for $e = 0.75$ we have $M_0 = 47.03$ degrees, and the number of steps needed to obtain E with an accuracy of 0.000 001 degree is as follows :

M	*iter.*	*M*	*iter.*
5°	51	60°	11
10°	37	70°	12
20°	23	90°	21
30°	15	110°	32
40°	9	130°	43
47°	5	150°	54
55°	8	170°	59

An interesting fact is that, when M is between M_0 and 180°, the results of the successive iterations *oscillate* while converging to the exact value: they do not constantly vary in the same direction as was the case in Example 29.a. For $e = 0.75$ and $M = 70°$, the results of the successive iterations are

70°.000 000	starting value
110.380 316	larger
110.281 870	smaller
110.307 524	larger
110.300 850	smaller
110.302 587	larger
110.302 135	smaller
etc...	

Second Method

When the orbital eccentricity e is larger than 0.4 or 0.5, the convergence of the method described above can be so slow that it may be advisable to use a better iteration formula. A better value E_1 for E is

$$E_1 = E_0 + \frac{M + e \sin E_0 - E_0}{1 - e \cos E_0} \tag{29.7}$$

where E_0 is the last obtained value for E. In this formula, the angles M, E_0 and E_1 are all expressed in *radians*. If one wishes to work in 'degree mode', then *in the numerator only* of the fraction the eccentricity e should be replaced by the 'modified' eccentricity $e_0 = 180\, e/\pi$.

Here, again, the process should be repeated as often as is necessary.

Note the difference between formulae (29.6) and (29.7). The first one directly gives a new approximation for E. While formula (29.7) too gives a new approximation E_1 for the eccentric anomaly, the fraction in the second member is actually a *correction* to the previous value E_0.

Example 29.b — Same problem as in Example 29.a, but now using formula (29.7).

We shall work in degree mode, so in this case formula (29.7) takes the following form :

$$E_1 = E_0 + \frac{5 + 5.729\,577\,95 \sin E_0 - E_0}{1 - 0.100 \cos E_0}$$

Starting with $E_0 = M = 5°$, we obtain the following values :

E_0	*correction*	E_1
5.000 000 000	+0.554 616 193	5.554 616 193
5.554 616 193	−0.000 026 939	5.554 589 254
5.554 589 254	−0.000 000 001	5.554 589 253

In this case, an accuracy of 0.000 000 001 degree is obtained after only three iterations.

We solved Kepler's equation for some values of e and M; see Table 29.A, where the successive columns give the orbital eccentricity e, the mean anomaly M, the corresponding value of E, and the number of steps needed by using the first (1) and the second (2) method, starting with $E = M$ as the first approximation. A computer working with twelve significant digits was used, and iterations were performed until the new value of E differed from the previous one by less than 0.000 001 degree.

It appears that, generally speaking, a larger value of e requires a larger number of iterations, for the first method as well as for the second one. But with the second method the number of these iterations is much smaller.

TABLE 29.A

e	M	E	*(1)*	*(2)*
0.1	5°	5°554589	6	2
0.2	5	6.246908	9	2
0.3	5	7.134960	12	2
0.4	5	8.313903	16	2
0.5	5	9.950063	21	2
0.6	5°	12.356653	28	3
0.7	5	16.167990	39	3
0.8	5	22.656579	52	4
0.9	5	33.344447	58	5
0.99	5	45.361023	50	11
0.99	1°	24.725822	150	8
0.99	33	89.722155	6	5

For small values of the eccentricity, say for $e < 0.3$, the first method still seems the best one: we may prefer to perform 5 or 10 *easy* iterations instead of two iterations with the more complicated formula (29.7). Only for larger values of the eccentricity is formula (29.7) to be preferred to the first method.

In some cases, the first method is disastrous. See the next-to-last line of the table: no less than 150 iterations are needed to obtain E for the values $e = 0.99$ and $M = 1°$.

Finally, Table 29.A shows that the number of steps needed to obtain a given accuracy does not only depend on the value of e, but on that of M too. See the last line of the table, where the first method requires only six iterations, in spite of the large value of the orbital eccentricity, namely $e = 0.99$.

Although for large values of the eccentricity formula (29.7) is superior to (29.6), there can still be problems. We performed some calculations with formula (29.7) on the HP-85 microcomputer, each time taking M as starting value for E. Table 29.B gives the successive 'better' values of E (in degrees) for three cases.

TABLE 29.B

$e = 0.99$ $M = 2°$	$e = 0.999$ $M = 6°$	$e = 0.999$ $M = 7°$
188.700250865	930.362114752	832.86912333
90.0043959725	418.384869795	275.954959759
58.7251974236	-345.064633754	-87.610596019
41.762008288	10182.3247508	-48.5623921307
34.1821261793	1840.68260539	-11.225108839
32.4485414136	-5573.41581953	340.962715254
32.361223124	-2776.37618814	-5996.93473678
32.3610074734	-478.97469399	-2079.96780001
32.3610074722	-185.902957505	511.49423506
32.3610074722	-86.6958017962	257.391360843
	-48.9711628749	5.969894505
	-14.7148241705	1094.05946279
	168.189220986	-33606.763133
	92.1098260913	-12599.3759885
	64.2252288664	11889243.763
	52.4123211568	3642203.90477
	49.7106850572	-432120.48862
	49.5699983807	-145379.711482
	49.5696248567	142691.415319
	49.5696248539	56806.8295471
		....

In the first example ($e = 0.99$, $M = 2°$) we start with $E = 2°$. The first iteration gives $E = 188°.7$, which is even farther away from the solution! But thereafter come the values 90°, 59°, 42°, and then the procedure converges rapidly: after the eighth iteration the result is reached with an accuracy of 0.000 004 arcsecond.

In the second case ($e = 0.999$, $M = 6°$), the first iterations give very bizarre values, almost as if by a random-number generator! There is no convergence at all, until after the 13th iteration the value 168° is obtained; seven more steps then give us the correct solution.

Third case: same eccentricity, but now $M = 7°$. Here too the successive results jump irregularly back and forth, and after 20 steps still nothing reasonable is reached. Not before the *47th iteration* (not given in the table) do we obtain the correct solution, namely 52°.270 2615.

It is truly remarkable that for the same eccentricity 0.999, but

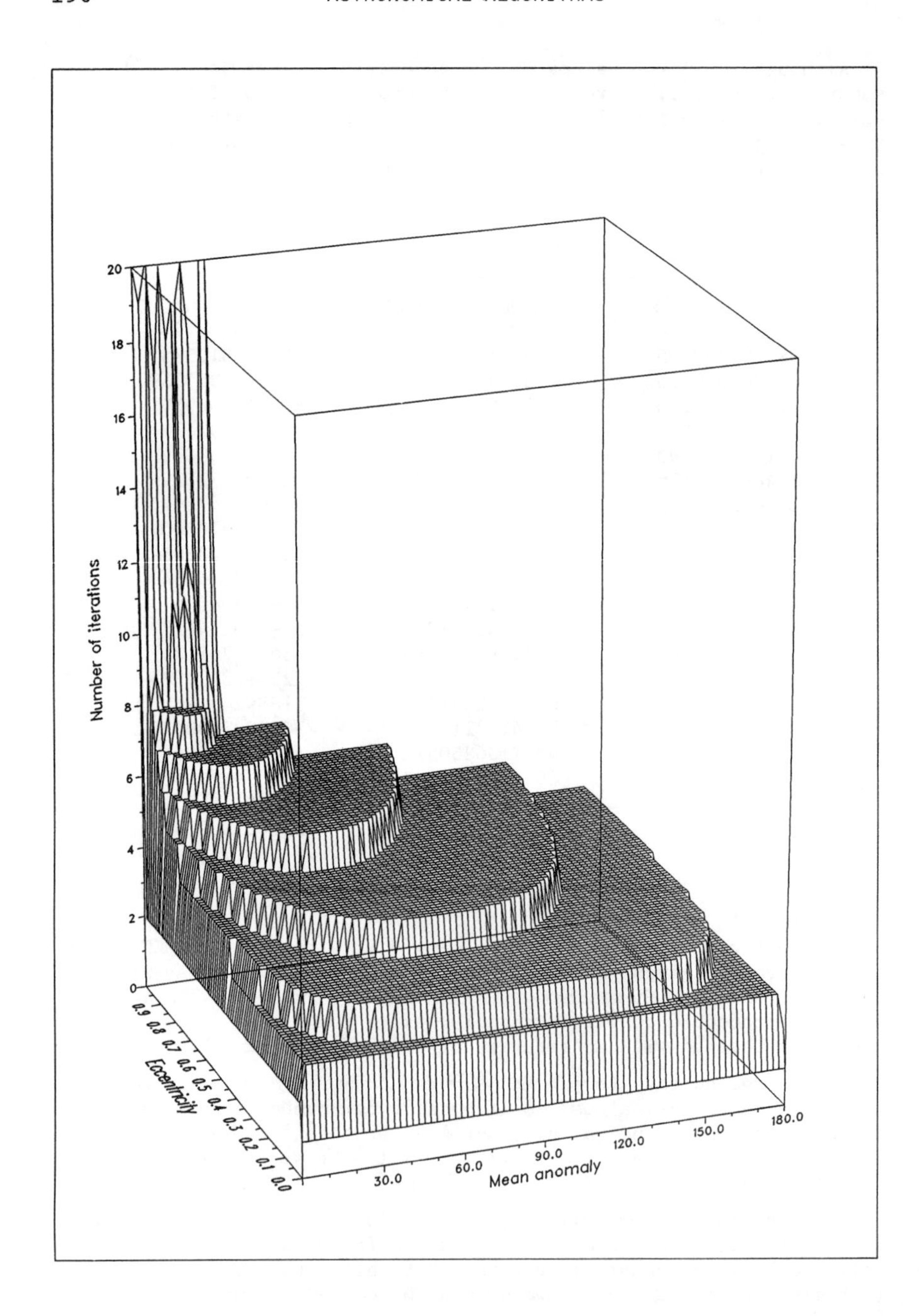
Number of iterations
20
18
16
14
12
10
8
6
4
2
0
Eccentricity
0.9
0.8
0.7
0.6
0.5
0.4
0.3
0.2
0.1
0.0
Mean anomaly
30.0
60.0
90.0
120.0
150.0
180.0

Figure 4

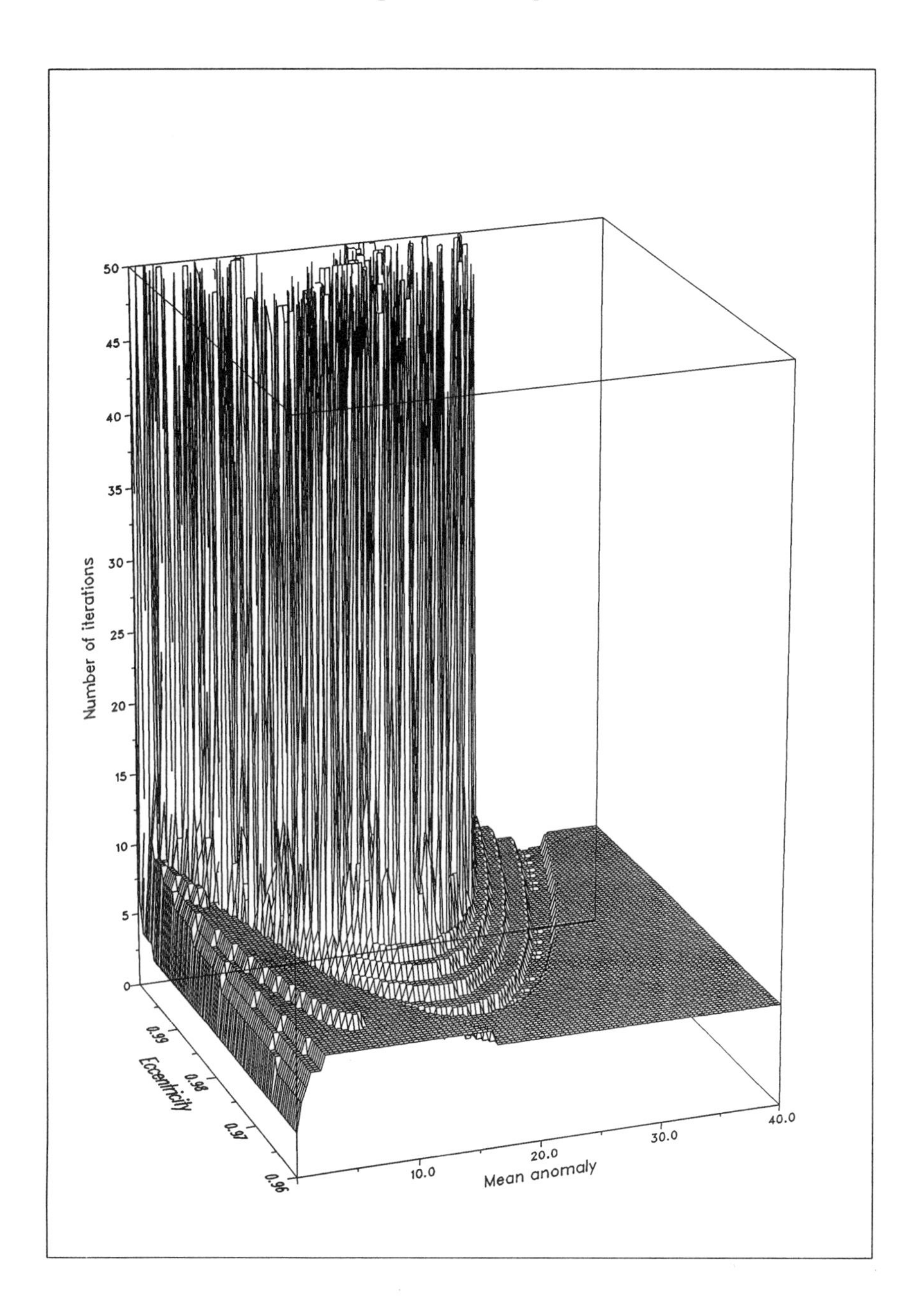
50
45
40
35
30
25
20
15
10
5
0
Number of iterations
0.99
0.98
0.97
0.96
Eccentricity
10.0
20.0
30.0
40.0
Mean anomaly

Figure 5

for $M = 7°.01$ instead of $7°.00$, the correct value of E is reached after only *twelve* iterations.

The HP-85 works with 12 significant digits. If you use another computer, the number of iterations can sometimes differ appreciably from those we mention here. When one calculates the second case ($e = 0.999$, $M = 6°$) with the HP-67 *pocket* calculator, which works with 10 significant digits, the successive results (in degrees) are

930.3621195
418.3848584
-345.0649049
10182.69391
1883.665232
-162.6729360
-85.06198931
-47.82386405
-13.18454655
211.0527629
84.65261970
60.76546811
51.35803706
49.62703439
49.56968687
49.56962485
49.56962485

It is interesting to compare these values with those of Table 29.B. After the third iteration, the difference with the value obtained with the HP-85 is still 0.00027 degree only; after the next iteration, the difference is $0°.37$, and after the next one it is 43 degrees! Nevertheless, convergence to the exact value is eventually achieved.

It is evident that, when e is large, formula (29.7) guarantees only a *local* convergence. The successive results jump irregularly back and forth, and only when by chance a result falls into the 'right domain' do the next results converge rapidly.

Figure 4, due to Edwin Goffin, is a three-dimentional representation of the number of steps needed to obtain E with an accuracy of 10^{-9} degree, as a function of the orbital eccentricity and the mean anomaly, when formula (29.7) is used. As before, M is used as the starting value for E. The left corner, near $e = 1$ and $M = 0°$, is the 'dangerous zone'. Figure 5 shows a magnification of that zone: we see a large number of peaks which are close together; the number of iterations needed to obtain the stated accuracy differs considerably even when e or M is changed very little.

Consequently, formula (29.7) is rather worrying for large values of e and small values of M. In some cases, the computer runs the risk of overflowing because the denominator of the fraction becomes almost zero.

This trouble can be avoided by choosing, as a starting value for E, a better value than just M. Mikkola [1] finds a good starting value as follows.

If M is expressed in radians, calculate

$$\alpha = \frac{1 - e}{4e + 0.5} \qquad \beta = \frac{M}{8e + 1}$$

$$z = \sqrt[3]{\beta \pm \sqrt{\beta^2 + \alpha^3}}$$

The sign of the square root is to be chosen as the sign of β. Attention: the number under the cubic root can be negative, resulting in a computer error!

Then calculate

$$s_o = z - \frac{\alpha}{2} \qquad s = s_o - \frac{0.078\, s_o^5}{1 + e}$$

Then a good starting value for formula (29.7) is

$$E = M + e\,(3s - 4s^3) \tag{29.8}$$

This procedure is useful only in the 'dangerous region', that is, when both $|M| < 30°$ and $0.975 < e < 1$. Otherwise, one just can use M as a starting value for E.

Figure 6, again due to Goffin, illustrates the number of iterations required to obtain an accuracy of 10^{-9} degree when formula (29.7) is used with the starting value given by (29.8).

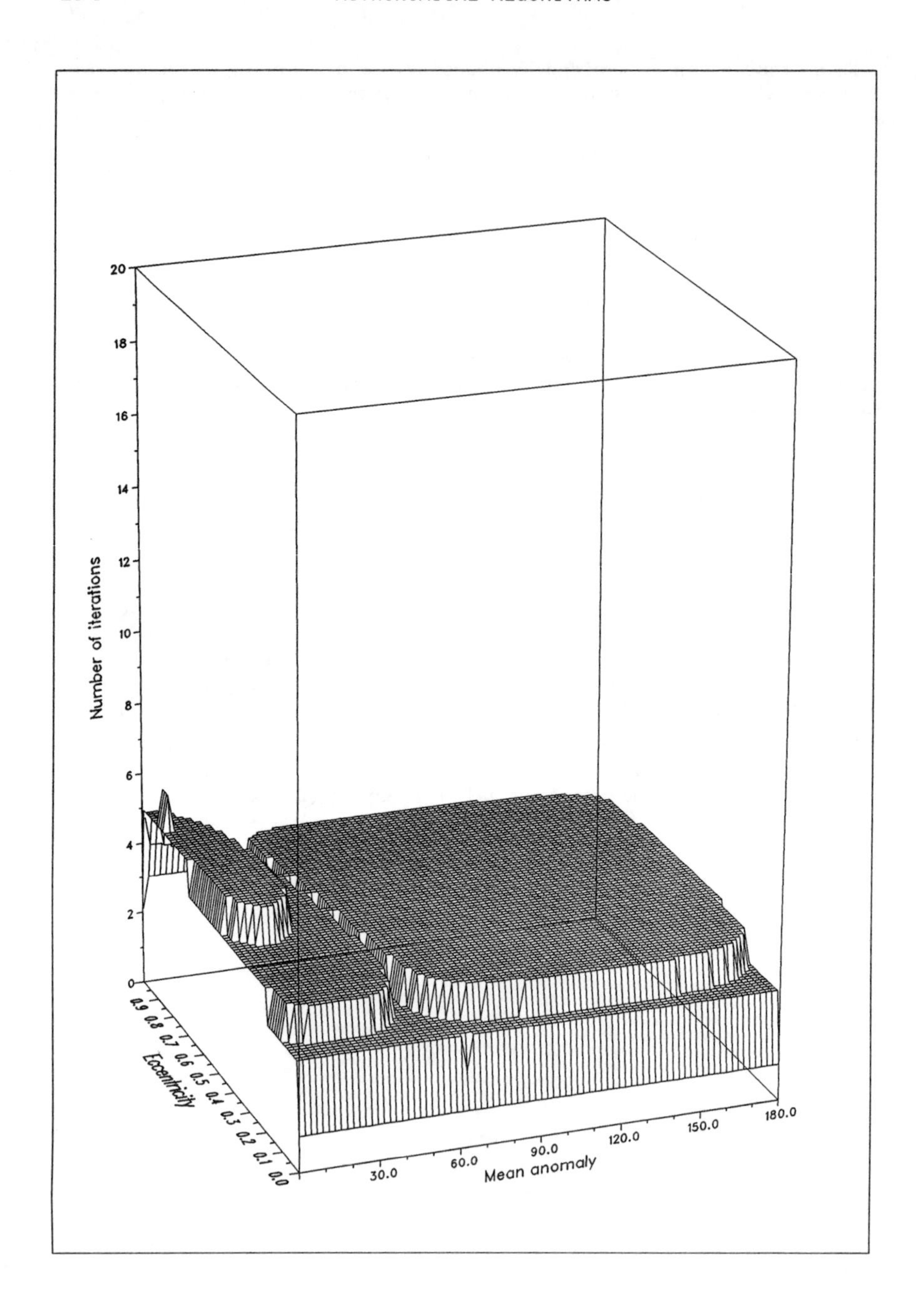
Number of iterations
20
18
16
14
12
10
8
6
4
2
0
0.9 0.8 0.7 0.6 0.5 0.4 0.3 0.2 0.1 0.0
Eccentricity
30.0
60.0
90.0
120.0
150.0
180.0
Mean anomaly

Figure 6

Third Method

Roger Sinnott [2] devised a method using a binary search to locate the correct value of E. The binary search has already been mentioned at the end of Chapter 5. The procedure is absolutely foolproof; it always converges to the most exact value of which the machine is capable, and it works for any eccentricity between 0 and 1. The relevant part of Sinnott's program, in BASIC, is given below. Here, E is the orbital eccentricity, and M the mean anomaly in radians. The result of the program is the eccentric anomaly $E0$ expressed in radians too.

For a computer language with 10-digit accuracy, 33 steps are needed in the binary search. The number of loops in line 180 should be increased to 53 if you are using a 16-digit BASIC. The number of steps needed is $3.32 \times$ the number of required digits, where 3.32 is equal to $1/\log_{10} 2$.

```
100  P1 = 3.14159265359
110  F = SGN (M)  :  M = ABS (M)/(2*P1)
120  M = (M - INT (M)) * 2 * P1 * F
130  IF M<0 THEN M = M+2*P1
140  F = 1
150  IF M>P1 THEN F = -1
160  IF M>P1 THEN M = 2*P1 -M
170  E0 = P1/2  :  D = P1/4
180  FOR J = 1 TO 33
190  M1 = E0 - E*SIN (E0)
200  E0 = E0 + D*SGN (M - M1)  :  D = D/2
210  NEXT J
220  E0 = E0*F
```

Fourth Method

The formula

$$\tan E = \frac{\sin M}{\cos M - e} \tag{29.9}$$

gives an *approximate* value for E, and is valid only for small values of the eccentricity.

For the same data as in Example 29.a, the formula (29.9) gives

$$\tan E = \frac{+0.087\,155\,74}{+0.896\,194\,70} = +0.097\,250\,90$$

whence $E = 5^\circ.554\,599$, the exact value being $5^\circ.554\,589$, so the error is only $0''.035$ in this case. But for the same eccentricity and $M = 82^\circ$, the error amounts to $35''$.

The greatest error due to the use of formula (29.9) is

$0^{\circ}.0327$	for $e = 0.15$
0.0783	for $e = 0.20$
0.1552	for $e = 0.25$
1.42	for $e = 0.50$
24.7	for $e = 0.99$

For the orbit of the Earth ($e = 0.0167$), the error will be less than $0''.2$. In that case, formula (29.9) can safely be used except when very high accuracy is needed.

References

1. Seppo Mikkola, 'A cubic approximation for Kepler's Equation', *Celestial Mechanics*, Vol. 40, pages 329 - 334 (1987).

2. Roger W. Sinnott, *Sky and Telescope*, Vol. 70, page 159 (August 1985).

Chapter 30

Elements of the Planetary Orbits

Although Appendix II mentions the principal periodic terms needed to calculate the heliocentric positions of the planets (with explanations given in Chapter 31), it may be of interest to have information about the *mean* orbits of these bodies.

The orbital elements of the major planets can be expressed as polynomials of the form

$$a_0 + a_1 T + a_2 T^2 + a_3 T^3$$

where T is the time measured in Julian centuries of 36525 ephemeris days from the epoch

$$\text{J2000.0} = \text{2000 January 1.5} = \text{JDE 2451 545.0}.$$

In other words,

$$T = \frac{\text{JDE} - 2451\,545.0}{36525} \qquad (30.1)$$

This quantity is negative before the beginning of the year 2000, positive afterwards. The orbital elements are :

L = mean longitude of the planet;
a = semimajor axis of the orbit ;
e = eccentricity of the orbit;
i = inclination on the plane of the ecliptic;
Ω = longitude of the ascending node;
π = longitude of the perihelion.

The longitude of the perihelion is often denoted by ϖ. But this is very confusing because the *argument* of the perihelion has the symbol ω. For this reason, we prefer the symbol π for the longitude of the perihelion, and we have $\pi = \Omega + \omega$.

It should be noted that the quantities L and π are measured in

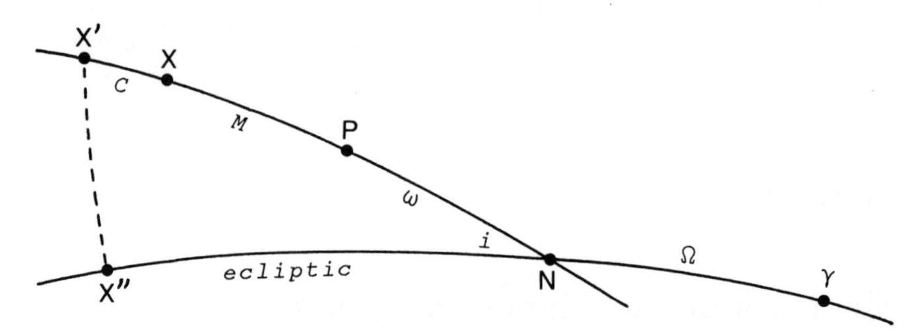

The arc γNX" *is a part of the ecliptic as seen from the Sun, and* NPXX' *is a part of the orbit of the planet (the intersection of the planet's orbital plane with the celestial sphere).* γ *is the vernal equinox (longitude 0°),* N *the ascending node of the orbit,* P *the planet's perihelion. At a given instant, the mean planet is at* X, *the true planet at* X'. *Then we have*

Ω = *arc* γN = *longitude of the ascending node,*
ω = *arc* NP = *argument of the perihelion,*
π = *arc* γN + *arc* NP = $\Omega + \omega$ = *longitude of the perihelion,*
L = *arc* γN + *arc* NX = $\Omega + \omega + M$ = *mean longitude of the planet,*
M = *arc* PX = *planet's mean anomaly,*
C = *arc* XX' = *equation of the center,*
v = *arc* PX' = $M + C$ = *planet's true anomaly,*
i = *inclination of the orbit* = *angle between arcs* NP *and* NX".

---

two different planes, namely from the vernal equinox along the ecliptic to the orbit's ascending node, and then from this node along the orbit. See the Figure above.

The planet's mean anomaly is given by

$$M = L - \pi$$

Table 30.A gives the coefficients a_0 to a_3 for the orbital elements of the planets Mercury to Neptune. The values for the semimajor axes are in astronomical units. Those for the angular quantities L, i, Ω and π are expressed in degrees and decimals; they are referred to the ecliptic and mean equinox of the date.

The tabular values have been deduced from the planetary theory VSOP82 of P. Bretagnon [1]. See Chapter 31 for more information about the theories VSOP82 and VSOP87. The elements L, i, Ω and π are actually referred to the mean *dynamical* ecliptic and equinox of the date, which differ very slightly from the FK5 system (see Chapter 24).

In some cases, it may be desirable to refer the elements L, i, Ω and π to a standard equinox. This is the case, for instance, when one wishes to calculate the least distance between the orbit of a comet and that of a major planet, when the elements of the first orbit are referred to a standard equinox.

By means of Table 30.B, it is possible to calculate these elements for the major planets, referred to the standard equinox of J2000.0. The elements a and e are not modified by a change of reference frame, of course. They should be calculated by means of Table 30.A.

For the Earth, in order to avoid a discontinuity in the variation of the inclination and a jump of 180° in the longitude of the ascending node at the epoch J2000.0, the inclination (on the ecliptic of 2000.0) is considered as negative before A.D. 2000.

Example 30.a — Calculate the mean orbital elements of Mercury on 2065 June 24 at 0^h TD.

We have (see Chapter 7)

2065 June 24.0 = JDE 2475 460.5

whence, by formula (30.1),

$$T = +0.654\,770\,704\,997$$

Consequently, from Table 30.A we find :

$$L = 252°.250\,906 + (149474°.072\,2491 \times 0.654\,770\,704\,997) + (0.000\,303\,97)(0.654\,770\,704\,997)^2 + (0.000\,000\,018)(0.654\,770\,704\,997)^3$$
$$= 98\,123°.494\,702 = 203°.494\,702$$

$a = 0.387\,098\,310$

$e = 0.205\,645\,10$

$i = 7°.006\,171$

$\Omega = 49°.107\,650$

$\pi = 78°.475\,382$

from which we deduce

$M = L - \pi = 125°.019\,320$

$\omega = \pi - \Omega = 29°.367\,732$

From Tables 30.A and 30.B, it appears that the inclination of the orbit of Mercury on the ecliptic of the date is increasing, but that it is decreasing with respect to the fixed ecliptic of 2000.0. The opposite occurs for Saturn and Neptune.

Between $T = -30$ and $T = +30$, Venus' orbital inclination on the ecliptic of the date is continuously increasing, but with respect to the fixed ecliptic of 2000.0 Venus' inclination reached a maximum about the year +690.

Uranus' orbital inclination on the ecliptic of the date reached a minimum about the year +1000, but with respect to the fixed equinox of 2000.0 its value is continuously decreasing during the time period considered here.

TABLE 30.A

Orbital Elements for the mean equinox of the date

	a_0	a_1	a_2	a_3
MERCURY				
L	252.250 906	+149 474.072 2491	+0.000 303 97	+0.000 000 018
a	0.387 098 310			
e	0.205 631 75	+0.000 020 406	−0.000 000 0284	−0.000 000 000 17
i	7.004 986	+0.001 8215	−0.000 018 09	+0.000 000 053
Ω	48.330 893	+1.186 1890	+0.000 175 87	+0.000 000 211
π	77.456 119	+1.556 4775	+0.000 295 89	+0.000 000 056
VENUS				
L	181.979 801	+58 519.213 0302	+0.000 310 60	+0.000 000 015
a	0.723 329 820			
e	0.006 771 88	−0.000 047 766	+0.000 000 0975	+0.000 000 000 44
i	3.394 662	+0.001 0037	−0.000 000 88	−0.000 000 007
Ω	76.679 920	+0.901 1190	+0.000 406 65	−0.000 000 080
π	131.563 707	+1.402 2188	−0.001 073 37	−0.000 005 315
EARTH				
L	100.466 449	+36 000.769 8231	+0.000 303 68	+0.000 000 021
a	1.000 001 018			
e	0.016 708 62	−0.000 042 037	−0.000 000 1236	+0.000 000 000 04
i	0			
π	102.937 348	+1.719 5269	+0.000 459 62	+0.000 000 499

TABLE 30.A (cont.)

	a_0	a_1	a_2	a_3
MARS				
L	355.433 275	+19 141.696 4746	+0.000 310 97	+0.000 000 015
a	1.523 679 342			
e	0.093 400 62	+0.000 090 483	−0.000 000 0806	−0.000 000 000 35
i	1.849 726	−0.000 6010	+0.000 012 76	−0.000 000 006
Ω	49.558 093	+0.772 0923	+0.000 016 05	+0.000 002 325
π	336.060 234	+1.841 0331	+0.000 135 15	+0.000 000 318
JUPITER				
L	34.351 484	+3036.302 7889	+0.000 223 74	+0.000 000 025
a	5.202 603 191	+0.000 000 1913		
e	0.048 494 85	+0.000 163 244	−0.000 000 4719	−0.000 000 001 97
i	1.303 270	−0.005 4966	+0.000 004 65	−0.000 000 004
Ω	100.464 441	+1.020 9550	+0.000 401 17	+0.000 000 569
π	14.331 309	+1.612 6668	+0.001 031 27	−0.000 004 569
SATURN				
L	50.077 471	+1223.511 0141	+0.000 519 52	−0.000 000 003
a	9.554 909 596	−0.000 002 1389		
e	0.055 508 62	−0.000 346 818	−0.000 000 6456	+0.000 000 003 38
i	2.488 878	−0.003 7363	−0.000 015 16	+0.000 000 089
Ω	113.665 524	+0.877 0979	−0.000 120 67	−0.000 002 380
π	93.056 787	+1.963 7694	+0.000 837 57	+0.000 004 899

TABLE 30.A (cont.)

	a_0	a_1	a_2	a_3
URANUS				
L	314.055 005	+429.864 0561	+0.000 304 34	+0.000 000 026
a	19.218 446 062	-0.000 000 0372	+0.000 000 000 98	
e	0.046 295 90	-0.000 027 337	+0.000 000 0790	+0.000 000 000 25
i	0.773 196	+0.000 7744	+0.000 037 49	-0.000 000 092
Ω	74.005 947	+0.521 1258	+0.001 339 82	+0.000 018 516
π	173.005 159	+1.486 3784	+0.000 214 50	+0.000 000 433
NEPTUNE				
L	304.348 665	+219.883 3092	+0.000 309 26	+0.000 000 018
a	30.110 386 869	-0.000 000 1663	+0.000 000 000 69	
e	0.008 988 09	+0.000 006 408	-0.000 000 0008	-0.000 000 000 05
i	1.769 952	-0.009 3082	-0.000 007 08	+0.000 000 028
Ω	131.784 057	+1.102 2057	+0.000 260 06	-0.000 000 636
π	48.123 691	+1.426 2677	+0.000 379 18	-0.000 000 003

The longitudes of the nodes, referred to the equinox of the date, are increasing for all planets. But with respect to the fixed equinox of 2000.0 these longitudes are decreasing, except for Jupiter and Saturn.

Reference

1. P. Bretagnon, 'Théorie du mouvement de l'ensemble des planètes. Solution VSOP82', *Astronomy and Astrophysics*, Vol. 114, pages 278 – 288 (1982).

TABLE 30.B

Orbital Elements for the standard equinox J2000.0

	a_0	a_1	a_2	a_3
MERCURY				
L	252.250 906	+149 472.674 6358	−0.000 005 35	+0.000 000 002
i	7.004 986	−0.005 9516	+0.000 000 81	+0.000 000 041
Ω	48.330 893	−0.125 4229	−0.000 088 33	−0.000 000 196
π	77.456 119	+0.158 8643	−0.000 013 43	+0.000 000 039
VENUS				
L	181.979 801	+58 517.815 6760	+0.000 001 65	−0.000 000 002
i	3.394 662	−0.000 8568	−0.000 032 44	+0.000 000 010
Ω	76.679 920	−0.278 0080	−0.000 142 56	−0.000 000 198
π	131.563 707	+0.004 8646	−0.001 382 32	−0.000 005 332
EARTH				
L	100.466 449	+35 999.372 8519	−0.000 005 68	+0.000 000 000
i	0	+0.013 0546	−0.000 009 31	−0.000 000 034
Ω	174.873 174	−0.241 0908	+0.000 040 67	−0.000 001 327
π	102.937 348	+0.322 5557	+0.000 150 26	+0.000 000 478
MARS				
L	355.433 275	+19 140.299 3313	+0.000 002 61	−0.000 000 003
i	1.849 726	−0.008 1479	−0.000 022 55	−0.000 000 027
Ω	49.558 093	−0.294 9846	−0.000 639 93	−0.000 002 143
π	336.060 234	+0.443 8898	−0.000 173 21	+0.000 000 300

TABLE 30.B (cont.)

	a_0	a_1	a_2	a_3
JUPITER				
L	34.351 484	+3034.905 6746	−0.000 085 01	+0.000 000 004
i	1.303 270	−0.001 9872	+0.000 033 18	+0.000 000 092
Ω	100.464 441	+0.176 6828	+0.000 903 87	−0.000 007 032
π	14.331 309	+0.215 5525	+0.000 722 52	−0.000 004 590
SATURN				
L	50.077 471	+1222.113 7943	+0.000 210 04	−0.000 000 019
i	2.488 878	+0.002 5515	−0.000 049 03	+0.000 000 018
Ω	113.665 524	−0.256 6649	−0.000 183 45	+0.000 000 357
π	93.056 787	+0.566 5496	+0.000 528 09	+0.000 004 882
URANUS				
L	314.055 005	+428.466 9983	−0.000 004 86	+0.000 000 006
i	0.773 196	−0.001 6869	+0.000 003 49	+0.000 000 016
Ω	74.005 947	+0.074 1461	+0.000 405 40	+0.000 000 104
π	173.005 159	+0.089 3206	−0.000 094 70	+0.000 000 413
NEPTUNE				
L	304.348 665	+218.486 2002	+0.000 000 59	−0.000 000 002
i	1.769 952	+0.000 2257	+0.000 000 23	−0.000 000 000
Ω	131.784 057	−0.006 1651	−0.000 002 19	−0.000 000 078
π	48.123 691	+0.029 1587	+0.000 070 51	−0.000 000 023

Chapter 31

Positions of the Planets

In 1982, P. Bretagnon of the Bureau des Longitudes of Paris published his planetary theory VSOP82. The acronym VSOP means 'Variations Séculaires des Orbites Planétaires'. The VSOP82 consists of long series of periodic terms for each of the major planets Mercury to Neptune. When, for a given planet, the sums of these series are evaluated for a given instant, one obtains the values of the following quantities for the osculating orbit. [The osculating orbit is the 'instantaneous' orbit of the planet; see more about this notion in Chapter 32.]

a = semimajor axis of the orbit
λ = mean longitude of the planet
$h = e \sin \pi$
$k = e \cos \pi$
$p = \sin \frac{1}{2} i \ \sin \Omega$
$q = \sin \frac{1}{2} i \ \cos \Omega$

where e is the orbital eccentricity, π the longitude of the perihelion, i the inclination, and Ω the longitude of the ascending node.

Once a, λ, e and π (from h and k), i and Ω (from p and q) are known, the true position of the planet in space can be obtained for the given instant.

The inconvenience of the VSOP82 solution is that one does not know where the several series should be truncated when no full accuracy is required. Fortunately, in 1987 Bretagnon and Francou constructed the version called VSOP87, which gives periodic terms for calculating the planets' heliocentric coordinates directly, namely

L, the ecliptical longitude
B, the ecliptical latitude
R, the radius vector (= distance to the Sun)

It should be noted that L is really the planet's ecliptical longitude, *not* the orbital longitude. In the figure of page 198, the *orbital* longitude of the planet is the sum of the arcs γN and NX' (in two different planes). Through the planet's position X', a great circle $X'X''$ is drawn perpendicularly to the ecliptic. Then the planet's *ecliptical* longitude is the measure of the arc $\gamma X''$.

Although the methods used for the construction of the VSOP82 and VSOP87 have been described in the astronomical literature (see the References 1 and 2), these theories themselves are available only on magnetic tape. By kind permission of Messrs. Bretagnon and Francou, we give in Appendix II the most important periodic terms from the VSOP87 theory. For each planet, series labelled L0, L1, L2, ..., B0, B1, ..., R0, R1, ... are provided.

The series L0, L1, ... are needed to calculate the planet's heliocentric ecliptical longitude L; the series B0, B1, ... are needed for the ecliptical latitude B; and the series R0, R1, ... are for the radius vector R.

Each horizontal line in the list represents one periodic term and contains four numbers :

- the current No. of the term in the series. It is *not* needed in the actual calculation and is given for reference purpose only;
- three numbers which we shall call here A, B, and C, respectively.

If JDE is the Julian Ephemeris Day corresponding to the given instant, calculate the time τ measured in Julian millennia from the epoch 2000.0

$$\tau = \frac{\text{JDE} - 2451\,545.0}{365\,250} \tag{31.1}$$

The value of each term is given by

$$A \cos(B + C\tau)$$

For example, the ninth term of the series L0 for Mercury is equal to $1803 \cos(4.1033 + 5661.3320\,\tau)$.

In the lists of Appendix II, the quantities B and C are expressed in *radians*. The coefficients A are in units of 10^{-8} radian in the case of the longitude and the latitude, in units of 10^{-8} astronomical unit for the radius vector.

When the coefficient A has less decimals, then less decimals too are given for B and C. This is merely done to avoid keypunching extraneous digits which do not influence the result.

To obtain the heliocentric ecliptical longitude L of a planet at a given instant, referred to the mean equinox of the date, proceed as follows. Calculate the sum $L0$ of the terms of series L0, the sum $L1$ of the terms of series L1, etc. Then the required longitude in radians is given by

$$L = (L0 + L1\,\tau + L2\,\tau^2 + L3\,\tau^3 + L4\,\tau^4 + L5\,\tau^5)/10^8 \tag{31.2}$$

Proceed in the same way for the heliocentric latitude B and for the radius vector R.

The planet's heliocentric longitude L and latitude B, obtained thus far, are referred to the mean *dynamical* ecliptic and equinox of the date defined by Bretagnon's VSOP planetary theory. This reference frame differs very slightly from the standard FK5 system mentioned in Chapter 20. The conversion of L and B to the FK5 system can be performed as follows, where T is the time in centuries from 2000.0, or $T = 10\,\tau$.

Calculate

$$L' = L - 1^\circ.397\,T - 0.00031\,T^2$$

Then the corrections to L and B are

$$\begin{aligned} \Delta L &= -0''.09033 + 0''.03916\,(\cos L' + \sin L')\,\tan B \\ \Delta B &= +0''.03916\,(\cos L' - \sin L') \end{aligned} \qquad (31.3)$$

These corrections are needed only for very accurate calculations. They may be dropped if use is made of the abridged version of the VSOP87 given in Appendix II.

How to obtain the *geocentric* positions of the planets will be explained in Chapter 32.

Example 31.a — Calculate the heliocentric coordinates of Venus on 1992 December 20 at 0^h TD.

This date corresponds to JDE 2448 976.5, from which

$$\tau = -0.007\,032\,169\,747.$$

For Venus, series L0 has 24 terms in Appendix II (there are many more in the original VSOP87 theory), L1 has 12 terms, L2 has 8 terms, L3 and L4 both have 3 terms, while L5 contains just a single term. For the sums of these series, we find

$$\begin{aligned} L0 &= +316\,402\,122 & L3 &= -56 \\ L1 &= +1\,021\,353\,038\,718 & L4 &= -109 \\ L2 &= +50\,055 & L5 &= -1 \end{aligned}$$

Hence, by formula (31.2), we find that the heliocentric longitude of Venus, for the given instant and referred to the mean equinox of the date, is

$$L = -68.659\,2582 \text{ radians} = -3933^\circ.88572 = +26^\circ.11428$$

We calculate the heliocentric latitude B and the radius vector R in the same way. It should be noted that, in the case of Venus, the series B5 and R5 do not exist. The results are

$$B = -0.045\,7399 \text{ radian} = -2^\circ.62070, \qquad R = 0.724\,603 \text{ AU}$$

Accuracy of the results

When high accuracy is desired, it appears that the periodic terms in the VSOP87 solution converge rather slowly. What is the magnitude of the errors in the coordinates if one truncates the list of terms at any point? The following empirical rule has been given by Bretagnon and Francou [3] :

> If n is the number of retained terms, and A the amplitude of the smallest retained term, the accuracy of the thus truncated series is about $\eta\sqrt{n} \times A$, where η is a number smaller than 2.

As an example, let us consider the heliocentric longitude of Mercury. In Appendix II, series L0 for this planet contains 38 terms, and the coefficient of the smallest retained term is 100×10^{-8} radian. Therefore, we may expect that the greatest possible error in Mercury's heliocentric longitude is approximately

$$2 \times \sqrt{38} \times 100 \times 10^{-8} \text{ radian} = 2''.54.$$

Of course, series L1, L2, etc., are truncated too, which gives rise to additional uncertainties of the order of $0''.41\ \tau$, $0''.08\ \tau^2$, etc.

References

1. P. Bretagnon, 'Théorie du mouvement de l'ensemble des planètes. Solution VSOP82', *Astronomy and Astrophysics*, Vol. 114, pages 278-288 (1982).

2. P. Bretagnon, G. Francou, 'Planetary theories in rectangular and spherical variables. VSOP87 solutions', *Astronomy and Astrophysics*, Vol. 202, pages 309-315 (1988).

3. *Ibid.*, page 314.

Chapter 32

Elliptic Motion

In this Chapter we will describe two methods for the calculation of geocentric positions in the case of an elliptic orbit. In the first method, the geocentric ecliptical longitude and latitude of a major planet (Mercury to Neptune) are obtained from the heliocentric ecliptical coordinates of the planet and of the Earth. In the second method, which is better suited for minor planets and periodic comets, the right ascension and declination of the body, referred to a standard equinox, are obtained directly; use is made of the geocentric rectangular coordinates of the Sun.

First Method

We will describe how the apparent right ascension and declination of a major planet can be calculated for a given instant.

For the given instant calculate, by means of the appropriate series given in Appendix II (using the method described in Chapter 31), the heliocentric coordinates L, B, R of the planet, and the heliocentric coordinates L_0, B_0, R_0 of the Earth. Do *not* convert from the dynamical ecliptic and equinox to the FK5 ecliptic and equinox at this stage.

Then find

$$\begin{aligned} x &= R \cos B \cos L - R_0 \cos B_0 \cos L_0 \\ y &= R \cos B \sin L - R_0 \cos B_0 \sin L_0 \\ z &= R \sin B - R_0 \sin B_0 \end{aligned} \qquad (32.1)$$

The geocentric longitude λ and latitude β of the planet are then given by

$$\tan \lambda = \frac{y}{x} \qquad \tan \beta = \frac{z}{\sqrt{x^2 + y^2}} \qquad (32.2)$$

Look out for the proper quadrant of λ. One may use the 'second' arctangent function, $\lambda = \text{ATN2}(y, x)$, or see the precepts given in Chapter 1 about 'the correct quadrant'.

However, the geocentric coordinates λ, β obtained in this manner are the planet's *geometric* coordinates referred to the mean equinox of the date. If high accuracy is needed, it is necessary to take into account the apparent displacement of the planet from its true position due to the finite velocity of light. This apparent displacement includes :

(a) the effect of light-time, the planet being seen where it was when light left it;

(b) the effect of the Earth's motion which, combined with the velocity of light, causes an apparent displacement of the object, just as the annual aberration in the case of a star.

The *combination* of the two effects is often called 'planetary aberration'. However, we prefer to reserve the term *aberration* to the effect (b) alone, because this effect is of the same nature as the aberration of the stars. Moreover, for some applications it is *not* necessary to take effect (b) into account. Suppose we want to calculate occultations of stars by planets. Then the effect of light-time *must* be taken into account in the calculation of the position of the planet; but we may drop effect (b) on the condition that the effect of aberration on the star's position is dropped too. Similarly, the effect of nutation can be neglected for *both* bodies in that particular case. The reason is evident: because the planet and the star are close together on the celestial sphere, the effects of aberration and of nutation will not change their *relative* positions.

(a) *effect of light-time*: at time t, the planet is seen where it was at time $t - \tau$, hence in the direction obtained by combining the Earth's position at time t with that of the planet at time $t - \tau$, where τ is the time taken by the light to reach the Earth from the planet. This time is given by

$$\tau = 0.005\,77\,55\,183\ \Delta \quad \text{day} \tag{32.3}$$

where Δ is the planet's distance to the Earth in astronomical units, given by

$$\Delta = \sqrt{x^2 + y^2 + z^2} \tag{32.4}$$

(b) the *effect of aberration* can be calculated as for the stars, namely by means of formulae (22.2), where $\odot$ is equal to $L_o \pm 180°$.

However, *both* effects can be calculated simultaneously. To the order of accuracy that the motion of the Earth during the light-time is rectilinear and uniform, the planet's apparent position at time t is the same as its geometric position at time $t - \tau$. In other words, in this method the Earth's position at time $t - \tau$ must be combined with the planet's position at the same time $t - \tau$.

Of course, the value of the light-time τ is not known in advance because the planet's distance Δ to the Earth is not known. But this distance can be found by iteration, using for instance the value $\Delta = 0$ (and hence $\tau = 0$) in the first calculation.

For very accurate calculations, the planet's geocentric longitude λ and latitude β can be converted from the dynamical ecliptic and equinox to the FK5 ecliptic and equinox by means of formulae (31.3), replacing L by λ, and B by β.

To complete the calculation of the planet's apparent position, the corrections for *nutation* should be applied. This is achieved by calculating the nutation in longitude ($\Delta\psi$) and in obliquity ($\Delta\varepsilon$), as explained in Chapter 21. Add $\Delta\psi$ to the planet's geocentric longitude, and $\Delta\varepsilon$ to the mean obliquity ε_0 of the ecliptic. The apparent right ascension and declination of the planet can then be deduced by means of formulae (12.3) and (12.4).

The *elongation* ψ of the planet, that is its angular distance to the Sun, can be calculated from

$$\cos\psi = \cos\beta \cos(\lambda - \lambda_0) \qquad (32.5)$$

where λ, β are the planet's apparent longitude and latitude, and λ_0 the Sun's apparent longitude. The Sun's latitude, which is always smaller than 1.2 arcsecond, may be neglected here.

Example 32.a — Calculate the apparent position of Venus on 1992 December 20 at 0^h TD = JDE 2448 976.5.

Because the planet's distance to the Earth is not known in advance, the value of the light-time is not known. Therefore, we start with the calculation of the true (geometric) position of the planet at the given time. We find the following values for the heliocentric coordinates (see Example 31.a) :

$L = 26°.11428 \qquad B = -2°.62070 \qquad R = 0.724\,603$

The coordinates of the Earth are calculated in the same way :

$L_0 = 88°.35704 \qquad B_0 = +0°.00014 \qquad R_0 = 0.983\,824 \qquad$ (A)

whence, by formulae (32.1), (32.4), and (32.3),

$x = +0.621\,746 \qquad \Delta = 0.910\,845$
$y = -0.664\,810 \qquad \tau = 0.005\,2606$ day
$z = -0.033\,134$

Δ is the true distance of Venus to the Earth on 1992 December 20.0. We now repeat the calculation of Venus' heliocentric coordinates for the instant $t - \tau$, that is for JDE = 2448 976.5 - 0.005 2606.

We find

$$L = 26°.10588, \qquad B = -2°.62102, \qquad R = 0.724\,604 \qquad (B)$$

Combining these new values with the values (A) of L_0, B_0, R_0, we find

$$x = +0.621\,794 \qquad \Delta = 0.910\,947$$
$$y = -0.664\,905 \qquad (C) \qquad \tau = 0.005\,2612 \text{ day}$$
$$z = -0.033\,138$$

If we repeat the calculation with this new value of τ, we find the same values (B) for L, B and R again, to the given accuracy.

Hence, the final value for the light-time is $\tau = 0.005\,2612$ day, and $\Delta = 0.910\,947$ is the *apparent* distance of the planet on 1992 December 20 at 0^h TD. It is the distance at which we 'see' the planet at that instant; in other words, it is the distance travelled by the light which left the planet at time $t - \tau$ to reach the Earth at time t.

Let us now calculate Venus' geocentric longitude and latitude. If we put the values (C) of x, y, z in formulae (32.2), we obtain

$$\lambda = 313°.08102 \qquad \beta = -2°.08474$$

which are corrected for light-time, but not yet for aberration.

From Chapter 22, we find

$$e = 0.016\,711\,573$$
$$\pi = 102°.88675$$

and formulae (22.2) give, for $\odot = 268°.35704$,

$$\Delta\lambda = -14''.868 = -0°.00413$$
$$\Delta\beta = -0''.531 = -0°.00015$$

and the apparent longitude and latitude of Venus (not yet corrected for nutation) are

$$\lambda = 313°.08102 - 0°.00413 = 313°.07689$$
$$\beta = -2°.08474 - 0°.00015 = -2°.08489$$

[Alternatively, we could have corrected for the light-time and the aberration *together* at once by calculating the coordinates of the Earth for the instant $t - \tau$, which gives

$$L_0 = 88°.35168, \qquad B_0 = +0°.00014, \qquad R_0 = 0.983\,825.$$

We now combine these values with Venus' coordinates (B). Formulae (32.1) and (32.2) then give

$$x = +0.621\,702 \qquad \lambda = 313°.07687$$
$$y = -0.664\,903 \qquad \beta = -2°.08489$$
$$z = -0.033\,138$$

or nearly the same values as before.]

The corrections for reduction to the FK5 system are, from (31.3),

$$\Delta\lambda = -0''.09027 = -0^\circ.00003$$
$$\Delta\beta = +0''.05535 = +0^\circ.00001$$

so the corrected values are

$$\lambda = 313^\circ.07689 - 0^\circ.00003 = 313^\circ.07686$$
$$\beta = -2^\circ.08489 + 0^\circ.00001 = -2^\circ.08488$$

From Chapter 21, we find

$$\Delta\psi = +16''.749, \qquad \Delta\varepsilon = -1''.933, \qquad \varepsilon = 23^\circ.439\,669$$

and the value of λ corrected for nutation is

$$\lambda = 313^\circ.07686 + 16''.749 = 313^\circ.08151$$

Finally, by (12.3) and (12.4),

apparent right ascension :

$$\alpha = 316^\circ.17291 = 21^h.078\,194 = 21^h04^m41^s.50$$

apparent declination :

$$\delta = -18^\circ.88802 = -18^\circ53'16''.9$$

The exact values, obtained by an accurate calculation using the complete VSOP87 theory, are $\alpha = 21^h04^m41^s.454$, $\delta = -18^\circ53'16''.84$, true distance = 0.910 845 96.

Second Method

Here we use the orbital elements referred to a standard equinox, for instance 2000.0, and the geocentric rectangular equatorial coordinates X, Y, Z of the Sun referred to that *same* equinox. These rectangular coordinates can be taken from an astronomical almanac, or they may be calculated by the method described in Chapter 25.

In this method, the heliocentric longitude and latitude of the body (minor planet or periodic comet) are not calculated. Instead, we calculate its heliocentric rectangular equatorial coordinates x, y, z, after which the right ascension, declination and other quantities are derived by means of simple formulae.

The following orbital elements are supposed to be known. They may be taken, for instance, from the *Circulars* of the I.A.U., from the *Minor Planet Circulars* of the Minor Planet Center, etc.

a = semimajor axis, in AU
e = eccentricity
i = inclination
ω = argument of perihelion
Ω = longitude of ascending node
n = mean motion, in degrees/day

where i, ω and Ω are referred to a standard equinox.

If a or n are not given, they can be calculated from

$$a = \frac{q}{1-e} \qquad n = \frac{0.985\,607\,6686}{a\sqrt{a}} \tag{32.6}$$

where q is the perihelion distance in AU. The numerator of the second fraction is the Gaussian gravitational constant 0.017 202 098 95 converted from radians to degrees.

Strictly speaking, all these elements are valid only for one given instant, called the *Epoch*. They vary with time under influence of planetary perturbations. (See, later in this Chapter, the note about *osculating elements*). Unless high accuracy is required, the elements may be considered as invariable during several weeks or even months, for example during the whole apparition of a comet.

Besides the above-mentioned orbital elements, either the value M_0 of the mean anomaly at the epoch, or the time T of passage through perihelion, is given. This allows the calculation of the mean anomaly M at any given instant. The mean anomaly increases by n degrees per day, and is zero at time T.

The orbital elements of a minor planet or of a periodic comet being given, the geocentric position for a given date can be calculated as follows. Firstly, we must calculate the quantities a, b, c and the angles A, B, C, which are constants for a given orbit.

Let ε be the obliquity of the ecliptic. If the orbital elements are referred to the standard equinox of 2000.0, one should use the value $\varepsilon_{2000} = 23°26'21''.448$, from which

$$\sin\varepsilon = 0.397\,777\,16$$
$$\cos\varepsilon = 0.917\,482\,06$$

Then calculate

$$\begin{array}{l|l} F = \cos\Omega & P = -\sin\Omega\cos i \\ G = \sin\Omega\cos\varepsilon & Q = \cos\Omega\cos i\cos\varepsilon - \sin i\sin\varepsilon \\ H = \sin\Omega\sin\varepsilon & R = \cos\Omega\cos i\sin\varepsilon + \sin i\cos\varepsilon \end{array} \tag{32.7}$$

As a check, we can use the relations

$$F^2 + G^2 + H^2 = 1, \qquad P^2 + Q^2 + R^2 = 1,$$

but of course this calculation is not needed in a program.

Then the quantities a, b, c, A, B, C are given by

$$\left.\begin{aligned} \tan A &= \frac{F}{P} & a &= \sqrt{F^2 + P^2} \\ \tan B &= \frac{G}{Q} & b &= \sqrt{G^2 + Q^2} \\ \tan C &= \frac{H}{R} & c &= \sqrt{H^2 + R^2} \end{aligned}\right\} \quad (32.8)$$

The quantities a, b, c should be taken *positive*, while the angles A, B, C should be placed in the correct quadrant, according to the following rules :

$\sin A$ has the same sign as $\cos \Omega$,

$\sin B$ and $\sin C$ have the same sign as $\sin \Omega$.

However, once again, one may use the 'second' arctangent function if it is available in the program language : $A = \text{ATN2}(F, P)$, etc.

Attention: do not confuse the quantity a with the semimajor axis a of the orbit!

For each required position, calculate the body's mean anomaly M, then the eccentric anomaly E (see Chapter 29), the true anomaly v by means of formula (29.1), and the radius vector r by means of (29.2). Then the heliocentric rectangular equatorial coordinates of the body are given by

$$\left.\begin{aligned} x &= r a \sin(A + \omega + v) \\ y &= r b \sin(B + \omega + v) \\ z &= r c \sin(C + \omega + v) \end{aligned}\right\} \quad (32.9)$$

The convenience of these formulae is seen when the rectangular coordinates are required for several positions of the body. The auxiliary quantities a, b, c, A, B, C are functions only of Ω, i and ε, and thus are constants for the whole ephemeris; for each position only the values of v and r must be calculated. However, it should be noted that Ω, i and ω are constant only if the body is in an unperturbed orbit.

For the same instant, calculate the Sun's rectangular coordinates X, Y, Z (Chapter 25), or take them from an astronomical almanac. The geocentric right ascension α and declination δ of the planet or comet are then calculated from

$$\left.\begin{aligned} \tan \alpha &= \frac{Y + y}{X + x} \\ \Delta^2 &= (X + x)^2 + (Y + y)^2 + (Z + z)^2 \\ \sin \delta &= \frac{Z + z}{\Delta} \end{aligned}\right\} \quad (32.10)$$

where Δ is the distance to the Earth and thus is positive. The correct quadrant of α is indicated by the fact that $\sin\alpha$ has the same sign as $(Y+y)$; however, once more, the second arctangent function can be used : $\alpha = \text{ATN2}\,(Y+y,\ X+x)$.

If α is negative, add 360 degrees. Then transform α from degrees into hours by dividing by 15.

The equatorial coordinates α and δ of the body will be referred to the same standard equinox as the orbital elements and the Sun's rectangular coordinates X, Y, Z. However, the values of α and δ obtained in the manner described above refer to the geometric (the true) position of the body in space. Just as in the 'First Method' in this Chapter, the *effect of light-time* should be taken into account. This is performed as follows.

For the given time t, calculate the distance Δ of the body to the Earth as described above, and then the light-time τ by means of (32.3). Then repeat the calculation of M, E, v, x, y, z for the time $t - \tau$, but *leave* the Sun's coordinates X, Y, Z unchanged. With the new values of x, y, z, formulae (32.10) will give the corrected values of α and δ.

When allowance is made for the light-time only, that is, if *no* correction is made for aberration nor for nutation, then the values obtained for α and δ are the so-called *astrometric* right ascension and declination of the body at the given instant. The astrometric position of a minor planet or a comet is directly comparable with the mean places of stars as given in star catalogues (corrected for proper motion and annual parallax, if significant). Of course, α and δ are geocentric.

The elongation ψ to the Sun, and the phase angle β (the angle Sun – body – Earth), can be calculated from

$$\cos\psi = \frac{(X+x)X + (Y+y)Y + (Z+z)Z}{R\,\Delta} = \frac{R^2 + \Delta^2 - r^2}{2R\Delta} \qquad (32.11)$$

$$\cos\beta = \frac{(X+x)x + (Y+y)y + (Z+z)z}{r\,\Delta} = \frac{r^2 + \Delta^2 - R^2}{2r\Delta} \qquad (32.12)$$

where $R = \sqrt{X^2 + Y^2 + Z^2}$ = the distance Earth-Sun. The angles ψ and β are both between 0 and +180 degrees.

Do not confuse this R with the quantity R of expressions (32.1), nor with that of (32.7).

The *magnitude* of the body is then calculated as follows. In the case of a *comet*, the 'total' magnitude is generally calculated from

$$m = g + 5\log\Delta + \kappa\log r \qquad (32.13)$$

where g is the absolute magnitude, and κ a constant which differs from one comet to another. In general, κ is a number between 5 and 15.

For the *minor planets*, a new magnitude system was adopted by Commission 20 of the International Astronomical Union (New Delhi, November 1985). The formula for the prediction of the apparent magnitude of a minor planet is

$$\text{magnitude} = H + 5 \log r\Delta - 2.5 \log \left[(1-G)\,\Phi_1 + G\,\Phi_2 \right] \qquad (32.14)$$

with

$$\Phi_1 = \exp \left[-3.33 \left(\tan \frac{\beta}{2}\right)^{0.63} \right]$$

$$\Phi_2 = \exp \left[-1.87 \left(\tan \frac{\beta}{2}\right)^{1.22} \right]$$

where β is the phase angle, and 'exp' is the exponential function, $\text{EXP}(x) = e^x$. Formula (32.14) is valid for $0° \leqq \beta \leqq 120°$. H and G are magnitude parameters, which are different for each minor planet. H is the mean absolute *visual* magnitude, while G is called the 'slope parameter'. Here are the values of H and G for the brightest minor planets and for some unusual objects [1] :

		H	G			H	G
1	Ceres	3.34	0.12	15	Eunomia	5.28	0.23
2	Pallas	4.13	0.11	18	Melpomene	6.51	0.25
3	Juno	5.33	0.32	20	Massalia	6.50	0.25
4	Vesta	3.20	0.32	433	Eros	11.16	0.46
5	Astraea	6.85	0.15	1566	Icarus	16.4	0.15
6	Hebe	5.71	0.24	1620	Geographos	15.60	0.15
7	Iris	5.51	0.15	1862	Apollo	16.25	0.09
8	Flora	6.49	0.28	2060	Chiron	6.0	0.15
9	Metis	6.28	0.17	2062	Aten	16.80	0.15

In formulae (32.13) and (32.14), the distance to the Sun (r) and the distance to the Earth (Δ) are in astronomical units, and all logarithms are to the base 10.

Example 32.b — Calculate the geocentric position of periodic comet Encke for 1990 October 6.0 TD, using the following orbital elements (see Example 23.b) :

T = 1990 Oct. 28.54502 TD
a = 2.209 1404 AU
e = 0.850 2196

i = 11°.94524
Ω = 334°.75006
ω = 186°.23352

(ecliptic and equinox 2000.0)

We first calculate the auxiliary constants of the orbit :

$$F = +0.904\,455\,59 \qquad P = +0.417\,330\,84$$
$$G = -0.391\,368\,30 \qquad Q = +0.729\,522\,09$$
$$H = -0.169\,678\,93 \qquad R = +0.541\,878\,67$$

whence, by formulae (32.8),

$$A = 65^\circ.230\,615 \qquad a = 0.996\,094\,85$$
$$B = 331^\circ.787\,680 \qquad b = 0.827\,871\,74$$
$$C = 342^\circ.613\,052 \qquad c = 0.567\,823\,42$$

From the value 2.209 1404 for the semimajor axis of the orbit, the second formula (32.6) yields $n = 0.300\,171\,252$ degree/day.

For the given date (1990 October 6.0), the time since the perihelion is -22.54502 days. Hence, the mean anomaly is

$$M = -22.54502 \times 0^\circ.300\,171\,252 = -6^\circ.767\,367$$

We then find

$$E = -34^\circ.026\,714 \qquad x = +0.250\,8066$$
$$v = -94^\circ.163\,310 \qquad y = +0.484\,9175$$
$$r = 0.652\,4867 \qquad z = +0.357\,3373$$

The Sun's geocentric rectangular equatorial coordinates for the same instant, and referred to the same standard equinox (2000.0), calculated by using the complete VSOP87 theory, are

$$X = -0.975\,6732, \qquad Y = -0.200\,3254, \qquad Z = -0.086\,8566$$

from which $\Delta = 0.824\,3689$, and the light-time is $\tau = 0.00476$ day.

Repeating the calculation of the comet's position for $t - \tau$, that is for 1990 October 5.99524, we find

$$M = -6^\circ.768\,796$$
$$E = -34^\circ.031\,552 \qquad x = +0.250\,9310$$
$$v = -94^\circ.171\,933 \qquad y = +0.484\,9477$$
$$r = 0.652\,5755 \qquad z = +0.357\,3712$$

$$X + x = -0.724\,7422$$
$$Y + y = +0.284\,6223$$
$$Z + z = +0.270\,5146$$
$$\Delta = 0.824\,2811$$

from which we deduce the astrometric right ascension and declination, and the elongation from the Sun :

$$\alpha_{2000} = 158^\circ.558\,965 = 10^h34^m14^s.2$$
$$\delta_{2000} = +19^\circ.158\,496 = +19^\circ09'31''$$
$$\psi = 40^\circ.51$$

Notes on the osculating elements

Mean orbital elements, such as those given in Chapter 30 for the major planets, represent the elements of a mean reference orbit. They refer to a slowly varying orbit.

For the periodic comets and the thousands of minor planets, however, no mean orbital elements are calculated. Instead, orbital elements are calculated for the 'instantaneous' orbit at a given instant (the Epoch); these are the so-called *osculating elements*, and the instant for which they are valid is the Epoch of osculation.

> "Osculating elements at a particular epoch are defined as the elements of an unperturbed elliptical orbit, referred to as the osculating orbit, in which the position and velocity of the planet at the epoch are identical with the actual position and velocity of the planet in its perturbed orbit at the same instant. The osculating elements therefore contain the effects of the perturbations due to the other planets, so that, unlike the mean elements, they are subject to periodic variations." [2]

While the *mean* elements vary slowly with time (for instance, the eccentricity of the mean orbit of Mars was 0.09331 in A.D. 1900 and will be 0.09340 in A.D. 2000), the osculating elements vary rather rapidly. These changes generally do *not* reflect the real changes of the mean orbit.

As an example, let us give the following osculating elements of minor planet Ceres for two epochs separated by only 200 days. These elements are taken from the yearly *Ephemerides of Minor Planets* (Leningrad); the elements i, ω and Ω are referred to the standard equinox of 1950.0.

Epoch (TD) :	1980 Dec. 27.0	1981 July 15.0
Semimajor axis (AU) :	a = 2.766 3951	a = 2.767 1238
Eccentricity :	e = 0.077 2343	e = 0.077 4937
Inclination (degrees) :	i = 10.598 78	i = 10.598 15
Argument of perihelion (deg.):	ω = 73.895 55	ω = 73.901 89
Long. of ascending node (deg.):	Ω = 80.102 59	Ω = 80.096 60
Mean anomaly (degrees) :	M = 319.239 14	M = 2.081 33
Mean motion (degrees/day) :	n = 0.214 206 55	n = 0.214 121 94

From 1980 December 27 to 1981 July 15, the semimajor axis of the 'instantaneous' orbit increased by 0.00073 AU; from this, we may not, however, deduce that during those 200 days the mean distance of Ceres to the Sun increased by 109 000 kilometers!

On 1980 December 27, the 'instantaneous' revolution period of Ceres was 1680.62 days (which is obtained by dividing 360° by n); 200 days later this had increased to 1681.28 days.

Neptune provides an even better illustration. While the eccentricity of its mean orbit is presently 0.0090, that of its osculating orbit reached a maximum of 0.0124 in November 1964, a minimum of 0.0039 in October 1970, another maximum (0.0122) in December 1976, and so on. These rather large variations are not surprising: the osculating orbit of Neptune refers to the instantaneous position and velocity of the Sun, which itself oscillates around the barycenter of the solar system, mainly due to the actions of the giant planets Jupiter and Saturn. Orbital elements of Neptune referred to that barycenter (instead of to the Sun) would show much smaller variations.

Accurate ephemerides of the periodic comets and the minor planets are obtained by numerical integration, and for these calculations the osculating orbital elements provide starting values.

Osculating elements may be used to give the actual position and motion of the body at the epoch of osculation, and they provide a good approximation to its actual orbit over short periods around the Epoch. They may *not*, however, be used as an unperturbed orbit over a long period!

In order to have an idea of the increasing error of an ephemeris calculated by using an osculating orbit as an unperturbed one, we used the above-mentioned osculating elements of Ceres valid for 1981 July 15. The heliocentric longitude of Ceres, calculated in this manner, was then compared with the exact one as deduced from the work of Duncombe [3]. It appeared that until 280 days after the Epoch the error was smaller than 9". During the first 40 days, the error was smaller than 1". The error in the calculated heliocentric longitude reached a maximum (+8") 180 days after the Epoch, but after a few months the error $\Delta\lambda$ quicky reached large negative values:

Number of days after 1981 July 15 :	0	40	80	120	160	200	240	280	320	360	400
$\Delta\lambda$ (arcseconds):	0	+1	+3	+5	+7	+7	+3	−8	−26	−52	−86

The further evolution of $\Delta\lambda$ is shown in the Figure on the next page. The oscillating curve represents the variation of the error as a function of the time. So, in this particular case, the error does not increase continually with time. We found the following extreme values for the error in Ceres' heliocentric longitude:

+8" in January 1982
−708" in mid-March 1984
+864" in mid-May 1986
−825" in July 1988
+1754" in August 1990

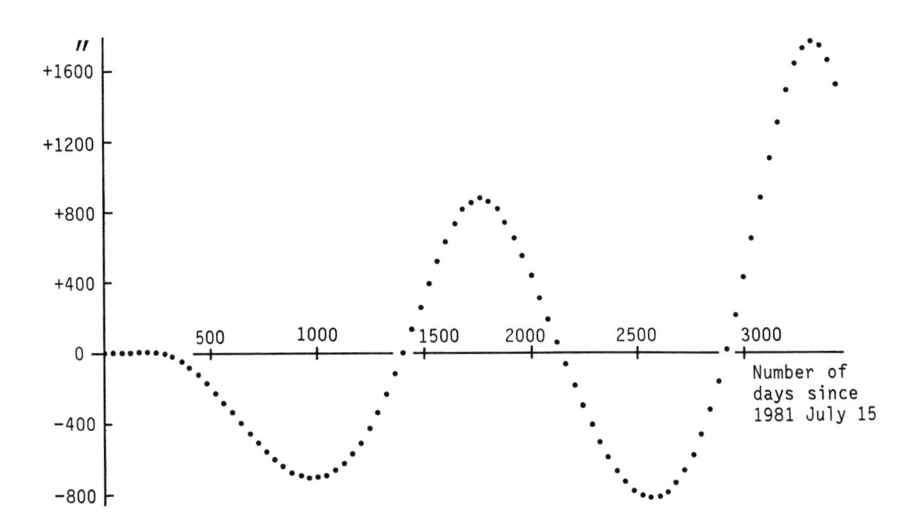

The error $\Delta\lambda$ in the calculated heliocentric longitude of Ceres, when osculating elements are used and the perturbations by the planets are neglected.

Vertically : $\Delta\lambda$ *in seconds of arc.*

Horizontally : the days elapsed since the Epoch, 1981 July 15.0.

The points are given at intervals of 40 days.

The Equation of the Center

If the orbital eccentricity is small, then instead of solving the equation of Kepler (Chapter 29) and then using formula (29.1), the equation of the center C, or the difference $v - M$, can be found directly in terms of e and M by means of the following formula.

$$C = \left(2e - \frac{e^3}{4} + \frac{5}{96}e^5\right)\sin M + \left(\frac{5}{4}e^2 - \frac{11}{24}e^4\right)\sin 2M$$
$$+ \left(\frac{13}{12}e^3 - \frac{43}{64}e^5\right)\sin 3M + \frac{103}{96}e^4 \sin 4M + \frac{1097}{960}e^5 \sin 5M$$

The result is expressed in radians, and thus should be multiplied by $180/\pi$ or 57.295 779 51 in order to be converted into degrees. The formula is derived from a series expansion [4], and has been truncated after the term in e^5. Therefore it is suitable only for small values of the eccentricity. If the eccentricity is *very* small, the terms in e^4 and e^5 may be neglected.

The greatest error is

	The formula up to terms in e^5	*The formula with terms e^4 and e^5 neglected*
for e = 0.03	0".0003	0".24
0.05	0.007	1.8
0.10	0.45	30
0.15	5	152
0.20	29	483
0.25	111	1183
0.30	331	2456

There exists a series expansion for the radius vector too. Its terms up to the fifth power of the eccentricity are as follows :

$$\frac{r}{a} = 1 + \frac{e^2}{2} - \left(e - \frac{3}{8}e^3 + \frac{5}{192}e^5\right)\cos M$$
$$- \left(\frac{e^2}{2} - \frac{e^4}{3}\right)\cos 2M$$
$$- \left(\frac{3}{8}e^3 - \frac{45}{128}e^5\right)\cos 3M$$
$$- \frac{e^4}{3}\cos 4M$$
$$- \frac{125}{384}e^5 \cos 5M$$

Velocity on an elliptic orbit

On an unperturbed elliptical orbit, the instantaneous velocity of the moving body, in kilometers per second, is given by

$$V = 42.1219\sqrt{\frac{1}{r} - \frac{1}{2a}}$$

where r is the distance of the body to the Sun, and a is the semimajor axis of the orbit, both expressed in astronomical units.

If e is the orbital eccentricity, then the velocities at perihelion and at aphelion, again in km/second, are respectively

$$V_p = \frac{29.7847}{\sqrt{a}}\sqrt{\frac{1+e}{1-e}}$$

$$V_a = \frac{29.7847}{\sqrt{a}}\sqrt{\frac{1-e}{1+e}}$$

Example 32.c — For the 1986 return of periodic comet Halley, we had [5]

$a = 17.940\,0782 \qquad e = 0.967\,274\,26$

these osculating values being valid strictly for the Epoch 1986 February 19.0 TD.

For this orbit, the velocities at perihelion and at aphelion are $V_p = 54.52$ km/second and $V_a = 0.91$ km/second, respectively.

At the distance $r = 1$ AU from the Sun, the comet's velocity was $V = 41.53$ km/second.

Length of the ellipse

While there is an exact formula giving the area of an ellipse (area = πab), there is no exact expression with a finite number of terms and ordinary functions for the length L (the perimeter) of an ellipse. In what follows, e is the eccentricity of the ellipse, a its semimajor axis, and b its semiminor axis given by $b = a\sqrt{1-e^2}$.

1. An approximate formula given by Ramanujan in 1914 is

$$L = \pi\left(3(a+b) - \sqrt{(a+3b)(3a+b)}\right)$$

The error is zero for $a = b$ (that is, for a circle), increasing to 0.4155% for $e = 1$, that is, for an infinitely flat ellipse.

2. Another interesting method for finding the length of an ellipse is as follows. Let A, G and H be the arithmetic, the geometric, and the harmonic means, respectively, of the semi-axes a and b of the ellipse. That is,

$$A = \frac{a+b}{2} \qquad G = \sqrt{ab} \qquad H = \frac{2ab}{a+b}$$

Then we have

$$L = \pi \left(\frac{21A - 2G - 3H}{8} \right)$$

with an error less than 0.001% if $e < 0.88$, and less than 0.01% if $e < 0.95$. But the error amounts to 1% for $e = 0.9997$, and to 3% for $e = 1$.

3. A formula with an infinite series expansion is

$$L = 2\pi a \left(1 - \left(\frac{1}{2}\right)^2 \frac{e^2}{1} - \left(\frac{1\times 3}{2\times 4}\right)^2 \frac{e^4}{3} - \left(\frac{1\times 3\times 5}{2\times 4\times 6}\right)^2 \frac{e^6}{5} - \cdots \right)$$

The expression between brackets takes the value 0.99937 for $e = 0.05$, the value 0.99750 for $e = 0.10$, and is equal to $0.63662 = 2/\pi$ for $e = 1$.

4. More rapid convergence is obtained with the following formula, where $m = (a - b) / (a + b)$,

$$L = \frac{2\pi a}{1+m} \left(1 + \left(\frac{1}{2}\right)^2 m^2 + \left(\frac{1}{2\times 4}\right)^2 m^4 + \left(\frac{1\times 3}{2\times 4\times 6}\right)^2 m^6 + \left(\frac{1\times 3\times 5}{2\times 4\times 6\times 8}\right)^2 m^8 + \cdots \right)$$

Example 32.d — Periodic comet Halley. Using the elements for the return of 1986 (see Example 32.c), we find that the length of the orbit is 77.07 AU = 11530 millions of kilometers.

References

1. *Minor Planet Circulars* 17256–17264 (1990 December 2).

2. *Explanatory Supplement to the Astronomical Ephemeris* (London, 1961); page 114.

3. R.L. Duncombe, 'Heliocentric Coordinates of Ceres, Pallas, Juno, Vesta, 1928-2000', *Astron. Papers*, XX, II (Washington, 1969).

4. *Annales de l'Observatoire de Paris*, Vol. I, pages 202-204.

5. *Minor Planet Circular* 10634 (1986 April 24).

Chapter 33

Parabolic Motion

In this Chapter we give formulae for the calculation of positions of a comet which moves around the Sun in a parabolic orbit. We will assume that the elements of this orbit are invariable (no planetary perturbations) and that they are referred to a standard equinox, for example that of 2000.0.

We assume that the following orbital elements are given :

T = time of passage in perihelion
q = perihelion distance, in AU
i = inclination
ω = argument of the perihelion
Ω = longitude of ascending node

First, calculate the auxiliary constants A, B, C, a, b, c as for an elliptic orbit: see formulae (32.7) and (32.8). Then, for each required position of the comet, proceed as follows.

Let $t - T$ be the time since perihelion, in days. This quantity is negative for an instant before the time of perihelion. Calculate

$$W = \frac{0.036\,491\,162\,45}{q\,\sqrt{q}}\,(t - T) \qquad (33.1)$$

The constant in the numerator is equal to $3k/\sqrt{2}$, where k is the Gaussian gravitational constant 0.017 202 098 95.

Then the true anomaly v and the radius vector r of the comet are given by

$$\tan\frac{v}{2} = s \qquad\qquad r = q\,(1 + s^2) \qquad (33.2)$$

where s is the root of the equation

$$s^3 + 3s - W = 0 \qquad (33.3)$$

It should be noted that, for an instant before the time of perihelion, s is negative and v is between $-180°$ and $0°$; after the perihelion, $s > 0$ and v is between $0°$ and $+180°$. At the instant of passage through perihelion, $s = 0$, $v = 0°$, and $r = q$.

There are several ways to solve equation (33.3), which is called *Barker's equation*.

1. The equation can easily be solved by iteration; this algorithm has the author's preference, because the iteration formula is simple, the convergence is rapid, no trigonometric functions or cubic roots are involved, and the procedure is valid for positive as well as negative values of $(t - T)$, and for $t = T$ (or $s = 0$) too.

One may start from *any* value for s; a good choice is $s = 0$. A better value for s is

$$\frac{2s^3 + W}{3\,(s^2 + 1)} \tag{33.4}$$

This calculation is repeated until the correct value of s is obtained. It should be noted that in expression (33.4) the cube of s must be calculated; if s is negative, this operation is not possible on some calculating machines; when this is the case, calculate $s \times s \times s$ instead of s^3.

2. Instead of solving equation (33.3) by iteration, s can be obtained directly as follows (J. Bauschinger, *Tafeln zur Theoretischen Astronomie*, page 9; Leipzig, 1934):

$$\left.\begin{aligned} \tan\beta &= \frac{2}{W} = 54.807\,791\,\frac{q\sqrt{q}}{t - T} \\ \tan\gamma &= \sqrt[3]{\tan\frac{\beta}{2}} \\ s &= \frac{2}{\tan 2\gamma} \end{aligned}\right\} \tag{33.5}$$

The constant 54.807 791 is equal to $\dfrac{2\sqrt{2}}{3k}$, where k is the Gaussian gravitational constant.

In this method, no iteration is performed, but two problems can occur:

— at the time of passage through perihelion, $t - T$ is zero, hence W is zero, and $2/W$ becomes infinite. However, in that case we have immediately $v = 0°$ and $r = q$, but the possible occurrence of this case must be anticipated in the computer program;

— before the perihelion we have $W < 0$, whence $\tan\beta$ is negative. But in that case, $\tan\beta/2$ is negative too, and computers cannot calculate the cubic root of a negative quantity. This difficulty can be avoided by replacing W by its absolute value in the first formula (33.5). At the end of the calculation, the sign of s should then be changed accordingly.

For instance, in BASIC the formulae (33.1) and (33.3) can be programmed as follows, where T stands for the number of days $t - T$ since perihelion :

```
IF T = 0 THEN ....
W = .03649116245 * T/(Q * SQR(Q))
B = ATN(2/ABS(W))
S = 2/TAN(2*ATN(TAN(B/2)^(1/3)))
IF T < 0 THEN S = -S
```

3. The following method is easier and does not use trigonometric functions. All expressions under the root signs are positive.

$$G = \frac{W}{2} \qquad Y = \sqrt[3]{G + \sqrt{G^2 + 1}} \qquad s = Y - \frac{1}{Y} \tag{33.6}$$

When s is obtained, v and r can be found by means of (33.2), after which the calculation continues as for the elliptic motion, formulae (32.9) and (32.10), with the same precept to take the effect of light-time into account.

The first formula (33.2) will give $v/2$ between -90 and +90 degrees, the range of the arctangent function of the computer languages. That will give v in the correct quadrant, between -180° and +180°, so no additional check will be required.

In the parabolic motion, $e = 1$ while a and the period of revolution are infinite; the mean daily motion is zero, and therefore the mean and eccentric anomalies do not exist (in fact, they are zero).

Example 33.a — Calculate the true anomaly and the distance to the Sun of comet Helin-Roman (1989*s* = 1989 IX) for 1989 October 31.0 TD, using the values

T = 1989 August 20.29104 TD
q = 1.324 5017

of a parabolic orbit calculated by B.G. Marsden (*Minor Planet Circular* No. 16001, 1990 March 11).

For the given date (1989 October 31.0), the time from perihelion is $t - T = +71.70896$ days. Hence, by formula (33.1),

$$W = +1.716\,652\,31.$$

Starting from the value $s = 0$, we obtain the following successive approximations by means of the iteration formula (33.4) :

0.000 0000
0.572 2174
0.525 1685
0.524 2029
0.524 2025

Hence, $s = +0.524\,2025$, and consequently

$$v = +55^\circ.32728 \qquad r = 1.688\,459$$

If, instead of the iteration procedure, formulae (33.6) are used, we obtain successively

$$G = 0.858\,326\,155$$
$$Y = 1.295\,879\,323$$
$s = Y - 1/Y = 0.524\,2025$, as before.

Chapter 34

Near-parabolic Motion

An eccentricity of exactly 1 means that the orbit is parabolic; in that case, it is easy to calculate the position of the body for a given instant (see Chapter 33). If the orbit has a high eccentricity (say, 0.98 to 1.1), but different from 1, it is more troublesome to deal with. An eccentricity greater than 1 means the orbit is hyperbolic.

The German astronomer Werner Landgraf has given an interesting program in BASIC [1], based on Karl Stumpff's work *Himmelsmechanik*, Vol. I (Berlin, 1959). Hereafter we give Landgraf's program, in a slightly modified form.

First, one calculates

$$Q = \frac{k}{2q}\sqrt{\frac{1+e}{q}} \qquad\qquad \gamma = \frac{1-e}{1+e}$$

where, as before, k is the Gaussian gravitational constant, e is the eccentricity of the orbit, and q is the perihelion distance in astronomical units.

Then solve the following equation iteratively for s :

$$s = Qt - (1-2\gamma)\frac{s^3}{3} + \gamma\,(2-3\gamma)\frac{s^5}{5} - \gamma^2(3-4\gamma)\frac{s^7}{7} + \ldots. \tag{34.1}$$

where t is the number of days before (-) or after (+) the perihelion. Begin by inserting into the right-hand side of the equation the value of s obtained for an orbit which would be precisely parabolic [with the value of W of formula (33.1) put equal to $Qt/3$]. This evaluation leads to an improved s, which is used in another iteration, and so on until the value of s ceases to change.

Once the final value of s is found, the true anomaly v and the distance r to the Sun are found from

$$\tan \frac{v}{2} = s \qquad\qquad r = \frac{q\,(1+e)}{1 + e \cos v}$$

The calculation of geocentric places can then be performed as for the elliptic and the parabolic motions.

Here is Landgraf's program in BASIC, slightly modified by us. It is valid for highly eccentric elliptical orbits (e slightly less than 1), for slightly hyperbolic orbits (e slightly larger than 1), as well as for an orbit which is exactly parabolic. The computer is assumed to be working in radians.

```
10  P1=4*ATN(1) : R1=180/P1
12  K=0.01720209895
14  D1 = 10000 : C = 1/3 : D = 1E-9
16  INPUT "PERIHELION DISTANCE = "; Q
18  INPUT "ECCENTRICITY = "; E0
20  Q1 = K*SQR((1+E0)/Q)/(2*Q) : G = (1-E0)/(1+E0)
22  INPUT "DAYS FROM PERIHELION = "; T
24  IF T<>0 THEN 28
26  R = Q : V = 0 : GOTO 72
28  Q2 = Q1*T
30  S = 2/(3*ABS(Q2))
32  S = 2/TAN(2*ATN(TAN(ATN(S)/2)^C))
34  IF T<0 THEN S = -S
36  IF E0 = 1 THEN 66
38  L = 0
40  S0 = S : Z = 1 : Y = S*S : G1 = -Y*S
42  Q3 = Q2 + 2*G*S*Y/3
44  Z = Z + 1
46  G1 = -G1*G*Y
48  Z1 = (Z-(Z+1)*G)/(2*Z+1)
50  F = Z1*G1
52  Q3 = Q3 + F
54  IF Z > 50 OR ABS(F) > D1 THEN 78
56  IF ABS(F) > D THEN 44
58  L = L+1 : IF L > 50 THEN 78
60  S1 = S : S = (2*S*S*S/3 + Q3)/(S*S+1)
62  IF ABS(S-S1) > D THEN 60
64  IF ABS(S-S0) > D THEN 40
66  V = 2*ATN(S)
68  R = Q*(1+E0)/(1 + E0*COS(V))
70  IF V<0 THEN V = V + 2*P1
72  PRINT "TRUE ANOMALY = "; V*R1
74  PRINT "RADIUS VECTOR (A.U.) = "; R
76  PRINT : GOTO 22
78  PRINT "NO CONVERGENCE"
80  PRINT : GOTO 22
```

Some comments about this program :

Line 10 : the first formula is a trick to obtain the number π.

Line 12 : the Gaussian gravitational constant k.

Line 14 : the number D = 10^{-9} adjusts to suit the computer's precision. If necessary, one may use 10^{-8} or 10^{-10}.

Line 26 : when $t = 0$ (the body being exactly in perihelion), we have $r = q$ and $v = 0°$.

Line 36 : if the orbit is exactly parabolic, the value of s has been found.

Line 54 : if in formula (34.1) more than 50 terms are needed, or if these terms become too large, there is no convergence.

Line 56 : as long as a term of formula (34.1) is not small enough, the next term should be calculated.

Line 58 : if after 50 iterations no result has still been found, the calculation must be halted.

Lines 60 and 62 : solving equation (34.1) by iteration. This is an iteration inside of an iteration!

As an exercise, try to calculate the following cases :

Data			Results	
perihelion distance q (AU)	*eccentricity* e	*days* t	*true anomaly* v (°)	*distance to the Sun* r (AU)
0.921 326	1.000 00	138.4783	102.744 26	2.364 192
0.100 000	0.987 00	254.9	164.500 29	4.063 777
0.123 456	0.999 97	-30.47	221.911 90	0.965 053
3.363 943	1.057 31	1237.1	109.405 98	10.668 551
0.587 1018	0.967 2746	20	52.853 31	0.729 116
0.587 1018	0.967 2746	0	0	0.587 1018

After having calculated some cases, you will notice that the calculation time is longer as $|t|$ is larger, that is, as the body is farther away from the perihelion. The calculation time is longer too as e differs more from unity. The table on the next page mentions some calculation times on the HP-85 microcomputer, together with a rounded value of the true anomaly v, and the number L of iterations.

q	e	t	Calculation time in seconds	v	L
0.1	0.9	10	14	126°	17
		20	47	142°	30
		30	no convergence	–	–
0.1	0.987	10	4	123°	7
		20	5	137°	8
		30	6	143°	10
		60	9	152°	12
		100	14	157°	16
		200	28	163°	23
		400	87	167°	38
		500	no convergence	–	–
0.1	0.999	100	3	156°	6
		200	4	161°	7
		500	5	166°	8
		1 000	7	169°	10
		5 000	18	174°	18
1	0.999 99	100 000	2	172°.5	4
		10 000 000	5	178°.41	8
		14 000 000	6	178°.58	9
		17 000 000	7	178°.68	9
		18 000 000	no convergence	–	–

For $q = 0.1$ and $e = 0.9$, the calculation takes 47 seconds for $t = 20$ days, and there is no convergence for $t = 30$ days.

For $q = 0.1$ and $e = 0.999$, there is no trouble up to $t = 5000$ days.

For $q = 1$ and $e = 0.999\,99$, there is no trouble even for $t =$ 17 million days. This is 465 centuries after the perihelion time; the object's distance from the Sun is then 7220 astronomical units – at least in theory!

Reference

1. *Sky and Telescope*, Vol. 73, pages 535-536 (May 1987).

Chapter 35

The Calculation of some Planetary Phenomena

There are two basically different methods for calculating planetary phenomena such as the greatest elongations of Venus, or the time of an opposition of Mars :

(i) either by comparing accurate positions of the planet with those of the Sun;

(ii) or by using formulae where a mean value is corrected by a sum of periodic terms.

The first method has the advantage of giving very accurate results, because use is made of very accurate positions of the bodies. It has the inconvenience, however, of requiring the availability or the calculation of these accurate ephemerides.

With the second method, the calculation can be performed easily and rapidly for any year. The results, while not so accurate as those of the first method, are still good enough for many applications, such as historical research, or even as a first approximation for a more accurate calculation.

In this Chapter, we provide formulae for calculating several configurations involving the planets Mercury to Neptune: oppositions and conjunctions with the Sun, and greatest elongations.

Oppositions and conjunctions with the Sun

From the proper line in Table 35.A, take the values of A, B, M_0 and M_1.

Let Y be an approximate time of the required phenomenon, expressed as *years and decimals*. For instance, 1993.0 means the beginning of the year 1993, 2028.5 denotes the middle of the year 2028, etc.

TABLE 35.A

Planet	Event	A	B	M_0	M_1
Mercury	Inferior conjunction	2451 612.023	115.877 4771	63.5867	114.208 8742
Mercury	Superior conjunction	2451 554.084	115.877 4771	6.4822	114.208 8742
Venus	Inferior conjunction	2451 996.706	583.921 361	82.7311	215.513 058
Venus	Superior conjunction	2451 704.746	583.921 361	154.9745	215.513 058
Mars	Opposition	2452 097.382	779.936 104	181.9573	48.705 244
Mars	Conjunction	2451 707.414	779.936 104	157.6047	48.705 244
Jupiter	Opposition	2451 870.628	398.884 046	318.4681	33.140 229
Jupiter	Conjunction	2451 671.186	398.884 046	121.8980	33.140 229
Saturn	Opposition	2451 870.170	378.091 904	318.0172	12.647 487
Saturn	Conjunction	2451 681.124	378.091 904	131.6934	12.647 487
Uranus	Opposition	2451 764.317	369.656 035	213.6884	4.333 093
Uranus	Conjunction	2451 579.489	369.656 035	31.5219	4.333 093
Neptune	Opposition	2451 753.122	367.486 703	202.6544	2.194 998
Neptune	Conjunction	2451 569.379	367.486 703	21.5569	2.194 998

$JDE_0 = A + kB$ $\qquad M = M_0 + kM_1$ (in degrees)

Then find the integer k nearest to

$$\frac{365.2425\,Y + 1721\,060 - A}{B} \tag{35.1}$$

It is important to note that k must be an *integer*. Non-integer values of k would yield meaningless results. Successive values of k will provide the data for the successive events, the value $k = 0$ corresponding to the first one after 2000 January 1. For years preceding A.D. 2000, k takes negative values.

Then calculate

$$\mathrm{JDE}_0 = A + kB, \qquad M = M_0 + kM_1$$

JDE_0 is the Julian Ephemeris Day corresponding to the time of the *mean* planetary configuration (that is, calculated from circular orbits and uniform planetary motions), and M is the mean anomaly of the Earth at that instant.

M is an angle expressed in *degrees* and decimals. Depending on the type of the calculating machine or programming language, it may be necessary or desirable to reduce the angle to the range 0-360 degrees, by adding or subtracting a convenient multiple of 360, and to convert the result into radians.

Find the time T, expressed in centuries from the beginning of the year 2000, from

$$T = \frac{\mathrm{JDE}_0 - 2451\,545}{36525}$$

T is positive after the beginning of A.D. 2000, negative before.

For the planets Jupiter to Neptune, additional angles are required. Expressed in degrees, these angles are :

for Jupiter : $a = 82.74 + 40.76\,T$

for Saturn :
$a = 82.74 + 40.76\,T$
$b = 29.86 + 1181.36\,T$
$c = 14.13 + 590.68\,T$
$d = 220.02 + 1262.87\,T$

for Uranus :
$e = 207.83 + 8.51\,T$
$f = 108.84 + 419.96\,T$

for Neptune :
$e = 207.83 + 8.51\,T$
$g = 276.74 + 209.98\,T$

The time JDE of the *true* configuration is obtained by adding to JDE_0 a correction which is given in Table 35.B as a sum of periodic terms which are functions of the angle M. By reason of the secular variations of the planetary orbits, the coefficients of these periodic terms are slowly varying with time, whence the presence of terms in T and T^2 in Table 35.B.

For instance, for an inferior conjunction of Mercury, the correction (in days) is

$$
\begin{aligned}
&+ 0.0545 + 0.0002\,T \\
&+ (-6.2008 + 0.0074\,T + 0.00003\,T^2)\ \sin M \\
&+ (-3.2750 - 0.0197\,T + 0.00001\,T^2)\ \cos M \\
&+ (0.4737 - 0.0052\,T - 0.00001\,T^2)\ \sin 2M \\
&+ \ \text{etc} \ldots.
\end{aligned}
$$

The corrected instant obtained in this way is expressed as a Julian Ephemeris Day (JDE), hence in the scale of Dynamical Time. This can be reduced to the standard Julian Day, JD, based on the Universal Time, by *subtracting* the quantity ΔT expressed in *days* (see Chapter 9). However, between the years 1500 and 2100, the correction $-\Delta T$ can be neglected for our purposes.

Finally, from the JD the corresponding calendar date can be obtained by means of standard procedures (see Chapter 7).

Example 35.a — Calculate Mercury's inferior conjunction that is nearest to 1993 October 1.

From Table 35.A, for Mercury, Inferior conjunction, we have

$$
\begin{aligned}
A &= 2451\,612.023 \\
B &= 115.877\,4771 \\
M_0 &= 63.5867 \\
M_1 &= 114.208\,8742
\end{aligned}
$$

October 1 is three quarters of a year since January 1, hence 1993 October 1 = 1993.75 = Y, and expression (35.1) yields the value -20.28, whence $k = -20$. (Remember that k must be an integer.) Then

$$
\begin{aligned}
\text{JDE}_0 &= 2449\,294.473 \\
M &= -2220^\circ.5908 = +299^\circ.4092 \\
T &= -0.06162
\end{aligned}
$$

The sum of the terms in the relevant part of Table 35.B (Mercury, Inferior conjunction) is +3.171, whence

$$\text{JDE} = \text{JDE}_0 + 3.171 = 2449\,297.644,$$

which corresponds to 1993 November 6, at 3^h TD.

Rounded to the nearest integer hour, this is indeed the correct instant.

Example 35.b — Find the instant of Saturn's conjunction with the Sun in the year 2125.

From Table 35.A, for Saturn, Conjunction, we have

$$A = 2451\,681.124$$
$$B = 378.091\,904$$
$$M_0 = 131.6934$$
$$M_1 = 12.647\,487$$

For $Y = 2125.0$ (i.e., the beginning of the year 2125), expression (35.1) gives the value +120.39. Since we are searching the first Saturn-Sun conjunction *after* the beginning of the year 2125, we take $k = +121$, not +120. Then

$$\text{JDE}_\circ = 2497\,430.244$$
$$M = 1662°.0393 = 222°.0393$$
$$T = +1.25627$$

and for Saturn we have to calculate the following additional angles:

$$a = 133°.95, \quad b = 73°.97, \quad c = 36°.18, \quad d = 6°.53$$

The sum of the terms in the relevant part of Table 35.B (Saturn, Conjunction with Sun) is +7.659, whence

$$\text{JDE} = \text{JDE}_\circ + 7.659 = 2497\,437.903,$$

which corresponds to 2125 August 26, at 10^h TD.

The correct instant, calculated with a more accurate method, is 2125 August 26, at 11^h TD.

Greatest elongations of Mercury and Venus

To calculate the times and the values of the greatest elongations of Mercury or Venus, *we start from the nearest inferior conjunction*. So we calculate k, $\text{JDE}_\circ$, M and T as explained before. We do *not* calculate the instant of the true inferior conjunction; instead, we use the periodic terms given in Table 35.C to find the correction (in days) to Mercury's or Venus' *mean* inferior conjunction, to obtain the time of greatest eastern or western elongation. In the same table, periodic terms are provided to find the value of this greatest elongation.

Do not forget that, if the planet is east from the Sun, it is visible in the evening in the west; if the elongation is west, the planet is visible in the morning in the east.

The value of the greatest elongation from the Sun is expressed in degrees and decimals. It concerns the maximum *angular distance* from the planet to the center of the Sun's disk, *not* the greatest difference between the geocentric ecliptical longitudes of the two bodies. There is no 'official' definition for the elongation of a planet to the Sun, and two different definitions could be considered :

(a) the *angular distance* between the object and the center of the solar disk;

(b) the difference between the *geocentric longitudes* of the object and the center of the solar disk.

Both definitions are used in the astronomical literature. Definition (a) has been used in the *Astronomical Ephemeris* since its beginning in 1960, and from 1981 onwards in its successor, the *Astronomical Almanac*. It is this definition we prefer. For example, for the visibility of Venus near its inferior conjunction, the important factor is not the longitude difference with the Sun, but the angular separation.

The French astronomers, however, use definition (b), for instance in their *Annuaire du Bureau des Longitudes*. On page 275 of the volume for 1990 we read :

> "Les plus grandes élongations des planètes inférieures : la différence des longitudes géocentriques de la planète et du Soleil est maximale."

As a consequence, the results will differ somewhat according as one uses definition (a) or (b). For example, for Mercury's greatest elongation of 1990 August 11: the difference between the geocentric ecliptical longitudes of the Sun and Mercury reached its maximum value (27°22') at 15^h UT, as mentioned on page 277 of the *Annuaire du Bureau des Longitudes* for 1990, but the maximum angular separation took place at 21^h and was equal to 27°25'.

Example 35.c — Find the instant and the value of Mercury's greatest western elongation in November 1993.

We start from the inferior conjunction of November 1993, for which we found in Example 35.a :

$$\text{JDE}_0 = 2449\,294.473, \qquad M = 299°.4092, \qquad T = -0.06162.$$

With these values of M and T, we find from the relevant part of Table 35.C (Mercury, greatest western elongation) :

correction = +19.665 days,   elongation = 19°.7506.

Hence, the time of Mercury's greatest western elongation is

$$\text{JDE} = \text{JDE}_0 + 19.665 = 2449\,314.14$$

which corresponds to 1993 November 22, at 15^h TD.

The value of this maximum elongation is 19°.7506 = 19°45'.

The accuracy of the results

It is evident that the expressions given in Tables 35.B and 35.C are valid only for a limited period of time, namely for a few millennia before and after A.D. 2000, and *not* for millions of years! Consequently, do not use the method given in this Chapter before the year -2000, nor after A.D. 4000.

For modern times, say between A.D. 1800 and 2200, the instants obtained for the phenomena involving Mercury and Venus will be less than 1 hour in error. The error can reach 2 hours in the case of Saturn, Uranus and Neptune, 3 hours for Mars, and 4 hours for Jupiter.

It is expected that the maximum possible error will be somewhat larger near the years -2000 and +4000. On the other hand, if the calculations are performed for epochs near A.D. 2000, say between 1900 and 2100, then the terms in T^2 may safely be ignored.

Exercises

Check your program with the following cases; all times are in TD.

Mercury	inferior conjunction	1631	Nov	7	7^h	(a)
Venus	inferior conjunction	1882	Dec	6	17^h	(b)
Mars	opposition	2729	Sep	9	3^h	(c)
Jupiter	opposition	-6	Sep	15	7^h	(d)
Saturn	opposition	-6	Sep	14	9^h	(d)
Uranus	opposition	1780	Dec	17	14^h	(e)
Neptune	opposition	1846	Aug	20	4^h	(f)

(a) the first observed transit of Mercury over the solar disk (by Gassendi, at Paris).

(b) the last transit of Venus before that of A.D. 2004.

(c) a perihelic opposition of Mars.

(d) because Jupiter and Saturn were in opposition with the Sun with a time difference less than one day, there occurred a triple conjunction between these two planets in that year.

(e) three months before Uranus' discovery by William Herschel.

(f) one month before Neptune's discovery.

TABLE 35.B

Periodic terms in days

	MERCURY Inferior conjunction	MERCURY Superior conjunction
	$+0.0545 + 0.0002\,T$	$-0.0548 - 0.0002\,T$
sin M	$-6.2008 + 0.0074\,T + 0.00003\,T^2$	$+7.3894 - 0.0100\,T - 0.00003\,T^2$
cos M	$-3.2750 - 0.0197\,T + 0.00001\,T^2$	$+3.2200 + 0.0197\,T - 0.00001\,T^2$
sin $2M$	$+0.4737 - 0.0052\,T - 0.00001\,T^2$	$+0.8383 - 0.0064\,T - 0.00001\,T^2$
cos $2M$	$+0.8111 + 0.0033\,T - 0.00002\,T^2$	$+0.9666 + 0.0039\,T - 0.00003\,T^2$
sin $3M$	$+0.0037 + 0.0018\,T$	$+0.0770 - 0.0026\,T$
cos $3M$	$-0.1768 + 0.00001\,T^2$	$+0.2758 + 0.0002\,T - 0.00002\,T^2$
sin $4M$	$-0.0211 - 0.0004\,T$	$-0.0128 - 0.0008\,T$
cos $4M$	$+0.0326 - 0.0003\,T$	$+0.0734 - 0.0004\,T - 0.00001\,T^2$
sin $5M$	$+0.0083 + 0.0001\,T$	$-0.0122 - 0.0002\,T$
cos $5M$	$-0.0040 + 0.0001\,T$	$+0.0173 - 0.0002\,T$

	VENUS Inferior conjunction	VENUS Superior conjunction
	$-0.0096 + 0.0002\,T - 0.00001\,T^2$	$+0.0099 - 0.0002\,T - 0.00001\,T^2$
sin M	$+2.0009 - 0.0033\,T - 0.00001\,T^2$	$+4.1991 - 0.0121\,T - 0.00003\,T^2$
cos M	$+0.5980 - 0.0104\,T + 0.00001\,T^2$	$-0.6095 + 0.0102\,T - 0.00002\,T^2$
sin $2M$	$+0.0967 - 0.0018\,T - 0.00003\,T^2$	$+0.2500 - 0.0028\,T - 0.00003\,T^2$
cos $2M$	$+0.0913 + 0.0009\,T - 0.00002\,T^2$	$+0.0063 + 0.0025\,T - 0.00002\,T^2$
sin $3M$	$+0.0046 - 0.0002\,T$	$+0.0232 - 0.0005\,T - 0.00001\,T^2$
cos $3M$	$+0.0079 + 0.0001\,T$	$+0.0031 + 0.0004\,T$

TABLE 35.B (cont.)

	MARS Opposition	MARS Conjunction with Sun
	$-0.3088 + 0.00002\,T^2$	$+0.3102 - 0.0001\,T + 0.00001\,T^2$
sin M	$-17.6965 + 0.0363\,T + 0.00005\,T^2$	$+9.7273 - 0.0156\,T + 0.00001\,T^2$
cos M	$+18.3131 + 0.0467\,T - 0.00006\,T^2$	$-18.3195 - 0.0467\,T + 0.00009\,T^2$
sin $2M$	$-0.2162 - 0.0198\,T - 0.00001\,T^2$	$-1.6488 - 0.0133\,T + 0.00001\,T^2$
cos $2M$	$-4.5028 - 0.0019\,T + 0.00007\,T^2$	$-2.6117 - 0.0020\,T + 0.00004\,T^2$
sin $3M$	$+0.8987 + 0.0058\,T - 0.00002\,T^2$	$-0.6827 - 0.0026\,T + 0.00001\,T^2$
cos $3M$	$+0.7666 - 0.0050\,T - 0.00003\,T^2$	$+0.0281 + 0.0035\,T + 0.00001\,T^2$
sin $4M$	$-0.3636 - 0.0001\,T + 0.00002\,T^2$	$-0.0823 + 0.0006\,T + 0.00001\,T^2$
cos $4M$	$+0.0402 + 0.0032\,T$	$+0.1584 + 0.0013\,T$
sin $5M$	$+0.0737 - 0.0008\,T$	$+0.0270 + 0.0005\,T$
cos $5M$	$-0.0980 - 0.0011\,T$	$+0.0433$

	JUPITER Opposition	JUPITER Conjunction with Sun
	$-0.1029 - 0.00009\,T^2$	$+0.1027 + 0.0002\,T - 0.00009\,T^2$
sin M	$-1.9658 - 0.0056\,T + 0.00007\,T^2$	$-2.2637 + 0.0163\,T - 0.00003\,T^2$
cos M	$+6.1537 + 0.0210\,T - 0.00006\,T^2$	$-6.1540 - 0.0210\,T + 0.00008\,T^2$
sin $2M$	$-0.2081 - 0.0013\,T$	$-0.2021 - 0.0017\,T + 0.00001\,T^2$
cos $2M$	$-0.1116 - 0.0010\,T$	$+0.1310 - 0.0008\,T$
sin $3M$	$+0.0074 + 0.0001\,T$	$+0.0086$
cos $3M$	$-0.0097 - 0.0001\,T$	$+0.0087 + 0.0002\,T$
sin a	$0 + 0.0144\,T - 0.00008\,T^2$	$0 + 0.0144\,T - 0.00008\,T^2$
cos a	$+0.3642 - 0.0019\,T - 0.00029\,T^2$	$+0.3642 - 0.0019\,T - 0.00029\,T^2$

TABLE 35.B (cont.)

	SATURN Opposition	SATURN Conjunction with Sun
	$-0.0209 + 0.0006T + 0.00023T^2$	$+0.0172 - 0.0006T + 0.00023T^2$
sin M	$+4.5795 - 0.0312T - 0.00017T^2$	$-8.5885 + 0.0411T + 0.00020T^2$
cos M	$+1.1462 - 0.0351T + 0.00011T^2$	$-1.1470 + 0.0352T - 0.00011T^2$
sin $2M$	$+0.0985 - 0.0015T$	$+0.3331 - 0.0034T - 0.00001T^2$
cos $2M$	$+0.0733 - 0.0031T + 0.00001T^2$	$+0.1145 - 0.0045T + 0.00002T^2$
sin $3M$	$+0.0025 - 0.0001T$	$-0.0169 + 0.0002T$
cos $3M$	$+0.0050 - 0.0002T$	$-0.0109 + 0.0004T$
sin a	$0 - 0.0337T + 0.00018T^2$	$0 - 0.0337T + 0.00018T^2$
cos a	$-0.8510 + 0.0044T + 0.00068T^2$	$-0.8510 + 0.0044T + 0.00068T^2$
sin b	$0 - 0.0064T + 0.00004T^2$	$0 - 0.0064T + 0.00004T^2$
cos b	$+0.2397 - 0.0012T - 0.00008T^2$	$+0.2397 - 0.0012T - 0.00008T^2$
sin c	$0 - 0.0010T$	$0 - 0.0010T$
cos c	$+0.1245 + 0.0006T$	$+0.1245 + 0.0006T$
sin d	$0 + 0.0024T - 0.00003T^2$	$0 + 0.0024T - 0.00003T^2$
cos d	$+0.0477 - 0.0005T - 0.00006T^2$	$+0.0477 - 0.0005T - 0.00006T^2$

TABLE 35.B (cont.)

	URANUS Opposition	URANUS Conjunction with Sun
	$+0.0844 - 0.0006T$	$-0.0859 + 0.0003T$
$\sin M$	$-0.1048 + 0.0246T$	$-3.8179 - 0.0148T + 0.00003T^2$
$\cos M$	$-5.1221 + 0.0104T + 0.00003T^2$	$+5.1228 - 0.0105T - 0.00002T^2$
$\sin 2M$	$-0.1428 + 0.0005T$	$-0.0803 + 0.0011T$
$\cos 2M$	$-0.0148 - 0.0013T$	$-0.1905 - 0.0006T$
$\sin 3M$	0	$+0.0088 + 0.0001T$
$\cos 3M$	+0.0055	0
$\cos e$	+0.8850	+0.8850
$\cos f$	+0.2153	+0.2153

	NEPTUNE Opposition	NEPTUNE Conjunction with Sun
	$-0.0140 + 0.00001T^2$	+0.0168
$\sin M$	$-1.3486 + 0.0010T + 0.00001T^2$	$-2.5606 + 0.0088T + 0.00002T^2$
$\cos M$	$+0.8597 + 0.0037T$	$-0.8611 - 0.0037T + 0.00002T^2$
$\sin 2M$	$-0.0082 - 0.0002T + 0.00001T^2$	$+0.0118 - 0.0004T + 0.00001T^2$
$\cos 2M$	$+0.0037 - 0.0003T$	$+0.0307 - 0.0003T$
$\cos e$	-0.5964	-0.5964
$\cos g$	+0.0728	+0.0728

TABLE 35.C

Periodic terms for greatest elongations

	MERCURY, greatest eastern elongation (evening visibility)	
	Correction (days) to the time of mean inferior conjunction	Elongation (degrees)
	$-21.6101 + 0.0002\,T$	22.4697
$\sin M$	$-1.9803 - 0.0060\,T + 0.00001\,T^2$	$-4.2666 + 0.0054\,T + 0.00002\,T^2$
$\cos M$	$+1.4151 - 0.0072\,T - 0.00001\,T^2$	$-1.8537 - 0.0137\,T$
$\sin 2M$	$+0.5528 - 0.0005\,T - 0.00001\,T^2$	$+0.3598 + 0.0008\,T - 0.00001\,T^2$
$\cos 2M$	$+0.2905 + 0.0034\,T + 0.00001\,T^2$	$-0.0680 + 0.0026\,T$
$\sin 3M$	$-0.1121 - 0.0001\,T + 0.00001\,T^2$	$-0.0524 - 0.0003\,T$
$\cos 3M$	$-0.0098 - 0.0015\,T$	$+0.0052 - 0.0006\,T$
$\sin 4M$	$+0.0192$	$+0.0107 + 0.0001\,T$
$\cos 4M$	$+0.0111 + 0.0004\,T$	$-0.0013 + 0.0001\,T$
$\sin 5M$	-0.0061	-0.0021
$\cos 5M$	$-0.0032 - 0.0001\,T$	$+0.0003$

TABLE 35.C (cont.)

	MERCURY, greatest western elongation (morning visibility)	
	Correction (days) to the time of mean inferior conjunction	Elongation (degrees)
	$+21.6249 - 0.0002\,T$	$22.4143 - 0.0001\,T$
sin M	$+0.1306 + 0.0065\,T$	$+4.3651 - 0.0048\,T - 0.00002\,T^2$
cos M	$-2.7661 - 0.0011\,T + 0.00001\,T^2$	$+2.3787 + 0.0121\,T - 0.00001\,T^2$
sin $2M$	$+0.2438 - 0.0024\,T - 0.00001\,T^2$	$+0.2674 + 0.0022\,T$
cos $2M$	$+0.5767 + 0.0023\,T$	$-0.3873 + 0.0008\,T + 0.00001\,T^2$
sin $3M$	$+0.1041$	$-0.0369 - 0.0001\,T$
cos $3M$	$-0.0184 + 0.0007\,T$	$+0.0017 - 0.0001\,T$
sin $4M$	$-0.0051 - 0.0001\,T$	$+0.0059$
cos $4M$	$+0.0048 + 0.0001\,T$	$+0.0061 + 0.0001\,T$
sin $5M$	$+0.0026$	$+0.0007$
cos $5M$	$+0.0037$	-0.0011

TABLE 35.C (cont.)

	VENUS, greatest eastern elongation (evening visibility)	
	Correction (days) to the time of mean inferior conjunction	Elongation (degrees)
	$-70.7600 + 0.0002\,T - 0.00001\,T^2$	$46.3173 + 0.0001\,T$
sin M	$+1.0282 - 0.0010\,T - 0.00001\,T^2$	$+0.6916 - 0.0024\,T$
cos M	$+0.2761 - 0.0060\,T$	$+0.6676 - 0.0045\,T$
sin $2M$	$-0.0438 - 0.0023\,T + 0.00002\,T^2$	$+0.0309 - 0.0002\,T$
cos $2M$	$+0.1660 - 0.0037\,T - 0.00004\,T^2$	$+0.0036 - 0.0001\,T$
sin $3M$	$+0.0036 + 0.0001\,T$	
cos $3M$	$-0.0011 + 0.00001\,T^2$	

	VENUS, greatest western elongation (morning visibility)	
	Correction (days) to the time of mean inferior conjunction	Elongation (degrees)
	$+70.7462 - 0.00001\,T^2$	46.3245
sin M	$+1.1218 - 0.0025\,T - 0.00001\,T^2$	$-0.5366 - 0.0003\,T + 0.00001\,T^2$
cos M	$+0.4538 - 0.0066\,T$	$+0.3097 + 0.0016\,T - 0.00001\,T^2$
sin $2M$	$+0.1320 + 0.0020\,T - 0.00003\,T^2$	-0.0163
cos $2M$	$-0.0702 + 0.0022\,T + 0.00004\,T^2$	$-0.0075 + 0.0001\,T$
sin $3M$	$+0.0062 - 0.0001\,T$	
cos $3M$	$+0.0015 - 0.00001\,T^2$	

Chapter 36

Pluto

As for the numerous minor planets (see Chapter 32), no analytical theory for the motion of Pluto is available. However, expressions for an accurate representation of the planet's motion (1950.0 coordinates) for the years 1885 to 2099 have been constructed by Goffin, Meeus and Steyaert [1]. The coefficients of the periodic terms were determined by the least-squares method, on the basis of a numerical integration of Pluto's heliocentric motion performed by E. Goffin. Perturbations by the first eight major planets were included. This integration itself was based on the osculating elements by Seidelmann *et al.* [2] which were obtained through a numerical integration fitted to all available observed positions of Pluto, spanning the years 1914 to 1979.

Using Goffin's numerical integration again, we have repeated the calculation of the periodic terms, but now referring Pluto's heliocentric longitude and latitude to the new standard equinox J2000.0 instead of B1950.0. The results are given in Table 36.A.

Method of calculation

Calculate, by means of formula (21.1), the time T in Julian centuries from the epoch J2000.0, and then the following angles (in degrees):

$$J = 34.35 + 3034.9057\,T$$
$$S = 50.08 + 1222.1138\,T$$
$$P = 238.96 + 144.9600\,T$$

Then calculate the periodic terms as given by Table 36.A. Here, each argument α is a linear combination of the angles J, S, P, namely

$$\alpha = iJ + jS + kP$$

and the contribution of each argument is

$$A \sin \alpha + B \cos \alpha$$

For instance, on the 13th line we read the numbers 0, 2, -1, so here the argument is $\alpha = 2S - P$, and for the latitude the contribution is $-94 \sin\alpha + 210 \cos\alpha$.

In Table 36.A, the numerical values of the coefficients A and B are given in units of the sixth decimal of a degree in the case of the longitude and the latitude, and in units of the seventh decimal (astronomical units) for the radius vector.

The heliocentric longitude l, latitude b (in degrees), and the radius vector r of Pluto are then given by

$$l = 238.956785 + 144.96\,T + \text{sum of periodic terms in longitude}$$
$$b = -3.908202 + \text{sum of periodic terms in latitude}$$
$$r = 40.7247248 + \text{sum of periodic terms in radius vector}$$

The longitude and latitude obtained by this method are heliocentric, not barycentric, and they are referred to the standard equinox of J2000.0.

Calculated in this way, l will be less than 0".6 in error, b less than 0".2, and the radius vector less than 0.000 02 AU, with respect to the numerical integration on which this representation of Pluto's motion is based. It is important to note, as has been said, that the method given here is not valid outside the period 1885 - 2099.

To find the *geocentric* astrometric 2000.0 equatorial coordinates α and δ of Pluto :

– find the geocentric 2000.0 rectangular equatorial coordinates X, Y, Z of the Sun (see Chapter 25);

– find those of Pluto by

$$\begin{aligned} x &= r \cos l \cos b \\ y &= r\,(\sin l \cos b \cos\varepsilon - \sin b \sin\varepsilon) \\ z &= r\,(\sin l \cos b \sin\varepsilon + \sin b \cos\varepsilon) \end{aligned} \qquad (36.1)$$

where ε is the mean obliquity of the ecliptic at epoch J2000.0. We have

$$\sin\varepsilon = 0.397777156$$
$$\cos\varepsilon = 0.917482062$$

– find α and δ, and Pluto's distance Δ to the Earth, by means of formulae (32.10).

However, the effect of light-time should be taken into account. See Chapter 32 and formula (32.3). Hence, to obtain the geocentric α and δ, the values of l, b, r should be calculated for an instant which is earlier than the given instant by the light-time τ.

It may seem strange that in our solution the mean longitudes of Uranus and Neptune are not needed. The reason is that the mean motion of Uranus is almost exactly twice that of Neptune, or three times that of Pluto. For this reason, the argument $2N - P$, for in-

TABLE 36.A

Periodic terms for the heliocentric coordinates of Pluto

No.	Argument			Longitude		Latitude		Radius vector	
	J	S	P	A	B	A	B	A	B
1	0	0	1	−19798886	19848454	−5453098	−14974876	66867334	68955876
2	0	0	2	897499	−4955707	3527363	1672673	−11826086	−333765
3	0	0	3	610820	1210521	−1050939	327763	1593657	−1439953
4	0	0	4	−341639	−189719	178691	−291925	−18948	482443
5	0	0	5	129027	−34863	18763	100448	−66634	−85576
6	0	0	6	−38215	31061	−30594	−25838	30841	−5765
7	0	1	−1	20349	−9886	4965	11263	−6140	22254
8	0	1	0	−4045	−4904	310	−132	4434	4443
9	0	1	1	−5885	−3238	2036	−947	−1518	641
10	0	1	2	−3812	3011	−2	−674	−5	792
11	0	1	3	−601	3468	−329	−563	518	518
12	0	2	−2	1237	463	−64	39	−13	−221
13	0	2	−1	1086	−911	−94	210	837	−494
14	0	2	0	595	−1229	−8	−160	−281	616
15	1	−1	0	2484	−485	−177	259	260	−395
16	1	−1	1	839	−1414	17	234	−191	−396
17	1	0	−3	−964	1059	582	−285	−3218	370
18	1	0	−2	−2303	−1038	−298	692	8019	−7869
19	1	0	−1	7049	747	157	201	105	45637
20	1	0	0	1179	−358	304	825	8623	8444
21	1	0	1	393	−63	−124	−29	−896	−801
22	1	0	2	111	−268	15	8	208	−122
23	1	0	3	−52	−154	7	15	−133	65
24	1	0	4	−78	−30	2	2	−16	1
25	1	1	−3	−34	−26	4	2	−22	7
26	1	1	−2	−43	1	3	0	−8	16
27	1	1	−1	−15	21	1	−1	2	9
28	1	1	0	−1	15	0	−2	12	5
29	1	1	1	4	7	1	0	1	−3
30	1	1	3	1	5	1	−1	1	0
31	2	0	−6	8	3	−2	−3	9	5
32	2	0	−5	−3	6	1	2	2	−1
33	2	0	−4	6	−13	−8	2	14	10
34	2	0	−3	10	22	10	−7	−65	12
35	2	0	−2	−57	−32	0	21	126	−233
36	2	0	−1	157	−46	8	5	270	1068
37	2	0	0	12	−18	13	16	254	155
38	2	0	1	−4	8	−2	−3	−26	−2
39	2	0	2	−5	0	0	0	7	0
40	2	0	3	3	4	0	1	−11	4
41	3	0	−2	−1	−1	0	1	4	−14
42	3	0	−1	6	−3	0	0	18	35
43	3	0	0	−1	−2	0	1	13	3

stance, where N is the mean longitude of Neptune, has almost exactly the same period as $2P$. The small difference could not have been detected by our investigation based on the rather short interval of 214 years. Therefore, Table 36.A does not contain the argument $2N - P$; the effects of the terms with this argument are included in the terms with argument $2P$. For the same reason, there are no terms in $S - 4P$, $S - 3P$, $S - 2P$, $J - 5P$, $J - 4P$, and $2S - 3P$: they have almost the same period as $4P$, $5P$, $6P$, $2S - P$, $2S$, and $J - S + P$, respectively.

Example 36.a — For 1992 October 13.0 TD = JDE 2448 908.5, find
(1) the geometric heliocentric coordinates of Pluto;
(2) its geocentric astrometric coordinates α and δ.

(1) We find

$$T = -0.072\,183\,4360$$
$$J = -184°.719\,921$$
$$S = -38°.136\,373$$
$$P = 228°.496\,289$$

Sum of periodic terms		
	in longitude :	+ 4 247 019
	in latitude :	+ 18 495 889
	in radius vector :	-110 133 423

from which

$$l = 228°.493\,074 + 4°.247\,019 = 232°.740\,09$$
$$b = -3°.908\,202 + 18°.495\,889 = +14°.587\,69$$
$$r = 40.724\,7248 - 11.013\,3423 = 29.711\,383 \text{ AU}$$

(2) For the given instant, the Sun's 2000.0 rectangular equatorial coordinates are (from Example 25.b)

$$X = -0.937\,3959$$
$$Y = -0.313\,1679$$
$$Z = -0.135\,7792$$

Using Pluto's coordinates l, b, r found above, formulae (36.1) give

$$x = -17.408\,3314$$
$$y = -23.973\,1135$$
$$z = -\ 2.237\,4336$$

whence, by formulae (32.10) and (32.3),

$$\Delta = 30.529\,024 \text{ AU} \qquad \text{and} \qquad \tau = 0.17632 \text{ day.}$$

(This value of Δ is Pluto's true distance to the Earth).

We now repeat the calculation of the planet's heliocentric coordinates for 1992 October 13.0 - 0.17632 = October 12.82368.

The results are

$$l = 232^\circ.73887$$
$$b = +14^\circ.58788$$
$$r = 29.711\,366$$

whence

$$x = -17.408\,7937 \qquad \Delta = 30.529\,017$$
$$y = -23.972\,7795 \qquad \tau = 0.17632 \text{ day}$$
$$z = -\ 2.237\,1895$$

We obtain for τ the same value as before, so no new iteration is needed.

The 2000.0 astrometric coordinates of Pluto for 1992 October 13.0 TD are then found by means of (32.10) :

$$\alpha = 232^\circ.93172 = 15^h31^m43^s.6$$
$$\delta = -4^\circ.45800 = -4^\circ27'29''$$

Mean orbital elements of the orbit of Pluto, near A.D. 2000 :

$$a = 39.543 \text{ AU}$$
$$e = 0.2490$$
$$\left.\begin{aligned} i &= 17^\circ.140 \\ \Omega &= 110^\circ.307 \\ \omega &= 113^\circ.768 \end{aligned}\right\} 2000.0$$

References

1. E. Goffin, J. Meeus, and C. Steyaert, 'An accurate representation of the motion of Pluto', *Astronomy and Astrophysics*, Vol. 155, pages 323 - 325 (1986).

2. P.K. Seidelmann, G.H. Kaplan, K.F. Pulkkinen, E.J. Santoro, and T.C. Van Flandern, *Icarus*, Vol. 44, page 20 (1980).

Chapter 37

Planets in Perihelion and Aphelion

The Julian Day corresponding to the time when a planet is in perihelion or in aphelion can be found by means of the following formulae :

Mercury	JDE = 2451 590.257 + 87.969 349 63 k − 0.000 000 0000 k^2
Venus	JDE = 2451 738.233 + 224.700 818 7 k − 0.000 000 0327 k^2
Earth	JDE = 2451 547.507 + 365.259 635 8 k + 0.000 000 0158 k^2
Mars	JDE = 2452 195.026 + 686.995 784 3 k − 0.000 000 1187 k^2
Jupiter	JDE = 2455 636.938 + 4332.897 090 k + 0.000 1368 k^2
Saturn	JDE = 2452 830.11 + 10764.217 31 k + 0.000 826 k^2
Uranus	JDE = 2470 213.5 + 30694.8767 k − 0.005 41 k^2
Neptune	JDE = 2468 895.7 + 60190.32 k + 0.031 75 k^2

where k is an integer for perihelion, and an integer increased by exactly 0.5 for aphelion.

Any other value for k would give a meaningless result !

A zero or a positive value of k will give a date after the beginning of the year 2000. If $k < 0$, one obtains a date earlier than A.D. 2000.

For example, $k = +14$ and $k = -222$ give passages through perihelion, while $k = +27.5$ and $k = -119.5$ give aphelion passages.

An *approximate* value for k can be found as follows, where the 'year' should be taken with decimals, if necessary :

Mercury	$k \cong 4.15201$ (year − 2000.12)
Venus	$k \cong 1.62549$ (year − 2000.53)
Earth	$k \cong 0.99997$ (year − 2000.01)
Mars	$k \cong 0.53166$ (year − 2001.78)

Jupiter	$k \cong 0.08430$ (year - 2011.20)
Saturn	$k \cong 0.03393$ (year - 2003.52)
Uranus	$k \cong 0.01190$ (year - 2051.1)
Neptune	$k \cong 0.00607$ (year - 2047.5)

Example 37.a — Find the time of passage of Venus at perihelion nearest to 1978 October 15, that is 1978.79.

An approximate value of k is

$$1.62549\ (1978.79 - 2000.53) = -35.34$$

and, since k must be an integer (perihelion!), we take $k = -35$. Putting this value in the formula for Venus, we find

$$\text{JDE} = 2443\,873.704,$$

which corresponds to 1978 December 31.204 = 1978 December 31, at 5^h Dynamical Time.

Example 37.b — Find the time of passage of Mars through aphelion in A.D. 2032.

Taking 'year' = 2032.0, we find $k \cong +16.07$. Since k must be an integer increased by 0.5 (aphelion!), the first aphelion of Mars after the beginning of the year 2032 occurs for $k = +16.5$.

Using the formula for Mars, this value of k gives

$$\text{JDE} = 2463\,530.456,$$

which corresponds to 2032 October 24.956 or 2032 October 24, at 23^h Dynamical Time.

It is important to note that the formulae for the calculation of JDE given above are based on unperturbed elliptic orbits. For this reason, the instants obtained for Mars can be a few hours in error.

Due to the mutual planetary perturbations, the instants for Jupiter, calculated by the method described here, may be up to half a month in error. For Saturn, the error may be larger than one month.

For instance, putting $k = -2.5$ in the formula for Jupiter gives 1981 July 19 as the date of an aphelion passage, while the correct date is 1981 July 28. For Saturn, $k = -2$ gives 1944 July 30, while the planet actually reached perihelion on 1944 September 8.

The error would be even larger for Uranus and Neptune. For these planets, the formulae are given merely for completeness.

Accurate times can be obtained by calculating the value of the planet's distance to the Sun for several instants near the expected time, and then finding when this distance reaches a maximum or a minimum. Here are the dates when Saturn (in the period 1920 - 2050) and Uranus (1750 - 2100) are in perihelion (P) or in aphelion (A). After the date, the distance to the Sun in astronomical units is given. These data have been calculated by means of P. Bretagnon's complete VSOP87 theory.

Saturn			Uranus		
A	1929 Nov 11	10.0468	A	1756 Nov 27	20.0893
P	1944 Sep 8	9.0288	P	1798 Mar 3	18.2890
A	1959 May 29	10.0664	A	1841 Mar 16	20.0976
P	1974 Jan 8	9.0153	P	1882 Mar 23	18.2807
A	1988 Sep 11	10.0444	A	1925 Apr 1	20.0973
P	2003 Jul 26	9.0309	P	1966 May 21	18.2848
A	2018 Apr 17	10.0656	A	2009 Feb 27	20.0989
P	2032 Nov 28	9.0149	P	2050 Aug 17	18.2830
A	2047 Jul 15	10.0462	A	2092 Nov 23	20.0994

The case of Neptune is peculiar. This planet has a slow motion and a small orbital eccentricity. On the other hand, the Sun is oscillating around the barycenter of the solar system, mainly due to the actions of Jupiter and Saturn. Consequently, the distance of Neptune to the Sun (not to the barycenter of the solar system) can reach a *double* maximum or minimum.

For example, we had the following extreme values for Neptune's radius vector :

minimum	1876 Aug. 28	r = 29.8148 AU
maximum	1881 Dec. 12	29.8213
minimum	1886 July 11	29.8174

Half a revolution later, near the aphelion part of the orbit, we had the following extrema :

maximum	1959 July 13	r = 30.3317 AU
minimum	1965 Oct. 6	30.3227
maximum	1968 Nov. 21	30.3241

The maximum of 1881 was *not* an aphelion, because Neptune was, at that time, near the perihelion of its orbit. Similarly, the minimum of 1965 did not correspond to a perihelion. The author has coined the new terms *apheloid* (= 'which resembles an aphelion') and *periheloid* for these odd maximum and minimum, respectively [1].

Figure 1 shows the variation of Neptune's distance to the Sun from 1954 to 1972. Note the principal aphelion (1), the periheloid

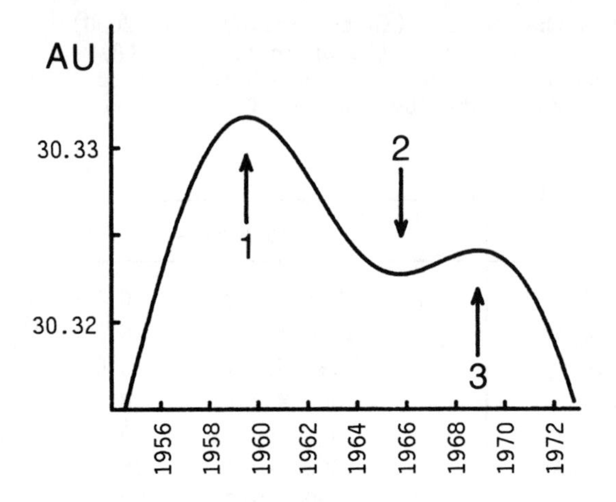

Figure 1

The variation of the distance of Neptune to the Sun, 1954 to 1972.

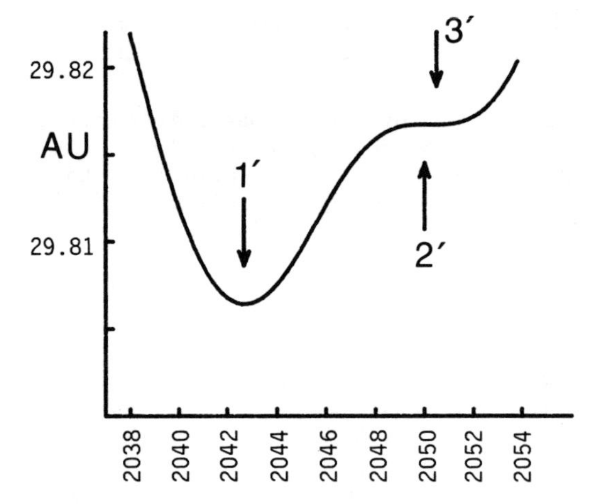

Figure 2

The variation of the distance of Neptune to the Sun, 2038 to 2054.

(2), and the secondary aphelion (3). Half a revolution later, we have the situation pictured in Figure 2; this will be almost a 'limiting case': the principal perihelion (1′) will occur in 2042, while in 2049-2050 the distance to the Sun will decrease only very slightly from the apheloid (2′) to the secondary perihelion (3′), as follows :

minimum	2042 Sep. 5	r = 29.8064 AU
maximum	2049 Oct. 24	29.816711
minimum	2050 June 25	29.816696

For the Earth, it is important to note that the formula given to calculate JDE is actually valid for the *barycenter* of the Earth-Moon system. Due to the action of the Moon, the time of least or greatest distance between the centers of Sun and Earth may differ from that for the barycenter by more than one day [2]. For instance, $k = -10$ in the formula for the Earth yields JDE = 2447 894.911, which corresponds to 1990 January 3.41, while the correct instant for the Earth is 1990 January 4, at 17^h TD.

The values obtained (for the Earth only) can be corrected as follows. Calculate the angles, in *degrees*,

$$A_1 = 328.41 + 132.788\,585\,k$$
$$A_2 = 316.13 + 584.903\,153\,k$$
$$A_3 = 346.20 + 450.380\,738\,k$$
$$A_4 = 136.95 + 659.306\,737\,k$$
$$A_5 = 249.52 + 329.653\,368\,k$$

Remember that k must be an integer for a perihelion, or an integer increased by 0.5 for an aphelion. Then we have the following correction terms, in days :

perihelion	*aphelion*		
+1.278	−1.352	×	sin A_1
−0.055	+0.061		sin A_2
−0.091	+0.062		sin A_3
−0.056	+0.029		sin A_4
−0.045	+0.031		sin A_5

Calculated in this way, the times for the years 1980-2019 have a mean error of 3 hours. Exceptionally, the error amounts to 6 hours.

For instance, for $k = -10$, we obtain a correction of +1.261 day, so the value JDE = 2447 894.911 mentioned above is corrected to 2447 896.172, which corresponds to 1990 January 4, at 16^h TD, which is much closer to the exact value.

Table 37.A gives the times of the passages of the Earth in perihelion and aphelion for the years 1991 to 2010, to the nearest 0.01 hour, together with the distance in AU between the centers of Sun and Earth. These data have been calculated accurately, using the complete VSOP87 theory, *not* the approximate method given above.

TABLE 37.A

Perihelion and Aphelion of the Earth, 1991 - 2010

Instants in Dynamical Time

Year	*Perihelion*			*Aphelion*		
		h			h	
1991	Jan. 3	3.00	0.983 281	July 6	15.46	1.016 703
1992	3	15.06	324	3	12.14	740
1993	4	3.08	283	4	22.37	666
1994	2	5.92	301	5	19.30	724
1995	4	11.10	302	4	2.29	742
1996	Jan. 4	7.43	0.983 223	July 5	19.02	1.016 717
1997	1	23.29	267	4	19.34	754
1998	4	21.28	300	3	23.86	696
1999	3	13.02	281	6	22.86	718
2000	3	5.31	321	3	23.84	741
2001	Jan. 4	8.89	0.983 286	July 4	13.65	1.016 643
2002	2	14.17	290	6	3.80	688
2003	4	5.04	320	4	5.67	728
2004	4	17.72	265	5	10.90	694
2005	2	0.61	297	5	4.98	742
2006	Jan. 4	15.52	0.983 327	July 3	23.18	1.016 697
2007	3	19.74	260	6	23.89	706
2008	2	23.87	280	4	7.71	754
2009	4	15.51	273	4	1.69	666
2010	3	0.18	290	6	11.52	702

References

1. J. Meeus, 'Le centre de gravité du système solaire et le mouvement de Neptune', *Ciel et Terre* (Belgium), Vol. 68, pages 288 - 292 (November - December 1952).

2. J. Meeus, 'A propos des passages de la Terre au périhélie', *l'Astronomie* (France), Vol. 97, pages 294 - 296 (June 1983).

Chapter 38

Passages through the Nodes

Given the orbital elements of a planet or comet, the times t of passages of that body through the nodes of its orbit can easily be calculated as follows.

We have

at the ascending node : $v = -\omega$ or $360° - \omega$

at the descending node : $v = 180° - \omega$

where, as before, v is the true anomaly, and ω the argument of the perihelion. Then, with these values of v, proceed as follows.

Case of an elliptic orbit

Calculate the eccentric anomaly E by

$$\tan \frac{E}{2} = \sqrt{\frac{1 - e}{1 + e}} \tan \frac{v}{2} \tag{38.1}$$

where e is the orbital eccentricity, and the mean anomaly M by

$$M = E - e \sin E \tag{38.2}$$

In formula (38.2), E should be expressed in radians; the resulting value for M is then in radians too. If, however, E is expressed in degrees and the computer is working in degree mode, then in formula (38.2) one should replace e by its value e_0 converted from radians into degrees, that is $e_0 = e \times 57°.295\,779\,51$.

Express M into degrees. Then, if T is the time of perihelion passage, and n is the mean motion in degrees/day, the required time of passage through the node is given by

$$t = T + \frac{M}{n} \text{ days} \tag{38.3}$$

The corresponding value of the radius vector r can be calculated from

$$r = a\,(1 - e \cos E) \tag{38.4}$$

where a is the semimajor axis of the orbit, expressed in astronomical units.

If a and n are not given, they can be calculated from (32.6).

Case of a parabolic orbit

Calculate

$$s = \tan \frac{v}{2}$$

Then

$$t = T + 27.403\,895\,(s^3 + 3s)\,q\sqrt{q} \text{ days}$$

where the perihelion distance q is expressed in AU. The corresponding value of the radius vector is

$$r = q\,(1 + s^2)$$

Note. — The nodes refer to the ecliptic of the same epoch as that of the equinox used for the orbital elements. For example, if the orbital elements are referred to the standard equinox of 1950.0, the above-mentioned formulae give the times of passage through the nodes on the ecliptic of 1950.0, *not* on the ecliptic of the date. The difference may generally be neglected, except when the inclination is very small or if the motion is very slow.

Example 38.a — For the 1986 return of periodic comet Halley, W. Landgraf [*Minor Planet Circular* No. 10634 (1986 April 24)] provided the following orbital elements :

T = 1986 February 9.45891 TD
ω = 111°.84644
e = 0.967 274 26
n = 0.012 970 82 degrees/day
a = 17.940 0782

the argument of perihelion ω being referred to the standard equinox of 1950.0.

For the passage at the ascending node, we have

$$v = 360° - \omega = 248°.15356$$

$$\tan \frac{E}{2} = -0.190\,6646$$

$E = -21°.589\,4332$

$M = -21°.589\,4332 - (0.967\,274\,26 \times 57°.295\,779\,51) \sin(-21°.589\,4332)$
$\quad = -1°.197\,2043$

$$t = T + \frac{-1.197\,2043}{0.012\,970\,82} = T - 92.2998 \text{ days}$$

Hence, the comet was at its ascending node (on the ecliptic of 1950.0) 92.2998 days before the perihelion passage, that is on 1985 November 9.16 TD.

Formula (38.4) then gives $r = 1.8045$ AU. So, at its ascending node the famous comet was a little outside of the orbit of Mars.

For the descending node, we find similarly :

$v = 180° - \omega = 68°.15356$
$E = +9°.972\,6067$
$M = +0°.374\,9928$
$t = T + 28.9105$ days $=$ 1986 March 10.37 TD
$r = 0.8493$ AU, between the orbits of Venus and Earth.

The fact that the comet's motion ($i = 162°$) is retrograde, is irrelevant here. Anyway, ω is measured from the ascending node in the direction of the motion of the body.

Example 38.b — For comet Helin-Roman (1989*s* = 1989 IX), Marsden (*Minor Planet Circular* No. 16001, of 1990 March 11) has calculated the following elements of a parabolic orbit :

T = 1989 August 20.29104 TD
$q = 1.324\,5017$ AU
$\omega = 154°.90425$ (1950.0)

For the ascending node, we have	For the descending node, we have
$v = -\omega = -154°.90425$	$v = 180° - \omega = +25°.09575$
$s = -4.492\,9389$	$s = +0.222\,5715$
$t = T - 4351.68$ days $=$ 1977 September 20	$t = T + 28.3527$ days $=$ 1989 September 17.644 TD
$r = 28.06$ AU	$r = 1.3901$ AU

Example 38.c — Calculate the time of passage of Venus at the ascending node nearest to the epoch 1979.0.

We use the elements given in Table 30.A. There we find

$$a = 0.723\,329\,820, \quad \text{whence} \quad n = 1.602\,137$$
$$e = 0.006\,771\,88 - 0.000\,047\,766\,T + 0.000\,000\,0975\,T^2$$
$$\omega = \pi - \Omega = 54°.883\,787 + 0°.501\,0998\,T - 0°.001\,4800\,T^2$$

The terms in T^3 can safely be dropped here. The elements e and ω vary (rather slowly) with time. We calculate their values for the epoch 1979.0, that is for $T = -0.21$. We find

$$e = 0.006\,781\,92 \qquad \omega = 54°.778\,491$$

We then find successively

$$v = -\omega = -54°.778\,491$$
$$E = -54°.461\,669$$
$$M = -54°.145\,475$$
$$t = T - 33.7958 \text{ days}$$

In Example 37.a, we have found T = 1978 December 31.204 for the time of passage of Venus in the perihelion. Therefore, we have

t = 1978 November 27.408 or 1978 November 27, at 10^h TD.

It is important to note that the algorithms given in this Chapter assume that the body moves on an unperturbed orbit. To obtain full accuracy, the heliocentric latitude of the body should be calculated for three or five instants near the expected time. At the node, we have, of course, latitude = zero.

Saturn reached the descending node (ecliptic of date) of its orbit on 1990 September 4, and will be at its ascending node on 2005 January 8.

Uranus was at the descending node on 1984 December 21, and will go through the ascending node on 2029 May 19.

For Neptune, we have

1920 June 3	ascending node
2003 Aug. 11	descending node
2084 Dec. 30	ascending node

Chapter 39

Correction for Parallax

Suppose we wish to calculate the topocentric coordinates of a body (Moon, Sun, planet, comet) when its geocentric coordinates are known. *Geocentric* = as seen from the center of the Earth; *topocentric* = as seen from the observer's place (Greek: *topos* = place; compare with the word 'Topology').

In other words, we wish to find the corrections $\Delta\alpha$ and $\Delta\delta$ (the parallaxes in right ascension and declination), in order to obtain the topocentric right ascension $\alpha' = \alpha + \Delta\alpha$ and the topocentric declination $\delta' = \delta + \Delta\delta$, when the geocentric values α and δ are known.

Let ρ be the geocentric radius and ϕ' the geocentric latitude of the observer. The expressions $\rho \sin \phi'$ and $\rho \cos \phi'$ can be calculated by the method described in Chapter 10.

Let π be the equatorial horizontal parallax of the body. For the Sun, planets and comets, it is frequently more convenient to use the distance Δ (in astronomical units) to the Earth instead of the parallax. We then have

$$\sin \pi = \frac{\sin 8''.794}{\Delta}$$

or, with sufficient accuracy,

$$\pi = \frac{8''.794}{\Delta} \tag{39.1}$$

Then, if H is the geocentric hour angle of the body, the rigorous formulae are :

$$\tan \Delta\alpha = \frac{-\rho \cos \phi' \sin \pi \sin H}{\cos \delta - \rho \cos \phi' \sin \pi \cos H} \tag{39.2}$$

In the case of the declination we may, instead of computing $\Delta\delta$, calculate δ' directly from

$$\tan \delta' = \frac{(\sin \delta - \rho \sin \phi' \sin \pi) \cos \Delta\alpha}{\cos \delta - \rho \cos \phi' \sin \pi \cos H} \qquad (39.3)$$

Except for the Moon, the following non-rigorous formulae may often be used instead of (39.2) and (39.3) :

$$\Delta\alpha = \frac{-\pi \rho \cos \phi' \sin H}{\cos \delta} \qquad (39.4)$$

$$\Delta\delta = -\pi(\rho \sin \phi' \cos \delta - \rho \cos \phi' \cos H \sin \delta) \qquad (39.5)$$

If π is expressed in seconds of a degree ("), then $\Delta\alpha$ and $\Delta\delta$ too are expressed in this unit. To express $\Delta\alpha$ in seconds of time, divide the result by 15.

It should be noted that $\Delta\alpha$ is a small angle, always lying between -2° and +2° in the case of the Moon; it is, of course, much less in the case of a planet.

An alternative method is as follows. Calculate

$$\left.\begin{aligned} A &= \cos \delta \sin H \\ B &= \cos \delta \cos H - \rho \cos \phi' \sin \pi \\ C &= \sin \delta - \rho \sin \phi' \sin \pi \end{aligned}\right\} \qquad (39.6)$$

$$q = \sqrt{A^2 + B^2 + C^2} \qquad (39.7)$$

Then the topocentric hour angle H' and declination δ' are given by

$$\tan H' = \frac{A}{B} \qquad \sin \delta' = \frac{C}{q}$$

Example 39.a — Calculate the topocentric right ascension and declination of Mars on 2003 August 28, at $3^h17^m00^s$ Universal Time at Palomar Observatory, for which (Example 10.a)

$$\rho \sin \phi' = +0.546\,861$$
$$\rho \cos \phi' = +0.836\,339$$
$$L = \text{longitude} = +7^h47^m27^s \text{ (West)}$$

Mars' geocentric apparent equatorial coordinates for the given instant, interpolated from an accurate ephemeris, are

$$\alpha = 22^h38^m07^s.25 = 339°.530\,208$$
$$\delta = -15°46'15''.9 = -15°.771\,083$$

The planet's distance at that time is 0.37276 AU. Hence, by formula (39.1), its equatorial horizontal parallax is $\pi = 23''.592$.

We still need the geocentric hour angle, which is equal to $H = \theta_0 - L - \alpha$, where θ_0, the apparent sidereal time at Greenwich, can be found as indicated in Chapter 11. For the given instant, we find $\theta_o = 1^h40^m45^s$. Consequently

$$H = 1^h40^m45^s - 7^h47^m27^s - 22^h38^m07^s$$
$$= -28^h44^m49^s = -431°.2042 = +288°.7958$$

Formula (39.2) then gives

$$\tan \Delta\alpha = \frac{+0.000\,090\,557}{+0.962\,324}$$

whence

$$\Delta\alpha = +0°.005\,3917 = +1^s.29$$
$$\alpha' = \alpha + \Delta\alpha = 22^h38^m08^s.54$$

Formula (39.3) gives

$$\tan \delta' = \frac{-0.271\,857\,13}{+0.962\,324\,47} \qquad \text{whence} \quad \delta' = -15°46'30''.0$$

If, instead of (39.2) and (39.3), we chose the non-rigorous formulae (39.4) and (39.5), we find

$$\Delta\alpha = +19''.409 = +1^s.29, \quad \text{as above;}$$
$$\Delta\delta = -14''.1, \quad \text{whence} \quad \delta' = \delta - 14''.1 = -15°46'30''.0, \text{ as above.}$$

As an exercise, perform the calculation for the Moon, again for Palomar Observatory, using fictive values, for example

$$\alpha = 1^h00^m00^s.00 = 15°.000\,000 \qquad H = 4^h00^m00^s.00 = +60°.000\,000$$
$$\delta = +5°.000\,000 \qquad \pi = 0°59'00''$$

First, use the formulae (39.2) and (39.3). Then do the calculation over again with (39.6) and (39.7). You should obtain the same results exactly. Compare the results with those obtained by means of the non-rigorous expressions (39.4) and (39.5).

We can consider the opposite problem: from the observed topocentric coordinates α' and δ', deduce the geocentric values α and δ. In the case of a planet or comet, the corrections $\Delta\alpha$ and $\Delta\delta$ are so small, that the formulae (39.4) and (39.5) can be used also for the reduction from topocentric to geocentric coordinates.

Parallax in horizontal coordinates

The parallax in azimuth is always very small. (It would be zero if the Earth were exactly a sphere). At the horizon, the parallax in azimuth is always less than $\pi/300$, where π is the equatorial horizontal parallax of the body.

Due to the parallax, the apparent altitude of a celestial body is smaller than its 'geocentric' altitude h. Except when high accuracy is needed, the parallax p in altitude may be calculated from $\sin p = \rho \sin \pi \cos h$.

Except in the case of the Moon, the parallax is so small that we may consider p and π to be proportional to their sines, and then we have $p = \rho\pi \cos h$.

The quantity ρ denotes the observer's distance to the center of the Earth, the equatorial radius being taken as unity — see Chapter 10. In many cases we may simply write $\rho = 1$.

Parallax in ecliptical coordinates

It is possible to calculate the topocentric coordinates of a celestial body (Moon or planet), from its geocentric values, directly in ecliptical coordinates. The following formulae are those given by Joseph Johann von Littrow (*Theoretische und Practische Astronomie*, Vol. I, p. 91; Wien, 1821), but in a slightly modified form. These expressions are rigorous.

Let λ = geocentric ecliptical longitude of the celestial body,
β = its geocentric ecliptical latitude,
s = its geocentric semidiameter,
λ', β', s' = the required topocentric values of the same quantities,
ϕ = the observer's latitude,
ε = the obliquity of the ecliptic,
θ = the local sidereal time,
π = the equatorial horizontal parallax of the body.

For the given place, calculate the quantities $\rho \sin \phi'$ and $\rho \cos \phi'$, as explained on page 78. For short, we shall call these quantities S and C, respectively. Then

$$N = \cos\lambda \cos\beta - C \sin\pi \cos\theta$$

$$\tan\lambda' = \frac{\sin\lambda \cos\beta - \sin\pi\,(S \sin\varepsilon + C \cos\varepsilon \sin\theta)}{N}$$

$$\tan\beta' = \frac{\cos\lambda' \left(\sin\beta - \sin\pi\,(S \cos\varepsilon - C \sin\varepsilon \sin\theta)\right)}{N}$$

$$\sin s' = \frac{\cos\lambda' \cos\beta' \sin s}{N}$$

As an exercise, calculate λ', β', s' from the following data :

$\lambda = 181°46'22''.5$ $\phi = +50°05'07''.8$, at sea level
$\beta = +2°17'26''.2$ $\varepsilon = 23°28'00''.8$
$\pi = 0°59'27''.7$ $\theta = 209°46'07''.9$
$s = 0°16'15''.5$

Answer :

$\lambda' = 181°48'05''.0$
$\beta' = +\ 1°29'07''.1$
$s' = 0°16'25''.5$

Chapter 40

Illuminated Fraction of the Disk and Magnitude of a Planet

The illuminated fraction k of the disk of a planet, as seen from the Earth, can be calculated from

$$k = \frac{1 + \cos i}{2} \tag{40.1}$$

where i is the phase angle, which can be found from

$$\cos i = \frac{r^2 + \Delta^2 - R^2}{2 r \Delta}$$

r being the planet's distance to the Sun, Δ its distance to the Earth, and R the distance Sun-Earth, all in astronomical units. Combining these two formulae, we find

$$k = \frac{(r + \Delta)^2 - R^2}{4 r \Delta} \tag{40.2}$$

If the planet's position has been obtained by the 'first method' of Chapter 32, then we have, using the notations used there,

$$\cos i = \frac{R - R_0 \cos B \cos (L - L_0)}{\Delta} \tag{40.3}$$

or

$$\cos i = \frac{x \cos B \cos L + y \cos B \sin L + z \sin B}{\Delta} \tag{40.4}$$

The position angle of the mid-point of the illuminated limb of a planet can be calculated in the same way as for the Moon — see Chapter 51.

Example 40.a — Find the illuminated fraction of the disk of Venus on 1992 December 20, at 0^h TD.

In Example 32.a we have found, for that instant,

$$r = 0.724\,604 \quad \text{(called } R \text{ there)}$$
$$R = 0.983\,824 \quad \text{(called } R_o \text{ there)}$$
$$\Delta = 0.910\,947$$

whence, by formula (40.2), $k = 0.647$.

Or, using, from the same Example 32.a, the values L_o and R_o from (A), L, B, R from (B), x, y, z from (C), and $\Delta = 0.910\,947$, formulae (40.3) and (40.4) both give $\cos i = 0.29312$, whence $k = 0.647$, as above.

For Mercury and Venus, k can take all values between 0 and 1. For Mars, k can never be less than approximately 0.838. In the case of Jupiter, the phase angle i is always less than 12°, whence k can vary only between 0.989 and 1. For Saturn, i is always less than $6\frac{1}{2}$ degrees, so for this planet k can vary only between 0.997 and 1, as seen from the Earth.

In the case of *Venus*, an *approximate* value of k can be found as follows.

Calculate T by means of formula (21.1). Then,

$$V = 261°.51 + 22\,518°.443\,T$$
$$M = 177°.53 + 35\,999°.050\,T$$
$$M' = 50°.42 + 58\,517°.811\,T$$
$$W = V + 1°.91 \sin M + 0°.78 \sin M'$$
$$\Delta^2 = 1.52321 + 1.44666 \cos W \qquad (\Delta > 0)$$

$$k = \frac{(0.72333 + \Delta)^2 - 1}{2.89332\,\Delta}$$

An *approximate* value of Venus' elongation ψ to the Sun is then given by

$$\cos \psi = \frac{\Delta^2 + 0.4768}{2\,\Delta}$$

Example 40.b — Same as in Example 40.a, but now using the approximate method described above.

We find successively

$$\begin{aligned} JD &= 2448\,976.5 \\ T &= -0.070\,321\,697 \\ V &= -1322°.025 = +117°.975 \\ M &= -2353°.984 = +166°.016 \\ M' &= -4064°.652 = +255°.348 \end{aligned} \qquad \begin{aligned} W &= V + 0°.462 - 0°.755 \\ W &= 117°.682 \\ \Delta^2 &= 0.851\,144 \\ \Delta &= 0.922\,575 \\ k &= 0.640 \end{aligned}$$

The correct value, found in Example 40.a, is 0.647.

Magnitude of the Planets

As seen from the Earth, the apparent (stellar) magnitude of a planet at a given instant depends of the planet's distance to the Earth (Δ), its distance to the Sun (r), and the phase angle (i). For Saturn, the magnitude depends also upon the aspect of the ring.

G. Müller's formulae, based on observations which he made from 1877 to 1891, are used since many years in astronomical almanacs. The numerical expressions for the visual magnitudes are as follows [1] :

Mercury : $+1.16 + 5 \log r\Delta + 0.02838\,(i - 50) + 0.000\,1023\,(i - 50)^2$

Venus : $-4.00 + 5 \log r\Delta + 0.01322\,i + 0.000\,000\,4247\,i^3$

Mars : $-1.30 + 5 \log r\Delta + 0.01486\,i$

Jupiter : $-8.93 + 5 \log r\Delta$

Saturn : $-8.68 + 5 \log r\Delta + 0.044\,|\Delta U| - 2.60 \sin |B| + 1.25 \sin^2 B$

Uranus : $-6.85 + 5 \log r\Delta$

Neptune : $-7.05 + 5 \log r\Delta$

in which i is expressed in degrees; r and Δ are in astronomical units, and the logarithms are to the base 10. For Saturn, the quantities ΔU and B, pertaining to the ring, are defined in Chapter 44; care must be taken to have ΔU and B positive, and to express ΔU in degrees. (As an approximation, the phase angle i might be used instead of ΔU).

Of course, Müller's expressions are not perfect. For instance, the effect of the phase is not taken into account in the case of Jupiter. In the formula for Saturn, the Sun's altitude B' above the plane of the ring is not considered; and when B and B' have opposite signs, the *dark* side of the ring is turned towards the Earth, but this case is not considered by Müller.

In any case, the calculated magnitudes should be rounded to the nearest tenth of a magnitude. Giving them to the nearest hundredth makes no sense. Mars, for instance, can differ by as much as 0.3 magnitude from the brightness it 'ought' to have. Some regions of Mars have more dark markings than others, so the planet's brightness depends on which face is turned towards us; and the varying polar caps and a major dust storm can add to its magnitude. In the case of Jupiter and Saturn, there are varying atmospheric phenomena, etc.

Example 40.c — Magnitude of Venus on 1992 December 20.0 TD.

From Example 40.a, we have

$$r = 0.724\,604, \qquad \Delta = 0.910\,947, \qquad \cos i = 0.29312,$$

whence $i = 72.96$ degrees.

Müller's formula for Venus then gives : magnitude = −3.8.

Example 40.d — Magnitude of Saturn on 1992 December 16.0 TD.

From Example 44.a, we have

$r = 9.867\,882$ $\quad B = 16°.442$

$\Delta = 10.464\,606$ $\quad \Delta U = 4°.198$

Müller's formula for Saturn then gives : magnitude = +0.9.

Since 1984, the American *Astronomical Almanac* uses other formulae for the calculation of the visual magnitudes of the planets. It has been stated [2] that these new expressions "are due to D.L. Harris". In fact, in his article [3], Harris does not provide new expressions at all. No expression is 'due' to Harris.

For Mercury and Venus, Harris (pages 277 and 278 of his article) just mentions expressions due to the French astronomer *A. Danjon*. For the outer planets, Harris discusses values of the absolute magnitude and of the phase coefficient made by others, but he himself does not propose or give new expressions.

If r and Δ (in astronomical units) and i (in degrees) have the same meanings as above, the new expressions used in the *Astronomical Almanac* since 1984 are :

Mercury : $-0.42 + 5 \log r\Delta + 0.0380\, i - 0.000\,273\, i^2 + 0.000\,002\, i^3$

Venus : $-4.40 + 5 \log r\Delta + 0.0009\, i + 0.000\,239\, i^2 - 0.000\,000\,65\, i^3$

Mars : $-1.52 + 5 \log r\Delta + 0.016\, i$

Jupiter : $-9.40 + 5 \log r\Delta + 0.005\, i$

Saturn : same as Müller's formula, except that for the absolute magnitude the value −8.88 is used instead of −8.68;

Uranus : $-7.19 + 5 \log r\Delta$

Neptune : $-6.87 + 5 \log r\Delta$

Pluto : $-1.00 + 5 \log r\Delta$

For the magnitudes of the minor planets, see Chapter 32.

References

1. *Explanatory Supplement to the Astronomical Ephemeris* (London, 1961), page 314.
2. *Astronomical Almanac* for 1984 (Washington, D.C.), page L8; and later volumes.
3. Daniel L. Harris, 'Photometry and Colorimetry of Planets and Satellites', Chapter 8 (pages 272*ff*) in *Planets and Satellites*, ed. G.P. Kuiper and B.L. Middlehurst (1961).

Chapter 41

Ephemeris for Physical Observations of Mars

In this Chapter, the following symbols will be used :

D_E = the planetocentric declination of the Earth. When it is positive, Mars' northern pole is tilted towards the Earth;

D_S = the planetocentric declination of the Sun. When it is positive, Mars' northern pole is illuminated;

P = the geocentric position angle of Mars' northern rotation pole, also called "position angle of axis". It is the angle that the Martian meridian from the center of the disk to the northern rotation pole form (on the geocentric celestial sphere) with the declination circle through the center, measured eastwards from the North Point of the disk. (By definition, position angle 0° means northwards on the sky, 90° east, 180° south, and 270° west);

q = the angular amount of the greatest defect of illumination; it is expressed in arcseconds;

Q = the position angle of this greatest defect of illumination;

ω = the (areographic) longitude of the central meridian.

The drawing on the next page shows the appearance of Mars on 1992 November 9. As seen from the Earth, the illuminated fraction of the planet's disk is 90% (k = 0.90). UV is the greatest defect of illumination. S is Mars' South Pole (just behind the limb, hence not visible), A is the northern extremity of the axis of rotation. AS is the central meridian. The arrow shows the direction of the northern celestial pole (on the celestial sphere of the *Earth*). N is the North Point of Mars' disk (not the planet's north pole!). The position angles are measured from N, towards the East. So we have

$$Q = \text{arc } NESV, \qquad P = \text{arc } NESVA.$$

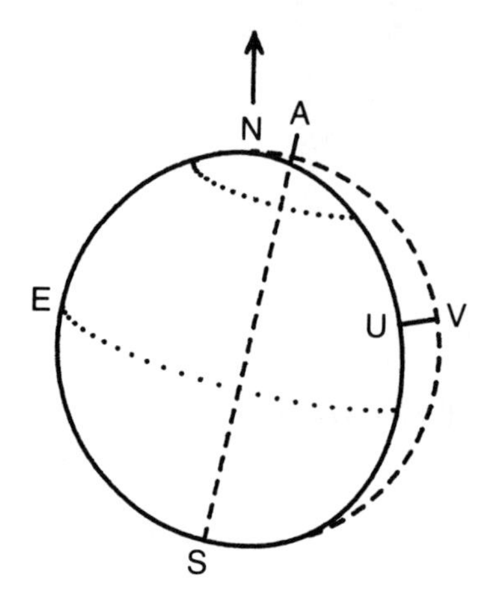

In the calculation of these quantities, the effect of light-time should be taken into account. Moreover, to obtain full accuracy, the aberration of the Sun as seen from *Mars* must be taken into account in the calculation of D_S; and in the calculation of P one should take into account the effects of nutation and aberration on Mars' position.

During the years, several positions for the north pole of Mars (that is, the coordinates of the point on the celestial sphere towards which the axis is directed) have been used in the astronomical almanacs.

According to Lowell and Crommelin [1], the right ascension α_0 and declination δ_0 of the north pole of Mars at the beginning of the year t, referred to the mean equinox of the date, are given by

$$\alpha_0 = 21^h10^m + 1^s.565\,(t - 1905.0)$$
$$\delta_0 = +54°30' + 12''.60\,(t - 1905.0)$$

This position of the north pole was adopted in 1909. But from 1968 to 1980, the *Astronomical Ephemeris* used the position obtained by G. de Vaucouleurs [3]: at the beginning of the year t

$$\alpha_0 = 316°.55 + 0°.006\,750\,(t - 1905.0)$$
$$\delta_0 = +52°.85 + 0°.003\,479\,(t - 1905.0)$$

Note the difference of 1°39' between the two values of δ_0, for the same epoch 1905.0.

Recently adopted values [4] are

$$\left.\begin{array}{l}\alpha_0 = 317°.342\\ \delta_0 = +52°.711\end{array}\right\}\ \text{equinox 1950.0 and epoch J1950.0}$$

$$\left.\begin{array}{l}\alpha_0 = 317°.681\\ \delta_0 = +52°.886\end{array}\right\}\ \text{equinox 2000.0 and epoch J2000.0}$$

From these values, we deduce the following expressions for the longitude and latitude of Mars' north pole, referred to the ecliptic and mean equinox of the date :

$$\begin{aligned}\lambda_0 &= 352°.9065 + 1°.17330\,T\\ \beta_0 &= +63°.2818 - 0°.00394\,T\end{aligned} \qquad (41.1)$$

where T is the time in Julian centuries from the epoch J2000.0; see formula (21.1). Formulae (41.1) take into account the precession of the rotational axis of both Earth and Mars.

For a given instant t, the values of D_E, D_S, etc., can be calculated as follows.

1. Calculate λ_0 and β_0 by means of (41.1).
2. Calculate the heliocentric longitude l_0, latitude b_0 and radius vector R of the Earth, referred to the ecliptic and mean equinox of the date, for instance by using the relevant data from Appendix II and the precepts given in Chapter 31.
3. Calculate the corresponding heliocentric coordinates l, b, r of Mars, but for the instant $t - \tau$, where τ is the light-time from Mars to the Earth, as given by (32.3). Because Mars' distance Δ is not known in advance, it should be found by iteration — see Step 4. One may use $\Delta = 0$ as a starting value.
4. Calculate

$$\begin{aligned} x &= r \cos b \cos l - R \cos l_0 \\ y &= r \cos b \sin l - R \sin l_0 \\ z &= r \sin b - R \sin b_0 \end{aligned} \qquad (41.2)$$

Then Mars' distance Δ to the Earth is

$$\Delta = \sqrt{x^2 + y^2 + z^2} > 0 \qquad (41.3)$$

5. Calculate Mars' geocentric longitude λ and latitude β from

$$\tan \lambda = \frac{y}{x} \qquad \tan \beta = \frac{z}{\sqrt{x^2 + y^2}}$$

6. $\sin D_E = -\sin \beta_0 \sin \beta - \cos \beta_0 \cos \beta \cos(\lambda_0 - \lambda)$
7. Calculate the longitude N of the ascending node of Mars' orbit from

$$N = 49^\circ.5581 + 0^\circ.7721\,T$$

Then correct l and b for the Sun's aberration as seen from Mars :

$$l' = l - 0^\circ.00697 / r$$

$$b' = b - 0^\circ.000\,225 \frac{\cos(l - N)}{r}$$

8. $\sin D_S = -\sin \beta_0 \sin b' - \cos \beta_0 \cos b' \cos(\lambda_0 - l')$
9. If JDE is the Julian Ephemeris Day corresponding to the given time, calculate the angle W, in degrees, from

$$W = 11.504 + 350.892\,000\,25\,(\text{JDE} - \tau - 2433\,282.5)$$

where τ is the light-time (in days) found in steps 3 and 4.

10. Calculate the mean obliquity of the ecliptic ε_0 by means of formula (21.2). Then use expressions (12.3) and (12.4) to find the pole's equatorial coordinates α_0, δ_0 from the ecliptical coordinates λ_0 and β_0.

11. Calculate

$$u = y \cos \varepsilon_0 - z \sin \varepsilon_0$$
$$v = y \sin \varepsilon_0 + z \cos \varepsilon_0$$

and the angles α, δ, ζ from

$$\tan \alpha = \frac{u}{x}$$

$$\tan \delta = \frac{v}{\sqrt{x^2 + u^2}}$$

$$\tan \zeta = \frac{\sin \delta_0 \cos \delta \cos (\alpha_0 - \alpha) - \sin \delta \cos \delta_0}{\cos \delta \sin (\alpha_0 - \alpha)}$$

Note that δ is between -90° and +90°. But α and ζ can take all values from 0° to 360°, and hence they should be taken in the proper quadrant!

12. Find $\omega = W - \zeta$, where ζ is expressed in degrees.

13. Calculate the nutations in longitude ($\Delta\psi$) and in obliquity ($\Delta\varepsilon$) as explained in Chapter 21. Only the most important terms may be used here; an accuracy of, say, 0".01 is not necessary.

14. Correct λ and β for the aberration of Mars :

$$\text{correction to } \lambda : \quad +0°.005\,693 \frac{\cos (l_0 - \lambda)}{\cos \beta}$$

$$\text{correction to } \beta : \quad +0°.005\,693 \sin (l_0 - \lambda) \sin \beta$$

15. Add $\Delta\psi$ to λ_0 and to λ. Add $\Delta\varepsilon$ to ε_0 to obtain the true obliquity of the ecliptic ε.

16. Transform (λ_0, β_0) and (λ, β) to the equatorial coordinates (α_0', δ_0') and (α', δ') by means of expressions (12.3) and (12.4), using for ε the true obliquity obtained above.

17. The position angle P is given by

$$\tan P = \frac{\cos \delta_0' \sin (\alpha_0' - \alpha')}{\sin \delta_0' \cos \delta' - \cos \delta_0' \sin \delta' \cos (\alpha_0' - \alpha')} \tag{41.4}$$

18. The position angle χ of the mid-point of the illuminated limb can be obtained as for the Moon — see Chapter 46. Then the position angle Q of the greatest defect of illumination is $\chi \pm 180°$.

19. Mars' apparent diameter d is given by

$$d = \frac{9''.36}{\Delta}$$

If k is the illuminated fraction of the planet (Chapter 40), then the greatest defect of illumination is $q = (1 - k) d$.

Example 41.a — Calculate the quantities concerning the appearance of Mars on 1992 November 9, at 0^h UT.

The instant corresponds to JD 2448 935.5. For the difference between Dynamical Time and Universal Time, we use the value $\Delta T = +59^s$, or +0.000 683 day, so that the given instant corresponds to

1992 November 9.000 683 TD = JDE 2448 935.500 683.

Step 1. $T = -0.071\,444\,1976$
$\lambda_0 = 352°.82267$
$\beta_0 = +63°.28208$

Step 2. From an accurate ephemeris, calculated by using the complete VSOP87 theory, we deduce

l_0 = 46°50'37''.90 = 46°.843 861
b_0 = −0''.60 = −0°.000 167
R = 0.990 413 01

Step 3. The following geometric heliocentric coordinates of Mars, referred to the ecliptic and mean equinox of the date, are taken from an accurate ephemeris :

TD	*l*	*b*	*r*
1992 Nov. 8.0	77°57'48''.45	+0°52'54''.74	1.540 3797
9.0	78 28 24.28	+0 53 46.72	1.541 6585
10.0	78 58 57.09	+0 54 38.36	1.542 9347

We use $\Delta = 0$ (hence $\tau = 0$) as a starting value. For 1992 November 9.000 683 TD we find, by interpolation,

$l = 78°.473\,759$, $b = +0°.896\,321$, $r = 1.541\,6594$ AU

Step 4. $x = -0.369\,4199$
$y = +0.787\,8856$ $\Delta = 0.870\,5266$
$z = +0.024\,1192$

Step 3. With this value of Δ, we obtain for the light-time the value $\tau = 0.005\,028$ day. Hence, $t - \tau$ is
1992 November 9.000 683 − 0.005 028 = November 8.995 655 TD.

For this instant we find, by interpolation of the tabulated values,

$l = 78°.471\,197$, $b = +0°.896\,249$, $r = 1.541\,6529$.

Step 4. $x = -0.369\,3536$
$y = +0.787\,8654$ $\Delta = 0.870\,4801$
$z = +0.024\,1172$

This new value of Δ yields for the light-time a value which differs by only 0.02 second from the preceding value, so a new iteration is not needed.

Step 5. $\lambda = 115°.117\,321$, $\beta = +1°.587\,619$

Step 6. $D_E = +12°.44$

Step 7. $N = 49^\circ.5029$ $l' = 78^\circ.466\,676$
$b' = +0^\circ.896\,121$

Step 8. $D_S = -2^\circ.76$

Step 9. $W = 5492522^\circ.4593 = 2^\circ.4593$

Step 10. $\varepsilon_0 = 23^\circ 26' 24''.793 = 23^\circ.440\,220$
$\alpha_0 = 317^\circ.632\,606$
$\delta_0 = +52^\circ.860\,916$

Step 11. $u = +0.713\,2537$ $\alpha = 117^\circ.377\,075$
$v = +0.335\,5335$ $\delta = +22^\circ.672\,176$
$\zeta = 250^\circ.9052$

Step 12. $\omega = -248^\circ.45 = 111^\circ.55$

Step 13. $\Delta\psi = +15''.42$ $\Delta\varepsilon = -1''.00$

Step 14. corrected $\lambda = 115^\circ.119\,429$
corrected $\beta = +1^\circ.587\,472$

Step 15. corrected $\lambda_0 = 352^\circ.826\,95$
corrected $\lambda = 115^\circ.123\,712$ $\varepsilon = 23^\circ.439\,942$

Step 16. $\alpha_0' = 317^\circ.63529$ $\alpha' = 117^\circ.38380$
$\delta_0' = +52^\circ.86236$ $\delta' = +22^\circ.67062$

Step 17. $P = 347^\circ.64$

Step 18. The right ascension and declination of the Sun can be obtained with sufficient accuracy from (24.6) and (24.7), with $\odot = l_0 + 180^\circ$. We find $224^\circ.378$ and $-16^\circ.869$.
The equatorial coordinates of Mars being α and δ, we find by means of formula (46.5) $\chi = 99^\circ.91$, whence $Q = 279^\circ.91$.

Step 19. Using the values of R, r and Δ found in Steps 2 to 4, formula (40.2) yields $k = 0.9012$. The greatest defect of illumination is $q = (1 - k) \times 9''.36 / \Delta = 1''.06$.
Mars' apparent diameter is $9''.36 / \Delta = 10''.75$.

References

1. *Monthly Notices of the Royal Astron. Soc.*, Vol. 66, page 56 (1905). Cited in [2].

2. *Explanatory Supplement to the Astronomical Ephemeris* (London, 1961), page 334.

3. *Icarus*, Vol. 3, page 243 (1964).

4. M.E. Davies e.a., 'Report of the IAU Working Group on Cartographic Coordinates and Rotational Elements of the Planets and Satellites: 1982', *Celestial Mechanics*, Vol. 29, pages 309-321 (1983).

Chapter 42

Ephemeris for Physical Observations of Jupiter

For Jupiter three rotational systems have been adopted. System I applies to features within about 10° of the planet's equator; it has an adopted sidereal rotation rate of exactly 877.90 degrees in 24 hours of mean solar time. System II, for use in higher latitudes, where the cloud features take about five minutes longer to circle the planet than those at the equator, rotates exactly 870.27 degrees per day. It follows that the planet's sidereal rotation period is $9^h50^m30^s.003$ in System I, and $9^h55^m40^s.632$ in System II.

System III, rooted deep in Jupiter's interior, applies to radio emissions of the planet. But in this Chapter we will consider only Systems I and II, which are of interest to the visual observer.

As for Mars (see Chapter 41), D_E and D_S will denote the planetocentric declinations of the Earth and the Sun, respectively, and P the position angle of Jupiter's northern rotation pole. The longitude of the central meridian will be denoted ω_1 for System I, and ω_2 for System II.

Because Jupiter's rotation axis is almost exactly perpendicular to the planet's orbital plane around the Sun, it is not needed to correct l and b for the Sun's aberration in the calculation of D_S. The error in D_S made by neglecting this aberration will never exceed 0".5.

For a given instant t, the values of D_E, D_S, ω_1, ω_2 and P can be obtained as follows.

1. Calculate

$$d = \text{JDE} - 2433\,282.5$$

$$T_1 = \frac{d}{36525}$$

and then the right ascension α_0 and declination δ_0 of the north pole of Jupiter, referred to the mean equinox of the date, by the following expressions given on page 725 of the Soviet almanac *Astronomicheskii Ezhegodnik* for 1985 :

$$\alpha_0 = 268^\circ.00 + 0^\circ.1061\ T_1$$
$$\delta_0 = +64^\circ.50 - 0^\circ.0164\ T_1$$

2. Calculate the angles W_1 and W_2 from

$$W_1 = 17^\circ.710 + 877^\circ.900\,035\,39\ d$$
$$W_2 = 16^\circ.838 + 870^\circ.270\,035\,39\ d$$

These can be large (positive or negative) angles; they should be reduced to less than 360 degrees. The angles W_1 and W_2 are related to the longitude Systems I and II, respectively. The constant terms 17°.710 and 16°.838 have been chosen in order to maintain consistency with the Jovian longitude systems established at the end of the 19th century. The other two constants are equal to the values 877°.90 and 870°.27 mentioned at the beginning of this Chapter, increased by 0°.000 035 39, the daily variation of the arc of the Jovian equator from its ascending node on the celestial equator to its ascending node on the orbit.

3. Calculate the heliocentric longitude l_0, latitude b_0 and radius vector R of the Earth, referred to the ecliptic and mean equinox of the date, for instance by using the relevant data of Appendix II and the precepts given in Chapter 31.

4. For the same instant, calculate the corresponding heliocentric coordinates l, b, r of Jupiter. Do *not* take the light-time into account here.

5. Calculate x, y, z by means of formulae (41.2), and then Jupiter's distance Δ by (41.3).

6. Correct Jupiter's heliocentric longitude l (in *degrees*) for the light-time :

$$\text{correction to } l = -0^\circ.012\,990\ \Delta / r^2$$

(The correction to the heliocentric latitude can be neglected here.)

7. Using the corrected value of l, calculate x, y, z, Δ again, as in Step 5.

8. Calculate the mean obliquity of the ecliptic ε_0 by means of formula (21.2).

9. Calculate α_s and δ_s from

$$\tan \alpha_s = \frac{\cos \varepsilon_0 \sin l - \sin \varepsilon_0 \tan b}{\cos l}$$

$$\sin \delta_s = \cos \varepsilon_0 \sin b + \sin \varepsilon_0 \cos b \sin l$$

The angle α_s should be taken in the proper quadrant.

10. $\sin D_S = -\sin\delta_\circ \sin\delta_s - \cos\delta_\circ \cos\delta_s \cos(\alpha_\circ - \alpha_s)$

    The extreme values of D_S are $+3^\circ.12$ and $-3^\circ.12$.

11. Calculate u, v, α, δ, and ζ as for Mars (see Step 11 of Chapter 41).

12. $\sin D_E = -\sin\delta_\circ \sin\delta - \cos\delta_\circ \cos\delta \cos(\alpha_\circ - \alpha)$

    The extreme values of D_E are $+3^\circ.4$ and $-3^\circ.4$.

13. If ζ is expressed in degrees, and Δ in astronomical units, then

$$\omega_1 = W_1 - \zeta - 5^\circ.07033\,\Delta$$
$$\omega_2 = W_2 - \zeta - 5^\circ.02626\,\Delta$$

    The last term in each formula is the amount of rotation during the light-time.

14. The values obtained for ω_1 and ω_2 should be reduced to the interval $0^\circ - 360^\circ$, by adding or subtracting a convenient multiple of 360 degrees. Moreover, it should be noted that the results refer to the geometric (the 'true') disk of Jupiter. The planet actually has a very small phase, and the longitudes of the 'central meridian' of the illuminated disk can be obtained by adding to ω_1 and ω_2 the *correction for phase* C which is equal to

$$C = \pm\, 57^\circ.2958 \times \frac{2r\Delta + R^2 - r^2 - \Delta^2}{4r\Delta}$$

    and has the same sign as $\sin(l - l_\circ)$. The angle C is always small, never exceeding $0^\circ.61$.

15. If an accuracy of 0.1 degree is sufficient for the position angle P, go to step 18.

    Otherwise, calculate the nutations in longitude ($\Delta\psi$) and in obliquity ($\Delta\varepsilon$), as explained in Chapter 21. Only the most important terms may be used; an accuracy of $0''.01$ is not needed. Add $\Delta\varepsilon$ to $\varepsilon_\circ$ to obtain ε.

16. Correct α and δ for Jupiter's aberration :

    correction to α :

$$+0^\circ.005693\;\frac{\cos\alpha\,\cos l_\circ\,\cos\varepsilon + \sin\alpha\,\sin l_\circ}{\cos\delta}$$

    correction to δ :

$$+0^\circ.005693\left[\cos l_\circ\,\cos\varepsilon\,(\tan\varepsilon\,\cos\delta - \sin\alpha\,\sin\delta) + \cos\alpha\,\sin\delta\,\sin l_\circ\right]$$

17. Correct α, δ, $\alpha_\circ$ and $\delta_\circ$ for the nutation, by means of expressions (22.1), giving α', δ', $\alpha_\circ'$ and $\delta_\circ'$.

18. Obtain P by means of formula (41.4).

Example 42.a — Calculate the quantities concerning the appearance of Jupiter on 1992 December 16, at 0^h UT.

This instant corresponds to JD 2448 972.5. For the difference between Dynamical Time and Universal Time, we shall use the value ΔT = +59 seconds = +0.00068 day, so that the given instant corresponds to 1992 December 16.00068 TD = JDE 2448 972.50068.

Step 1. d = 15690.00068 $\qquad$ α_0 = 268°.04558
T_1 = +0.429 569 $\qquad$ δ_0 = +64°.49296
(keeping extra decimals to minimize rounding errors)

Step 2. W_1 = 13 774 269°.8622 = 309°.8622
W_2 = 13 654 554°.2851 = 114°.2851

Steps 3-4. From accurate ephemerides, calculated by using the complete VSOP87 theory, we deduce :

l_0 = 84°.285 703 $\qquad$ l = 181°.882 168
b_0 = +0°.000 197 $\qquad$ b = +1°.290 464
R = 0.984 123 16 $\qquad$ r = 5.446 423 20

Step 5. x = −5.540 0914
y = −1.158 0704 $\qquad$ Δ = 5.661 1645
z = +0.122 6552

Step 6. l = 181°.882 168 − 0°.002 479 = 181°.879 689

Step 7. x = −5.540 0991
y = −1.157 8350 $\qquad$ Δ = 5.661 1239
z = +0.122 6552

Step 8. ε_0 = 23°26'24".745 = 23°.440 2069

Step 9. α_s = 182°.237 749
δ_s = +0°.436 472

Step 10. D_S = −2°.20

Step 11. u = −1.111 0767 $\qquad$ α = 191°.340 327
v = −0.348 0441 $\qquad$ δ = −3°.524 749
ζ = 13°.5238

Step 12. D_E = −2°.48

Step 13. ω_1 = 267°.63 $\qquad$ ω_2 = 72°.31

These are the longitudes of the Central Meridian of the *geometric* disk in Systems I and II, respectively.

Step 14. C = +0°.43. Since $\sin(l - l_0)$ is positive, so is C.

The longitudes of the Central Meridian of the *illuminated* disk are :

System I : ω_1 = 267°.63 + 0°.43 = 268°.06
System II : ω_2 = 72°.31 + 0°.43 = 72°.74

Step 15. $\Delta\psi$ = +16".86 $\qquad$ $\Delta\varepsilon$ = −1".79 $\qquad$ ε = 23°.439 710

Step 16. correction to α : $-0°.001\,627$ $\alpha = 191°.338\,700$
correction to δ : $+0°.000\,560$ $\delta = -3°.524\,189$

Step 17. $\alpha' = 191°.34305$ $\alpha_o' = 268°.04594$
$\delta' = -3°.52592$ $\delta_o' = +64°.49339$

Step 18. $P = 24°.80$

Lower accuracy

The following, shorter method may be used when high accuracy is not needed.

For the given instant (TD !), calculate the JDE (see Chapter 7), and then proceed as follows.

Number of days (and decimals of a day) since 2000 January 1, at 12^h TD :

$$d = \text{JDE} - 2451\,545.0$$

Argument for the long-period term in the motion of Jupiter :

$$V = 172°.74 + 0°.001\,115\,88\,d$$

Mean anomalies of Earth and Jupiter :

$$M = 357°.529 + 0°.985\,6003\,d$$
$$N = 20°.020 + 0°.083\,0853\,d + 0°.329 \sin V$$

Difference between the mean heliocentric longitudes of Earth and Jupiter :

$$J = 66°.115 + 0°.902\,5179\,d - 0°.329 \sin V$$

The angles V, M, N and J are expressed in degrees and decimals. If necessary, they should be reduced to the interval 0-360 degrees ; this depends on your computer language.

Equations of the center of Earth and Jupiter, in degrees :

$$A = 1.915 \sin M + 0.020 \sin 2M$$
$$B = 5.555 \sin N + 0.168 \sin 2N$$

and then

$$K = J + A - B$$

Radius vector of the Earth :

$$R = 1.00014 - 0.01671 \cos M - 0.00014 \cos 2M$$

Radius vector of Jupiter :

$$r = 5.20872 - 0.25208 \cos N - 0.00611 \cos 2N$$

Distance Earth-Jupiter :

$$\Delta = \sqrt{r^2 + R^2 - 2rR \cos K}$$

The distances R, r and Δ are expressed in astronomical units, and Δ should of course be taken positive. The phase angle of Jupiter (that is, the angle Earth-Jupiter-Sun) is then given by

$$\sin \psi = \frac{R}{\Delta} \sin K$$

The angle ψ always lies between $-12°$ and $+12°$. Because R and Δ are always positive, the angle ψ has the same sign as $\sin K$.

The longitudes of the central meridian in Systems I and II are then, respectively,

$$\omega_1 = 210°.98 + 877°.816\,9088\,(d - \frac{\Delta}{173}) + \psi - B$$

$$\omega_2 = 187°.23 + 870°.186\,9088\,(d - \frac{\Delta}{173}) + \psi - B$$

where $-\Delta/173$ is the correction for the light-time, expressed in days. The denominator 173 results from the fact that the light-time for unit distance is 1/173 day.

The values obtained for ω_1 and ω_2 should be reduced to the interval $0° - 360°$, by adding or subtracting a convenient multiple of 360 degrees. The results refer to the geometric disk of Jupiter. The longitudes of the 'central meridian' of the illuminated disk can be obtained by adding to ω_1 and ω_2 the *correction for phase* which is equal to

$$\pm\, 57°.3 \sin^2 \frac{\psi}{2}$$

and the sign is opposite the sign of $\sin K$.

Calculated in this manner, ω_1 and ω_2 can be up to 0.1 or 0.2 degree in error.

Find Jupiter's heliocentric longitude λ referred to the equinox of 2000.0 by the formula

$$\lambda = 34°.35 + 0°.083\,091\, d + 0°.329 \sin V + B$$

Then we obtain, in degrees and decimals,

$$D_S = 3.12 \sin(\lambda + 42°.8)$$

$$D_E = D_S - 2.22 \sin\psi \cos(\lambda + 22°) - 1.30 \frac{r - \Delta}{\Delta} \sin(\lambda - 100°.5)$$

In these expressions, 3°.12 is the inclination of the equator of Jupiter on the orbital plane, 2°.22 its inclination on the ecliptic, and 1°.30 the inclination of the orbital plane on the ecliptic.

Example 42.b — Let us take the same instant as in Example 42.a, namely 1992 December 16, 0^h UT
= JD 2448 972.5
= JDE 2448 972.50068.

We find successively

$$d = -2572.49932$$
$$V = 169^\circ.87$$
$$M = -2177^\circ.927 = +342^\circ.073$$
$$N = -193^\circ.659$$
$$J = -2255^\circ.670 = +264^\circ.330$$
$$A = -0^\circ.601$$
$$B = +1^\circ.235$$
$$K = 262^\circ.494$$
$$R = 0.98413$$
$$r = 5.44824$$
$$\Delta = 5.66151$$
$$\sin\psi = -0.17234$$
$$\psi = -9^\circ.924$$
$$d - \frac{\Delta}{173} = -2572.53205$$

From this we deduce, for the geometric disk of Jupiter :

$$\omega_1 = -2258\,012^\circ.31 = 267^\circ.69$$
$$\omega_2 = -2238\,407^\circ.64 = 72^\circ.36$$

The correct values are 267°.63 and 72°.31 (see Step 13 of Example 42.a).

For the correction for phase, we find +0°.43, exactly as in Example 42.a, Step 14.

$$\lambda = -178^\circ.11$$
$$D_S = -2^\circ.194$$
$$D_E = -2^\circ.194 - 0^\circ.350 + 0^\circ.048 = -2^\circ.50$$

Chapter 43

Positions of the Satellites of Jupiter

This Chapter gives two methods to calculate, for any given instant, the positions of the four great satellites of Jupiter with respect to the planet, as seen from the Earth. These apparent rectangular coordinates X and Y of the satellites will be measured from the center of the disk of Jupiter, in units of the planet's equatorial radius. X is measured positively to the west of Jupiter, negatively to the east, the X-axis coinciding with the equator of the planet. Y is positive to the north, negative to the south, the Y-axis coinciding with the planet's rotation axis (see the drawing).

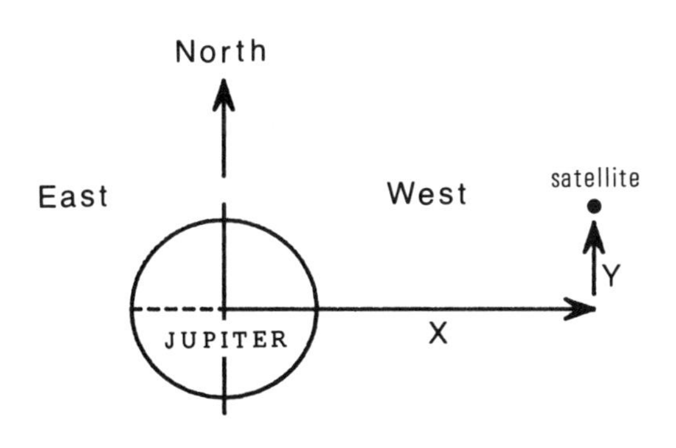

The accuracy of the first method ('low accuracy') is sufficient for identifying the satellites at the telescope, or for drawing a wavy-line diagram showing their positions with respect to Jupiter, as given in several astronomical almanacs and magazines. The high-accuracy method is needed, for instance, to calculate the classical phenomena of the satellites (eclipses, transits, etc.) and their mutual phenomena.

Low accuracy

First, convert the date and the instant (TD) into the Julian Day, using the method described in Chapter 7. Then, obtain the following quantities as explained in Chapter 42 ('lower accuracy'): d, V, M, N, J, A, B, K, R, r, Δ, ψ, and the planetocentric declination D_E of the Earth.

For each of the four satellites, we now calculate an angle u which is measured from the inferior conjunction with Jupiter, so that $u = 0°$ corresponds to the satellite's inferior conjunction, $u = 90°$ to its greatest western elongation, $u = 180°$ to the superior conjunction, and $u = 270°$ to the greatest eastern elongation.

$$u_1 = 163°.8067 + 203°.405\,8643 \left(d - \frac{\Delta}{173}\right) + \psi - B$$

$$u_2 = 358°.4108 + 101°.291\,6334 \left(d - \frac{\Delta}{173}\right) + \psi - B$$

$$u_3 = 5°.7129 + 50°.234\,5179 \left(d - \frac{\Delta}{173}\right) + \psi - B$$

$$u_4 = 224°.8151 + 21°.487\,9801 \left(d - \frac{\Delta}{173}\right) + \psi - B$$

If necessary, these angles u should be reduced to the interval 0° - 360°. In order to obtain more accurate values, the results can be corrected as follows. Calculate the angles G and H by means of the formulae

$$G = 331°.18 + 50°.310\,482 \left(d - \frac{\Delta}{173}\right)$$

$$H = 87°.40 + 21°.569\,231 \left(d - \frac{\Delta}{173}\right)$$

Then we have the following corrections, in degrees :

correction to u_1 : $+0.473 \sin 2(u_1 - u_2)$

correction to u_2 : $+1.065 \sin 2(u_2 - u_3)$

correction to u_3 : $+0.165 \sin G$

correction to u_4 : $+0.841 \sin H$

The first correction is due to a periodic perturbation of satellite I by satellite II. The second correction is a perturbation of II by III. The two last corrections are due to the eccentricities of the orbits of satellites III and IV. (The orbits of I and II are almost exactly circular.)

It should be noted that we take into account only the largest periodic terms in the motions of the satellites. There are many other (but smaller) periodic terms. For instance, satellite I is perturbed by satellite III too, satellite III by II and by IV, etc. — see further the 'high accuracy' method in this Chapter.

The distances of the satellites to the center of Jupiter, in units of Jupiter's equatorial radius, are given by

$$r_1 = 5.9073 - 0.0244 \cos 2(u_1 - u_2)$$
$$r_2 = 9.3991 - 0.0882 \cos 2(u_2 - u_3)$$
$$r_3 = 14.9924 - 0.0216 \cos G$$
$$r_4 = 26.3699 - 0.1935 \cos H$$

where the *un*corrected values of u_1, etc., should be used. In these expressions, the periodic terms are again due to mutual perturbations of the satellites or to their orbital eccentricities.

The apparent rectangular coordinates X and Y of the satellites are then given by

$$X_1 = r_1 \sin u_1 \qquad \text{and} \qquad Y_1 = -r_1 \cos u_1 \sin D_E$$

with similar expressions for the other three satellites.

Example 43.a — Calculate the configuration of the satellites of Jupiter on 1992 December 16, at 0^h UT = JD 2448 972.5 = JDE 2448 972.50068. (The value ΔT = +59 seconds is used).

For this instant we have found, in Example 42.b,

$$d = -2572.49932 \qquad d - \frac{\Delta}{173} = -2572.53205$$
$$B = +1°.235$$
$$\psi = -9°.924 \qquad D_E = -2°.50$$

By means of the formulae given in the present Chapter, we then find successively :

u_1 = −523 115°.457 = 324°.543 $\qquad$ $2(u_1 - u_2)$ = 546°.53 = 186°.53
u_2 = −260 228°.722 = 51°.278 $\qquad$ $2(u_2 - u_3)$ = 93°.26
u_3 = −129 235°.353 = 4°.647 $\qquad$ G = −129 094°.15 = 145°.85
u_4 = − 55 064°.861 = 15°.139 $\qquad$ H = − 55 400°.14 = 39°.86

correction to u_1 : −0°.054 $\qquad$ corrected u_1 = 324°.489
correction to u_2 : +1°.063 $\qquad$ corrected u_2 = 52°.341
correction to u_3 : +0°.093 $\qquad$ corrected u_3 = 4°.740
correction to u_4 : +0°.539 $\qquad$ corrected u_4 = 15°.678

r_1 = 5.9073 + 0.0242 = 5.9315 $\qquad$ X_1 = −3.45 $\qquad$ Y_1 = +0.21
r_2 = 9.3991 + 0.0050 = 9.4041 $\qquad$ X_2 = +7.44 $\qquad$ Y_2 = +0.25
r_3 = 14.9924 + 0.0179 = 15.0103 $\qquad$ X_3 = +1.24 $\qquad$ Y_3 = +0.65
r_4 = 26.3699 − 0.1485 = 26.2214 $\qquad$ X_4 = +7.09 $\qquad$ Y_4 = +1.10

(It is just a coincidence that all four Y-values are positive !)

With these values of X and Y we can draw the following figure which shows the configuration of the satellites at the given time. In this figure the South is up, and the West to the left, as in the field of an inverting telescope in the northern hemisphere.

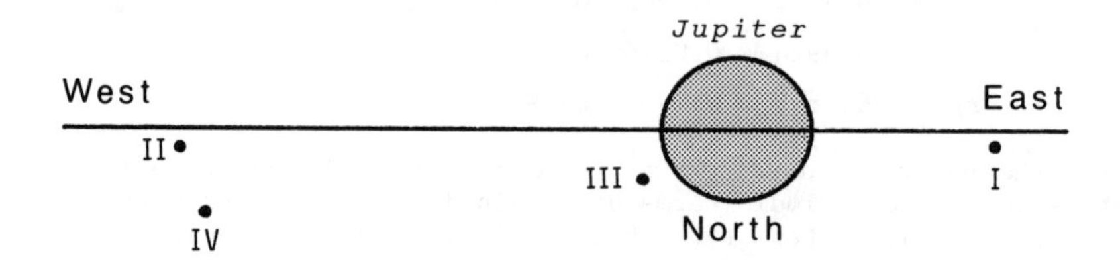

The X- and Y-values, resulting from an accurate calculation, are mentioned in Example 43.b. The discrepancies between the Y-values are mainly due to the fact that, in this simplified method, the inclinations of the orbits of the satellites on the equatorial plane of Jupiter have been neglected. Actually, the four satellites can reach extreme latitudes of 0°03', 0°31', 0°20', and 0°44', respectively, with respect to the equatorial plane of the planet. As a consequence, mutual occultations cannot be calculated with certainty by means of the simplified method described above. In the case of a very close conjunction, it is even not possible to deduce which of the two satellites passes to the north of the other.

High accuracy

The following method is based on the theory E2 of the satellites due to Lieske [1], with improvements known as E2x3 [2].

For the given instant, calculate the following quantities (see Chapter 24) :

$\odot$ = geocentric geometric longitude of the Sun,
β = geocentric geometric latitude of the Sun,
R = radius vector of the Sun in astronomical units.

Let τ be the light-time from Jupiter to the Earth. Because the distance of Jupiter to the Earth is not known in advance, so is τ not known. The distance Δ should be found by iteration. A good starting value is $\Delta = 5$, since the extreme values of Jupiter's distance to the Earth are 3.95 and 6.5 astronomical units. The light-time is given by (32.3); a better value for Δ will be provided by formula (43.2).

Calculate the following values for the given time decreased by the light-time τ (see Chapter 31) :

l = heliocentric longitude of Jupiter,
b = heliocentric latitude of Jupiter,
r = radius vector of Jupiter, in AU.

In the above, the longitudes and latitudes are referred to the ecliptic and mean equinox of the date.

Calculate the rectangular geocentric ecliptical coordinates of Jupiter

$$\begin{aligned} x &= r \cos b \cos l + R \cos \odot \\ y &= r \cos b \sin l + R \sin \odot \\ z &= r \sin b + R \sin \beta \end{aligned} \qquad (43.1)$$

and its distance to the Earth

$$\Delta = \sqrt{x^2 + y^2 + z^2} \qquad (43.2)$$

Calculate

$$\Lambda = \text{ATN2}\,(y, x) \qquad \text{and} \qquad \alpha = \text{ATN}\left(\frac{z}{\sqrt{x^2 + y^2}}\right)$$

where, as mentioned earlier in this book, ATN2 is the 'second' arc-tangent function. In other words, Λ is equal to $\text{ATN}(y/x)$ taken in the proper quadrant.

Let t be the time measured in ephemeris days from 1976 August 10 at 0^h TD = JDE 2443 000.5, decreased by the light-time τ. In other words, if JDE is the Julian Ephemeris Day corresponding to the given instant,

$$t = \text{JDE} - 2443\,000.5 - \tau$$

In the following expressions, all numerical values are expressed in degrees and decimals. The longitudes are referred to the standard equinox of 1950.0.

Mean longitudes of the satellites :

$$\begin{aligned} \ell_1 &= 106.07947 + 203.488\,955\,432\,t \\ \ell_2 &= 175.72938 + 101.374\,724\,550\,t \\ \ell_3 &= 120.55434 + 50.317\,609\,110\,t \\ \ell_4 &= 84.44868 + 21.571\,071\,314\,t \end{aligned}$$

Longitudes of the perijoves :

$$\begin{aligned} \pi_1 &= 58.3329 + 0.161\,039\,36\,t \\ \pi_2 &= 132.8959 + 0.046\,479\,85\,t \\ \pi_3 &= 187.2887 + 0.007\,127\,40\,t \\ \pi_4 &= 335.3418 + 0.001\,839\,98\,t \end{aligned}$$

Longitudes of the nodes on the equatorial plane of Jupiter :

$$\omega_1 = 311.0793 - 0.132\,794\,30\,t$$
$$\omega_2 = 100.5099 - 0.032\,630\,47\,t$$
$$\omega_3 = 119.1688 - 0.007\,177\,04\,t$$
$$\omega_4 = 322.5729 - 0.001\,759\,34\,t$$

Principal inequality in the longitude of Jupiter :

$$\Gamma = 0.33033 \sin(163°.679 + 0°.001\,0512\,t) + 0.03439 \sin(34°.486 - 0°.016\,1731\,t)$$

There is a small libration, with a period of 2070 days, in the longitudes of the three inner satellites: when satellite II decelerates, I and III accelerate. To take this into account, we need the phase of free libration :

$$\Phi_\lambda = 191.8132 + 0.173\,900\,23\,t$$

Longitude of the node of the equator of Jupiter on the ecliptic :

$$\psi = 316.5182 - 0.000\,002\,08\,t$$

Mean anomalies of Jupiter and Saturn :

$$G = 30.23756 + 0.083\,092\,5701\,t + \Gamma$$
$$G' = 31.97853 + 0.033\,459\,7339\,t$$

Longitude of the perihelion of Jupiter :

$\Pi = 13.469\,942$ (considered as a constant in the E2 theory)

PERIODIC TERMS IN THE LONGITUDES OF THE SATELLITES

Satellite I

$+0°.47\,259 \sin 2(\ell_1 - \ell_2)$
$-0.03\,480 \sin(\pi_3 - \pi_4)$
$-0.01\,756 \sin(\pi_1 + \pi_3 - 2\Pi - 2G)$
$+0.01\,080 \sin(\ell_2 - 2\ell_3 + \pi_3)$
$+0.00\,757 \sin \Phi_\lambda$
$+0.00\,663 \sin(\ell_2 - 2\ell_3 + \pi_4)$
$+0.00\,453 \sin(\ell_1 - \pi_3)$
$+0.00\,453 \sin(\ell_2 - 2\ell_3 + \pi_2)$
$-0.00\,354 \sin(\ell_1 - \ell_2)$
$-0.00\,317 \sin(2\psi - 2\Pi)$
$-0.00\,269 \sin(\ell_2 - 2\ell_3 + \pi_1)$
$+0.00\,263 \sin(\ell_1 - \pi_4)$
$+0°.00\,186 \sin(\ell_1 - \pi_1)$
$-0.00\,186 \sin G$
$+0.00\,167 \sin(\pi_2 - \pi_3)$
$+0.00\,158 \sin 4(\ell_1 - \ell_2)$
$-0.00\,155 \sin(\ell_1 - \ell_3)$
$-0.00\,142 \sin(\psi + \omega_3 - 2\Pi - 2G)$
$-0.00\,115 \sin 2(\ell_1 - 2\ell_2 + \omega_2)$
$+0.00\,089 \sin(\pi_2 - \pi_4)$
$+0.00\,084 \sin(\omega_2 - \omega_3)$
$+0.00\,084 \sin(\ell_1 + \pi_3 - 2\Pi - 2G)$
$+0.00\,053 \sin(\psi - \omega_2)$

Call $\Sigma 1$ the sum of these terms.

Satellite II

$+1^{\circ}.06\,476 \sin 2(\ell_2 - \ell_3)$
$+0.04\,253 \sin (\ell_1 - 2\ell_2 + \pi_3)$
$+0.03\,579 \sin (\ell_2 - \pi_3)$
$+0.02\,383 \sin (\ell_1 - 2\ell_2 + \pi_4)$
$+0.01\,977 \sin (\ell_2 - \pi_4)$
$-0.01\,843 \sin \Phi_\lambda$
$+0.01\,299 \sin (\pi_3 - \pi_4)$
$-0.01\,142 \sin (\ell_2 - \ell_3)$
$+0.01\,078 \sin (\ell_2 - \pi_2)$
$-0.01\,058 \sin G$
$+0.00\,870 \sin (\ell_2 - 2\ell_3 + \pi_2)$
$-0.00\,775 \sin 2(\psi - \Pi)$
$+0.00\,524 \sin 2(\ell_1 - \ell_2)$
$-0.00\,460 \sin (\ell_1 - \ell_3)$
$+0.00\,450 \sin (\ell_2 - 2\ell_3 + \pi_1)$
$+0.00\,327 \sin (\psi - 2G + \omega_3 - 2\Pi)$
$-0.00\,296 \sin (\pi_1 + \pi_3 - 2\Pi - 2G)$
$-0.00\,151 \sin 2G$
$+0.00\,146 \sin (\psi - \omega_3)$
$+0.00\,125 \sin (\psi - \omega_4)$
$-0.00\,117 \sin (\ell_1 - 2\ell_3 + \pi_3)$
$-0^{\circ}.00\,095 \sin 2(\ell_2 - \omega_2)$
$+0.00\,086 \sin 2(\ell_1 - 2\ell_2 + \omega_2)$
$-0.00\,086 \sin (5G' - 2G + 52^{\circ}.225)$
$-0.00\,078 \sin (\ell_2 - \ell_4)$
$-0.00\,064 \sin (\ell_1 - 2\ell_3 + \pi_4)$
$-0.00\,063 \sin (3\ell_3 - 7\ell_4 + 4\pi_4)$
$+0.00\,061 \sin (\pi_1 - \pi_4)$
$+0.00\,058 \sin 2(\psi - \Pi - G)$
$+0.00\,058 \sin (\omega_3 - \omega_4)$
$+0.00\,056 \sin 2(\ell_2 - \ell_4)$
$+0.00\,055 \sin 2(\ell_1 - \ell_3)$
$+0.00\,052 \sin (3\ell_3 - 7\ell_4 + \pi_3 + 3\pi_4)$
$-0.00\,043 \sin (\ell_1 - \pi_3)$
$+0.00\,042 \sin (\pi_3 - \pi_2)$
$+0.00\,041 \sin 5(\ell_2 - \ell_3)$
$+0.00\,041 \sin (\pi_4 - \Pi)$
$+0.00\,038 \sin (\ell_2 - \pi_1)$
$+0.00\,032 \sin (\omega_2 - \omega_3)$
$+0.00\,032 \sin 2(\ell_3 - G - \Pi)$
$+0.00\,029 \sin (\pi_1 - \pi_3)$

Call $\Sigma 2$ the sum of these terms.

Satellite III

$+0^{\circ}.16\,477 \sin (\ell_3 - \pi_3)$
$+0.09\,062 \sin (\ell_3 - \pi_4)$
$-0.06\,907 \sin (\ell_2 - \ell_3)$
$+0.03\,786 \sin (\pi_3 - \pi_4)$
$+0.01\,844 \sin 2(\ell_3 - \ell_4)$
$-0.01\,340 \sin G$
$+0.00\,703 \sin (\ell_2 - 2\ell_3 + \pi_3)$
$-0.00\,670 \sin 2(\psi - \Pi)$
$-0.00\,540 \sin (\ell_3 - \ell_4)$
$+0.00\,481 \sin (\pi_1 + \pi_3 - 2\Pi - 2G)$
$-0.00\,409 \sin (\ell_2 - 2\ell_3 + \pi_2)$
$+0.00\,379 \sin (\ell_2 - 2\ell_3 + \pi_4)$
$+0.00\,235 \sin (\psi - \omega_3)$
$+0.00\,198 \sin (\psi - \omega_4)$
$+0.00\,180 \sin \Phi_\lambda$
$+0.00\,129 \sin 3(\ell_3 - \ell_4)$
$+0^{\circ}.00\,124 \sin (\ell_1 - \ell_3)$
$-0.00\,119 \sin (5G' - 2G + 52^{\circ}.225)$
$+0.00\,109 \sin (\ell_1 - \ell_2)$
$-0.00\,099 \sin (3\ell_3 - 7\ell_4 + 4\pi_4)$
$+0.00\,091 \sin (\omega_3 - \omega_4)$
$+0.00\,081 \sin (3\ell_3 - 7\ell_4 + \pi_3 + 3\pi_4)$
$-0.00\,076 \sin (2\ell_2 - 3\ell_3 + \pi_3)$
$+0.00\,069 \sin (\pi_4 - \Pi)$
$-0.00\,058 \sin (2\ell_3 - 3\ell_4 + \pi_4)$
$+0.00\,057 \sin (\ell_3 + \pi_3 - 2\Pi - 2G)$
$-0.00\,057 \sin (\ell_3 - 2\ell_4 + \pi_4)$
$-0.00\,052 \sin (\pi_2 - \pi_3)$
$-0.00\,052 \sin (\ell_2 - 2\ell_3 + \pi_1)$

$+0^\circ.00048 \sin(\ell_3 - 2\ell_4 + \pi_3)$
$-0.00045 \sin(2\ell_2 - 3\ell_3 + \pi_4)$
$-0.00041 \sin(\pi_2 - \pi_4)$
$-0.00038 \sin 2G$
$-0.00033 \sin(\pi_3 - \pi_4 + \omega_3 - \omega_4)$
$-0.00032 \sin(3\ell_3 - 7\ell_4 + 2\pi_3 + 2\pi_4)$
$+0.00030 \sin 4(\ell_3 - \ell_4)$
$-0^\circ.00029 \sin(\omega_3 + \psi - 2\Pi - 2G)$
$+0.00029 \sin(\ell_3 + \pi_4 - 2\Pi - 2G)$
$+0.00026 \sin(\ell_3 - \Pi - G)$
$+0.00024 \sin(\ell_2 - 3\ell_3 + 2\ell_4)$
$+0.00021 \sin 2(\ell_3 - \Pi - G)$
$-0.00021 \sin(\ell_3 - \pi_2)$
$+0.00017 \sin 2(\ell_3 - \pi_3)$

Call $\Sigma 3$ the sum of these terms.

Satellite IV

$+0^\circ.84109 \sin(\ell_4 - \pi_4)$
$+0.03429 \sin(\pi_4 - \pi_3)$
$-0.03305 \sin 2(\psi - \Pi)$
$-0.03211 \sin G$
$-0.01860 \sin(\ell_4 - \pi_3)$
$+0.01182 \sin(\psi - \omega_4)$
$+0.00622 \sin(\ell_4 + \pi_4 - 2G - 2\Pi)$
$+0.00385 \sin 2(\ell_4 - \pi_4)$
$-0.00284 \sin(5G' - 2G + 52^\circ.225)$
$-0.00233 \sin 2(\psi - \pi_4)$
$-0.00223 \sin(\ell_3 - \ell_4)$
$-0.00208 \sin(\ell_4 - \Pi)$
$+0.00177 \sin(\psi + \omega_4 - 2\pi_4)$
$+0.00134 \sin(\pi_4 - \Pi)$
$+0.00125 \sin 2(\ell_4 - G - \Pi)$
$-0.00117 \sin 2G$
$-0.00112 \sin 2(\ell_3 - \ell_4)$
$+0.00106 \sin(3\ell_3 - 7\ell_4 + 4\pi_4)$
$+0.00102 \sin(\ell_4 - G - \Pi)$
$+0.00096 \sin(2\ell_4 - \psi - \omega_4)$
$+0.00087 \sin 2(\psi - \omega_4)$
$-0.00087 \sin(3\ell_3 - 7\ell_4 + \pi_3 + 3\pi_4)$
$+0.00085 \sin(\ell_3 - 2\ell_4 + \pi_4)$
$-0.00081 \sin 2(\ell_4 - \psi)$
$+0.00071 \sin(\ell_4 + \pi_4 - 2\Pi - 3G)$
$+0.00060 \sin(\ell_1 - \ell_4)$
$-0.00056 \sin(\psi - \omega_3)$
$-0.00055 \sin(\ell_3 - 2\ell_4 + \pi_3)$
$+0^\circ.00051 \sin(\ell_2 - \ell_4)$
$+0.00042 \sin 2(\psi - G - \Pi)$
$+0.00039 \sin 2(\pi_4 - \omega_4)$
$+0.00036 \sin(\psi + \Pi - \pi_4 - \omega_4)$
$+0.00035 \sin(2G' - G + 188^\circ.37)$
$-0.00035 \sin(\ell_4 - \pi_4 + 2\Pi - 2\psi)$
$-0.00032 \sin(\ell_4 + \pi_4 - 2\Pi - G)$
$+0.00030 \sin(3\ell_3 - 7\ell_4 + 2\pi_3 + 2\pi_4)$
$+0.00030 \sin(2G' - 2G + 149^\circ.15)$
$+0.00028 \sin(\ell_4 - \pi_4 + 2\psi - 2\Pi)$
$-0.00028 \sin 2(\ell_4 - \omega_4)$
$-0.00027 \sin(\pi_3 - \pi_4 + \omega_3 - \omega_4)$
$-0.00026 \sin(5G' - 3G + 188^\circ.37)$
$+0.00025 \sin(\omega_4 - \omega_3)$
$-0.00025 \sin(\ell_2 - 3\ell_3 + 2\ell_4)$
$-0.00023 \sin 3(\ell_3 - \ell_4)$
$+0.00021 \sin(2\ell_4 - 2\Pi - 3G)$
$-0.00021 \sin(2\ell_3 - 3\ell_4 + \pi_4)$
$+0.00019 \sin(\ell_4 - \pi_4 - G)$
$-0.00019 \sin(2\ell_4 - \pi_3 - \pi_4)$
$-0.00018 \sin(\ell_4 - \pi_4 + G)$
$-0.00016 \sin(\ell_4 + \pi_3 - 2\Pi - 2G)$

Call $\Sigma 4$ the sum of these terms.

The true longitudes of the satellites are

$$L_1 = \ell_1 + \Sigma 1, \qquad L_2 = \ell_2 + \Sigma 2, \qquad L_3 = \ell_3 + \Sigma 3, \qquad L_4 = \ell_4 + \Sigma 4$$

PERIODIC TERMS FOR THE LATITUDES OF THE SATELLITES

The sum of the following terms gives the *tangent* of the satellite's latitude B_i with respect to Jupiter's equatorial plane.

Satellite I

$$+0.000\,6502 \sin (L_1 - \omega_1)$$
$$+0.000\,1835 \sin (L_1 - \omega_2)$$
$$+0.000\,0329 \sin (L_1 - \omega_3)$$
$$-0.000\,0311 \sin (L_1 - \psi)$$
$$+0.000\,0093 \sin (L_1 - \omega_4)$$
$$+0.000\,0075 \sin (3L_1 - 4\ell_2 - 1.9927\,\Sigma 1 + \omega_2)$$
$$+0.000\,0046 \sin (L_1 + \psi - 2\,\Pi - 2G)$$

Satellite II

$$+0.008\,1275 \sin (L_2 - \omega_2)$$
$$+0.000\,4512 \sin (L_2 - \omega_3)$$
$$-0.000\,3286 \sin (L_2 - \psi)$$
$$+0.000\,1164 \sin (L_2 - \omega_4)$$
$$+0.000\,0273 \sin (\ell_1 - 2\ell_3 + 1.0146\,\Sigma 2 + \omega_2)$$
$$+0.000\,0143 \sin (L_2 + \psi - 2\,\Pi - 2G)$$
$$-0.000\,0143 \sin (L_2 - \omega_1)$$
$$+0.000\,0035 \sin (L_2 - \psi + G)$$
$$-0.000\,0028 \sin (\ell_1 - 2\ell_3 + 1.0146\,\Sigma 2 + \omega_3)$$

Satellite III

$$+0.003\,2364 \sin (L_3 - \omega_3)$$
$$-0.001\,6911 \sin (L_3 - \psi)$$
$$+0.000\,6849 \sin (L_3 - \omega_4)$$
$$-0.000\,2806 \sin (L_3 - \omega_2)$$
$$+0.000\,0321 \sin (L_3 + \psi - 2\Pi - 2G)$$
$$+0.000\,0051 \sin (L_3 - \psi + G)$$
$$-0.000\,0045 \sin (L_3 - \psi - G)$$
$$-0.000\,0045 \sin (L_3 + \psi - 2\Pi)$$
$$+0.000\,0037 \sin (L_3 + \psi - 2\Pi - 3G)$$
$$+0.000\,0030 \sin (2\ell_2 - 3L_3 + 4.03\,\Sigma 3 + \omega_2)$$
$$-0.000\,0021 \sin (2\ell_2 - 3L_3 + 4.03\,\Sigma 3 + \omega_3)$$

Satellite IV

$$-0.007\,6579 \sin (L_4 - \psi)$$
$$+0.004\,4148 \sin (L_4 - \omega_4)$$
$$-0.000\,5106 \sin (L_4 - \omega_3)$$
$$+0.000\,0773 \sin (L_4 + \psi - 2\,\Pi - 2G)$$
$$+0.000\,0104 \sin (L_4 - \psi + G)$$
$$-0.000\,0102 \sin (L_4 - \psi - G)$$
$$+0.000\,0088 \sin (L_4 + \psi - 2\,\Pi - 3G)$$
$$-0.000\,0038 \sin (L_4 + \psi - 2\,\Pi - G)$$

PERIODIC TERMS FOR THE RADIUS VECTOR

Satellite I	−0.004 1339	$\cos 2(\ell_1 - \ell_2)$
	−0.000 0395	$\cos(\ell_1 - \pi_3)$
	−0.000 0214	$\cos(\ell_1 - \pi_4)$
	+0.000 0170	$\cos(\ell_1 - \ell_2)$
	−0.000 0162	$\cos(\ell_1 - \pi_1)$
	−0.000 0130	$\cos 4(\ell_1 - \ell_2)$
	+0.000 0106	$\cos(\ell_1 - \ell_3)$
	−0.000 0063	$\cos(\ell_1 + \pi_3 - 2\Pi - 2G)$
Satellite II	+0.009 3847	$\cos(\ell_1 - \ell_2)$
	−0.000 3114	$\cos(\ell_2 - \pi_3)$
	−0.000 1738	$\cos(\ell_2 - \pi_4)$
	−0.000 0941	$\cos(\ell_2 - \pi_2)$
	+0.000 0553	$\cos(\ell_2 - \ell_3)$
	+0.000 0523	$\cos(\ell_1 - \ell_3)$
	−0.000 0290	$\cos 2(\ell_1 - \ell_2)$
	+0.000 0166	$\cos 2(\ell_2 - \omega_2)$
	+0.000 0107	$\cos(\ell_1 - 2\ell_3 + \pi_3)$
	−0.000 0102	$\cos(\ell_2 - \pi_1)$
	−0.000 0091	$\cos 2(\ell_1 - \ell_3)$
Satellite III	−0.001 4377	$\cos(\ell_3 - \pi_3)$
	−0.000 7904	$\cos(\ell_3 - \pi_4)$
	+0.000 6342	$\cos(\ell_2 - \ell_3)$
	−0.000 1758	$\cos 2(\ell_3 - \ell_4)$
	+0.000 0294	$\cos(\ell_3 - \ell_4)$
	−0.000 0156	$\cos 3(\ell_3 - \ell_4)$
	+0.000 0155	$\cos(\ell_1 - \ell_3)$
	−0.000 0153	$\cos(\ell_1 - \ell_2)$
	+0.000 0070	$\cos(2\ell_2 - 3\ell_3 + \pi_3)$
	−0.000 0051	$\cos(\ell_3 + \pi_3 - 2\Pi - 2G)$
Satellite IV	−0.007 3391	$\cos(\ell_4 - \pi_4)$
	+0.000 1620	$\cos(\ell_4 - \pi_3)$
	+0.000 0974	$\cos(\ell_3 - \ell_4)$
	−0.000 0541	$\cos(\ell_4 + \pi_4 - 2\Pi - 2G)$
	−0.000 0269	$\cos 2(\ell_4 - \pi_4)$
	+0.000 0182	$\cos(\ell_4 - \Pi)$
	+0.000 0177	$\cos 2(\ell_3 - \ell_4)$
	−0.000 0167	$\cos(2\ell_4 - \psi - \omega_4)$
	+0.000 0167	$\cos(\psi - \omega_4)$
	−0.000 0155	$\cos 2(\ell_4 - \Pi - G)$
	+0.000 0142	$\cos 2(\ell_4 - \psi)$
	+0.000 0104	$\cos(\ell_1 - \ell_4)$

Satellite IV (cont.)

$+0.000\,0092 \cos(\ell_2 - \ell_4)$
$-0.000\,0089 \cos(\ell_4 - \Pi - G)$
$-0.000\,0062 \cos(\ell_4 + \pi_4 - 2\Pi - 3G)$
$+0.000\,0048 \cos 2(\ell_4 - \omega_4)$

The radius vector R_i of satellite No. i, in equatorial radii of Jupiter, is given by

$$R_i = a_i \times (1 + \text{sum of periodic terms})$$

with the following values for the mean distances :

satellite		
I	a_1 =	5.90730
II	a_2 =	9.39912
III	a_3 =	14.99240
IV	a_4 =	26.36990

If JDE is the Julian Ephemeris Day corresponding to the given instant, calculate

$$T_\circ = \frac{\text{JDE} - 2433\,282.423}{36525}$$

Then the precession in longitude from the epoch B1950.0 to the date, in degrees, is given by

$$P = 1.396\,6626\,T_\circ + 0.000\,3088\,T_\circ^2$$

Add P to the four longitudes L_i and to ψ.

Inclination of Jupiter's axis of rotation on the orbital plane :

$$I = 3°.120\,262 + 0°.0006\,T$$

where T is the time in centuries since 1900.0.

For each of the four (i = 1 to 4) satellites, we have found the tropical longitude L_i, the equatorial latitude B_i, and the radius vector R_i (in equatorial Jupiter radii).

For each of them, calculate

$$X_i = R_i \cos(L_i - \psi) \cos B_i$$
$$Y_i = R_i \sin(L_i - \psi) \cos B_i$$
$$Z_i = R_i \sin B_i$$

Now consider a 'fifth, fictitious satellite', situated at unit distance from the center of Jupiter, above the planet's north pole :

$$X_5 = 0, \qquad Y_5 = 0, \qquad Z_5 = 1.$$

This fictitious satellite will be needed later.

To obtain the apparent rectangular coordinates of the satellites as they appear on the celestial sphere, as defined at the beginning of this Chapter, several rotations must be performed. So, calculate the following for all five satellites (the four real ones and the fifth, fictitious satellite):

Rotation towards Jupiter's orbital plane:

$$A_1 = X$$
$$B_1 = Y \cos I - Z \sin I$$
$$C_1 = Y \sin I + Z \cos I$$

Rotation towards the ascending node of the orbit of Jupiter:

$$A_2 = A_1 \cos \Phi - B_1 \sin \Phi$$
$$B_2 = A_1 \sin \Phi + B_1 \cos \Phi$$
$$C_2 = C_1$$

where $\Phi = \psi - \Omega$, Ω being the longitude of the node of Jupiter, referred to the mean equinox of the date. See in Table 30.A, under 'Jupiter', the formula for Ω.

Rotation towards the plane of the ecliptic:

$$A_3 = A_2$$
$$B_3 = B_2 \cos i - C_2 \sin i$$
$$C_3 = B_2 \sin i + C_2 \cos i$$

where i is the inclination of the orbit of Jupiter on the ecliptic. See in Table 30.A the expression for i.

Rotation towards the vernal equinox:

$$A_4 = A_3 \cos \Omega - B_3 \sin \Omega$$
$$B_4 = A_3 \sin \Omega + B_3 \cos \Omega$$
$$C_4 = C_3$$

Then calculate

$$A_5 = A_4 \sin \Lambda - B_4 \cos \Lambda$$
$$B_5 = A_4 \cos \Lambda + B_4 \sin \Lambda$$
$$C_5 = C_4 = C_3$$

$$A_6 = A_5$$
$$B_6 = C_5 \sin \alpha + B_5 \cos \alpha$$
$$C_6 = C_5 \cos \alpha - B_5 \sin \alpha$$

If ξ, η are the values of A_6 and C_6 for the 'fifth satellite', that is, $\xi = A_6(5)$, $\eta = C_6(5)$, then calculate the angle

$$D = \text{ATN2}(\xi, \eta)$$

where, as mentioned earlier in this book, ATN2 is the 'second' arctangent function, giving the angle D in the proper quadrant.

Calculate

$$X = A_6 \cos D - C_6 \sin D$$
$$Y = A_6 \sin D + C_6 \cos D \qquad (43.3)$$
$$Z = B_6$$

X and Y are the required apparent rectangular coordinates of the satellite, as defined at the beginning of this Chapter. The quantity Z is negative if the satellite is closer to the Earth than Jupiter, positive if it is more distant than Jupiter.

However, to obtain full accuracy, the apparent coordinates X and Y just obtained should be corrected for two effects :

1. differential light-time : if a satellite is on the nearer half of its orbit, its light-time is smaller than that of Jupiter; if on the far half, its light-time is larger. The correction to be *added* to X is

$$\frac{|Z|}{K} \sqrt{1 - (X/R)^2}$$

where

K = 17295	for satellite	I
21819	–	II
27558	–	III
36548	–	IV

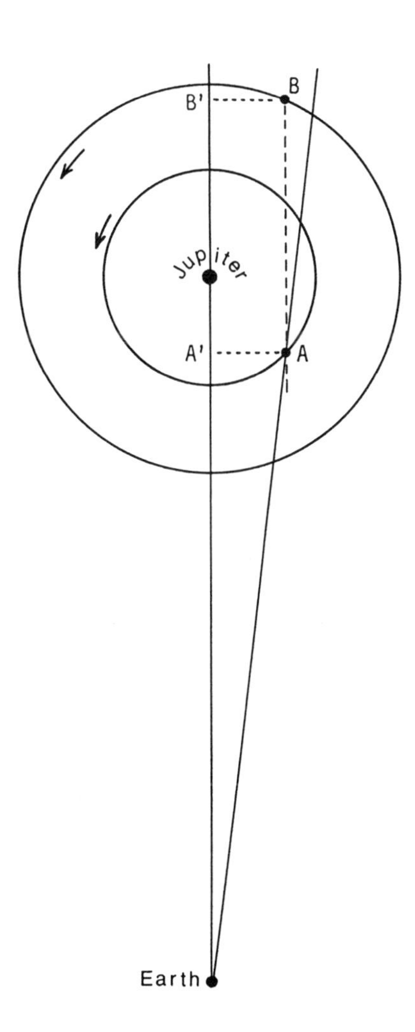

This correction is zero at the greatest elongations, and *positive* in all other cases. It is always very small, being at most 0.0003 for satellite I, or 0.0007 for satellite IV. The correction to Y is negligible. In the formula above, R is the radius vector of the satellite, while X and Z are the values given by (43.3).

2. the perspective effect, which is due to the fact that Jupiter is not situated at an infinite distance from the Earth. This is illustrated by the figure at the right, showing the orbits of two satellites around Jupiter (not to scale!). Although satellites A and B have equal X-coordinates *in space* (distances AA' and BB' are equal), they are not exactly in conjunction as seen from the Earth: their *apparent* X-coordinates are not equal.

To correct for this perspective effect, the X and Y values obtained thus far should be *multiplied* by the factor

$$W = \frac{\Delta}{\Delta + Z/2095}$$

where Δ is Jupiter's distance to Earth in astronomical units as given by (43.2), while Z is in Jupiter radii (43.3). The constant 2095 is the number of equatorial radii of Jupiter in one astronomical unit.

Example 43.b — Same instant as in Example 43.a.

We shall not give the details of the calculation. Let us just mention the values of the sums

$\Sigma 1 = -0°.01171$, $\Sigma 2 = +1°.09596$, $\Sigma 3 = +0°.03879$, $\Sigma 4 = +0°.58932$,

and the final results :

	Satell. I	Satell. II	Satell. III	Satell. IV
X	-3.4515	+7.4435	+1.1996	+7.0754
Y	+0.2138	+0.2756	+0.5903	+1.0294

Mutual conjunctions — Two satellites are in conjunction when their X-coordinates are equal. The difference between the Y-coordinates then corresponds to the separation of the satellites. Of course, if one satellite (or both) is eclipsed or occulted by Jupiter, the conjunction is inobservable.

Conjunctions with Jupiter — A satellite is in inferior conjunction with Jupiter when its X-coordinate is zero and changing from negative to positive; its Z-coordinate is then negative.

Similarly, a satellite is in superior conjunction with Jupiter when its X-coordinate, passing from positive to negative, becomes zero. Its Z-coordinate is then positive.

Exercise. — On 1988 November 23, satellites III and IV were almost simultaneously in conjunction with Jupiter. Confirm this with your program. Take ΔT from Table 9.A.

Answer: Satellite III was in inferior conjunction with Jupiter on 1988 November 23, at 7^h28^m UT; at that instant, its Y-value was -0.8045; the satellite was in transit over the planet's disk.

Satellite IV was in superior conjunction that same day, at 5^h15^m. Its Y-value was then +1.3995. Since this is larger than the polar radius of Jupiter (0.933), the satellite was not occulted, but was visible above the planet's northern polar regions.

Satellite phenomena — The X and Y coordinates are the basic data for the calculation of the satellite phenomena: occultations behind Jupiter, and transits across the planet's disk. If the calculations are made for the center of the satellite, then an occultation or a transit begins or ends when the distance d of the satellite to the center of Jupiter's disk, given by $d^2 = X^2 + Y^2$, is equal to the planet's radius ρ at the point of contact. Due to Jupiter's flattening, ρ varies between 1 (at the equator) and 0.933 (at the poles). One can avoid working with an elliptical disk by 'stretching' the scale vertically: multiply the Y-values by the factor 1.071374, leaving the X-values unchanged :

$$Y_1 = 1.071374\,Y$$

Jupiter's disk then becomes exactly circular, and the condition for the beginning or end of an occultation or of a transit becomes $X^2 + Y_1^2 = 1$.

In the case of an occultation, it remains to be checked whether the satellite is visible at the time of its immersion or emersion, because it could be eclipsed in the shadow of the planet.

Eclipses and shadow transits can be calculated in the same way, except that one should replace X and Y by the apparent coordinates X_0 and Y_0 as seen from the *Sun*. These coordinates are obtained by putting $R = 0$ in expressions (43.1). Moreover, the light-time τ to the *Earth* should be *added* to the true times of the eclipses or to those of the shadow transits, because we on Earth see these events later by the amount τ. Finally, in the case of an eclipse it remains to be checked whether the disappearance or the reappearance is visible from Earth: indeed, the satellite could be occulted by Jupiter at that instant.

References

1. J.H. Lieske, *Astronomy and Astrophysics*, Vol. 82, pages 340 - 348 (1980).

2. J.H. Lieske, *Astronomy and Astrophysics*, Vol. 176, pages 146 - 158 (1987).

radius of Jupiter [illegible], the satellite was [illegible] but was visible above the [illegible] north polar regions.

[illegible] The [illegible] coordinates are the [illegible] of the satellite [illegible] the planet [illegible] of the satellite [illegible] of the satellite [illegible]

[illegible]

[illegible]

[illegible] in the [illegible] of the time.

[illegible] can be calculated [illegible] by the apparent coordinates [illegible] coordinates are obtained [illegible] however [illegible] of the [illegible]

[illegible]

References

1. [illegible] Astronomy and Astrophysics, Vol. 2[illegible], pages [illegible] (19[illegible])

2. [illegible] Astronomy and Astrophysics, Vol. 1[illegible], pages 1[illegible] (19[illegible])

Chapter 44

The Ring of Saturn

In this Chapter, the following symbols will be used with respect to the ring of Saturn. (Of course, we know that Saturn has *many* rings. But they form one single, compact, planar system. We shall use the word *ring*, in singular form, to denote the ring system.)

B = the Saturnicentric latitude of the Earth referred to the plane of the ring, positive towards the north; when B is positive, the visible surface of the ring is the northern one;

B' = the Saturnicentric latitude of the Sun referred to the plane of the ring, positive towards the north; when B' is positive, the illuminated surface of the ring is the northern one;

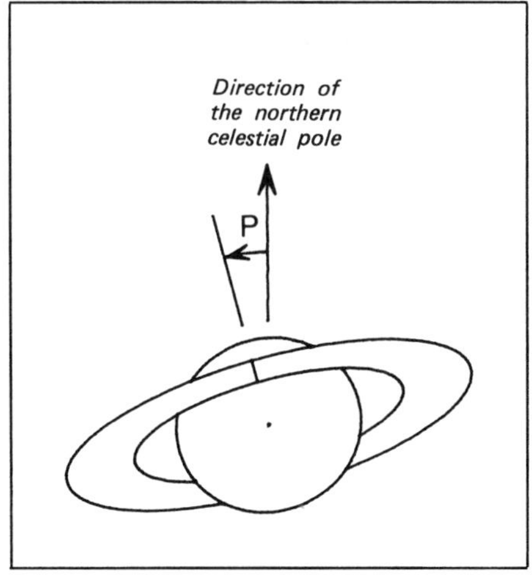

P = the geocentric position angle of the northern semiminor axis of the apparent ellipse of the ring, measured from the North towards the East (see the Figure). Because the ring is situated exactly in Saturn's equator plane, P is also the position angle of the north pole of rotation of the planet;

a, b = the major and the minor axes of the outer edge of the outer ring, in arcseconds.

In the calculation of these quantities, the effect of light-time should be taken into account. Moreover, to obtain full accuracy, the aberration of the Sun as seen from *Saturn* must be taken into account in the calculation of B'; and in the calculation of P one should take into account the effect of the nutation and Saturn's aberration.

G. Dourneau [1] gives the following values for the inclination of the plane of the ring and the longitude of the ascending node referred to the ecliptic and mean equinox of B1950.0 :

$$i = 28°.0817 \pm 0°.0035$$
$$\Omega = 168°.8112 \pm 0°.0089$$

From these values, we deduce the following expressions to calculate i and Ω referred to the ecliptic and mean equinox of the date:

$$\begin{aligned} i &= 28°.075\,216 - 0°.012\,998\,T + 0°.000\,004\,T^2 \\ \Omega &= 169°.508\,470 + 1°.394\,681\,T + 0°.000\,412\,T^2 \end{aligned} \qquad (44.1)$$

where T is the time from J2000.0 in Julian centuries, as given by formula (21.1). In expressions (44.1), we retained extra decimals in order to avoid loss in accuracy.

For a given instant t, the values of B, B', etc., can be calculated as follows.

1. Calculate i and Ω by means of (44.1).
2. Calculate the heliocentric longitude l_0, latitude b_0 and radius vector R of the Earth, referred to the ecliptic and mean equinox of the date, FK5 system, for instance by using the relevant data of Appendix II and the precepts given in Chapter 31.
3. Calculate the corresponding coordinates l, b, r for Saturn, but for the instant $t - \tau$, where τ is the light-time from Saturn to the Earth, as given by (32.3). Because Saturn's distance Δ is not known in advance, it should be found by iteration — see Step 4. One may use $\Delta = 9$ as a starting value, since Saturn's distance to the Earth is always between 8.0 and 11.1 AU.
4. Calculate

$$\begin{aligned} x &= r \cos b \cos l - R \cos l_0 \\ y &= r \cos b \sin l - R \sin l_0 \\ z &= r \sin b - R \sin b_0 \end{aligned}$$

Then Saturn's distance Δ to the Earth is

$$\Delta = \sqrt{x^2 + y^2 + z^2} > 0$$

5. Calculate the geocentric longitude λ and latitude β of Saturn from

$$\tan \lambda = \frac{y}{x} \qquad \tan \beta = \frac{z}{\sqrt{x^2 + y^2}}$$

6. $$\sin B = \sin i \cos \beta \sin(\lambda - \Omega) - \cos i \sin \beta$$

$$a = \frac{375''.35}{\Delta} \qquad b = a \sin |B|$$

Factor by which the axes a and b of the outer edge of the outer ring are to be multiplied to obtain the axes of :

Inner edge of outer ring	0.8801
Outer edge of inner ring	0.8599
Inner edge of inner ring	0.6650
Inner edge of dusky ring	0.5486

7. Calculate the longitude N of the ascending node of Saturn's orbit from

$$N = 113°.6655 + 0°.8771\ T$$

Then correct l and b for the Sun's aberration as seen from Saturn :

$$l' = l - 0°.01759/r$$

$$b' = b - 0°.000\,764\ \frac{\cos(l - N)}{r}$$

8. $$\sin B' = \sin i \cos b' \sin(l' - \Omega) - \cos i \sin b'$$

9. For the calculation of Saturn's magnitude (see Chapter 40), we need the quantity ΔU, the difference between the Saturnicentric longitudes of the Sun and the Earth, measured in the plane of the ring.

$$\tan U_1 = \frac{\sin i \sin b' + \cos i \cos b' \sin(l' - \Omega)}{\cos b' \cos(l' - \Omega)}$$

$$\tan U_2 = \frac{\sin i \sin \beta + \cos i \cos \beta \sin(\lambda - \Omega)}{\cos \beta \cos(\lambda - \Omega)}$$

$$\Delta U = |U_1 - U_2|, \text{ to be expressed in degrees.}$$

ΔU is a small angle, equal to at most 7°.

10. Calculate the nutations in longitude ($\Delta\psi$) and in obliquity ($\Delta\varepsilon$) and then the true obliquity of the ecliptic ε (see Chapter 21). For the nutation, only the most important terms may be used; an accuracy of, say, 0".01, is unnecessary.

11. Find the ecliptical longitude λ_0 and latitude β_0 of the northern pole of the ringplane from

$$\lambda_0 = \Omega - 90°, \qquad \beta_0 = 90° - i$$

12. Correct λ and β for the aberration of Saturn :

correction to λ : $\quad +0^\circ.005\,693 \dfrac{\cos(l_0 - \lambda)}{\cos\beta}$

correction to β : $\quad +0^\circ.005\,693 \sin(l_0 - \lambda) \sin\beta$

13. Add $\Delta\psi$ to λ_0 and to λ.
14. Transform (λ_0, β_0) and (λ, β) to the equatorial coordinates (α_0, δ_0) and (α, δ), by means of formulae (12.3) and (12.4), using for ε the true obliquity obtained in Step 10.
15. The position angle P is given by

$$\tan P = \frac{\cos\delta_0 \sin(\alpha_0 - \alpha)}{\sin\delta_0 \cos\delta - \cos\delta_0 \sin\delta \cos(\alpha_0 - \alpha)}$$

Example 44.a — Calculate the quantities concerning the appearance of Saturn's ring on 1992 December 16, at 0^h UT.

The instant corresponds to JD = 2448 972.5. For the difference between Dynamical Time and Universal Time, we use the value ΔT = +59 seconds = +0.00068 day, so that the instant corresponds to 1992 December 16.00068 TD = JDE 2448 972.50068.

Step 1. T = −0.070 431 193
i = 28°.076 131
Ω = 169°.410 243

Step 2. From an accurate ephemeris, calculated by using the complete VSOP87 theory, we deduce

l_0 = 84°17'08".53 = 84°.285 703
b_0 = +0".71 = +0°.000 197
R = 0.984 123 16

Step 3. The following geometric heliocentric coordinates of Saturn, referred to the ecliptic and mean equinox of the date, are taken from an accurate ephemeris :

TD	l	b	r
1992 Dec. 15.0	319°09'44".23	−1°04'26".52	9.868 0846
16.0	319 11 36.61	−1 04 30.92	9.867 8690
17.0	319 13 28.99	−1 04 35.31	9.867 6534

Using $\Delta = 9$ as a first approximation for Saturn's distance, formula (32.3) yields $\tau = 0.05198$. Hence,

$t - \tau$ = 1992 December 16.00068 − 0.05198
= 1992 December 15.94870 TD.

For this instant we find, by interpolation of the values tabulated above,

l = 319°.191 900, $\quad b$ = −1°.075 192, $\quad r$ = 9.867 8801.

Step 4. $x = +7.369\,7225$ $\Delta = 10.464\,6006$
$y = -7.427\,0295$
$z = -0.185\,1696$

Step 3. With this value for Δ, we obtain the new value $\tau = 0.06044$ day for the light-time; hence,
$t - \tau$ = 1992 December 16.00068 - 0.06044
= 1992 December 15.94024 TD.

For this instant we find, by interpolation of the tabulated values,

$l = 319°.191\,636$, $b = -1°.075\,183$, $r = 9.867\,8819$.

Step 4. $x = +7.369\,6942$ $\Delta = 10.464\,6059$
$y = -7.427\,0651$
$z = -0.185\,1681$

This new value of Δ gives $\tau = 0.06044$ again, so no new iteration is needed.

Step 5. $\lambda = 314°.777\,850$
$\beta = -1°.013\,885$

Step 6. $B = +16°.442$
$a = 35''.87$
$b = 10''.15$

Step 7. $N = 113°.6037$
$l' = 319°.189\,853$
$b' = -1°.075\,113$

Step 8. $B' = +14°.679$

Step 9. $U_1 = 153°.2645$
$U_2 = 149°.0663$
$\Delta U = 4°.198$

Step 10. $\Delta\psi = +16''.86$
$\Delta\varepsilon = -1''.79$
$\varepsilon = 23°26'22''.96 = 23°.43971$

Step 11. $\lambda_0 = 79°.410\,243$
$\beta_0 = 61°.923\,869$

Step 12. corrected $\lambda = 314°.774\,228$
corrected $\beta = -1°.013\,963$

Step 13. corrected $\lambda_0 = 79°.414\,926$
corrected $\lambda = 314°.778\,911$

Step 14. $\alpha_0 = 40°.36365$ $\alpha = 317°.55421$
$\delta_0 = +83°.48486$ $\delta = -17°.37056$

Step 15. $P = +6°.741$

Reference

1. Gérard Dourneau, 'Observations et étude du mouvement des huit premiers satellites de Saturne', Thèse de doctorat d'État, Université de Bordeaux I (1987).

Chapter 45

Position of the Moon

In order to calculate accurately the position of the Moon for a given instant, it is necessary to take into account *hundreds* of periodic terms in the Moon's longitude, latitude and distance. Because this is outside the scope of this book, we shall limit ourselves to the most important periodic terms; the accuracy of the results will be approximately 10" in the longitude of the Moon, and 4" in its latitude.

Using the algorithm described in this Chapter, one obtains the geocentric longitude λ and latitude β of the center of the Moon, referred to the mean equinox of the date, and the distance Δ, in kilometers, between the centers of Earth and Moon. The equatorial horizontal parallax π of the Moon can then be obtained from

$$\sin \pi = \frac{6378.14}{\Delta}$$

The periodic terms given in this Chapter are based on the Chapront ELP-2000/82 lunar theory [1]. However, for the mean arguments L', D, M, M', F the improved expressions given later by Chapront [2] have been used.

For the given instant (Dynamical Time), calculate T by means of formula (21.1). Remember that T is expressed in centuries, and thus should be taken with a sufficient number of decimals (at least nine, since during 0.000 000 001 century the Moon moves over an arc of 1.7 arcseconds).

Then calculate the angles L', D, M, M', and F by means of the following expressions. The angles so calculated will be expressed in *degrees*. In order to avoid working with large angles, reduce them to less than 360°.

Moon's mean longitude, referred to the mean equinox of the date, and including the constant term of the effect of light-time :

$$L' = 218.316\,4591 + 481\,267.881\,342\,36\,T - 0.001\,3268\,T^2 + T^3/538\,841 - T^4/65\,194\,000 \qquad (45.1)$$

Mean elongation of the Moon :

$$D = 297.850\,2042 + 445\,267.111\,5168\,T - 0.001\,6300\,T^2 + T^3/545\,868 - T^4/113\,065\,000 \qquad (45.2)$$

Sun's mean anomaly :

$$M = 357.529\,1092 + 35\,999.050\,2909\,T - 0.000\,1536\,T^2 + T^3/24\,490\,000 \qquad (45.3)$$

Moon's mean anomaly :

$$M' = 134.963\,4114 + 477\,198.867\,6313\,T + 0.008\,9970\,T^2 + T^3/69\,699 - T^4/14\,712\,000 \qquad (45.4)$$

Moon's argument of latitude (mean distance of the Moon from its ascending node) :

$$F = 93.272\,0993 + 483\,202.017\,5273\,T - 0.003\,4029\,T^2 - T^3/3\,526\,000 + T^4/863\,310\,000 \qquad (45.5)$$

Three further arguments (again, in degrees) are needed :

$$A_1 = 119°.75 + 131°.849\,T$$
$$A_2 = 53°.09 + 479\,264°.290\,T$$
$$A_3 = 313°.45 + 481\,266°.484\,T$$

Calculate the sums Σl and Σr of the terms given in Table 45.A, and the sum Σb of the terms given in Table 45.B. The argument of each sine (for Σl and Σb) and cosine (for Σr) is a linear combination of the four fundamental arguments D, M, M' and F. For example, the argument on the eight line of Table 45.A is $2D - M - M'$, and the contributions to Σl and Σr are $+57\,066 \sin(2D - M - M')$ and $-152\,138 \cos(2D - M - M')$, respectively.

However, the terms whose argument contains the angle M depend on the eccentricity of the Earth's orbit around the Sun, which presently is decreasing with time. For this reason, the amplitude of these terms is actually variable. To take this effect into account, multiply the terms whose argument contains M (or $-M$) by E, and those containing $2M$ (or $-2M$) by E^2, where

$$E = 1 - 0.002\,516\,T - 0.000\,0074\,T^2 \qquad (45.6)$$

Moreover, add the following additive terms to Σl and to Σb. The terms involving A_1 are due to the action of Venus, the term involving A_2 is due to Jupiter, while those involving L' are due to the flattening of the Earth.

TABLE 45.A

Periodic terms for the longitude (Σl) and distance (Σr) of the Moon. The unit is 0.000 001 degree for Σl, and 0.001 km for Σr.

Argument				Σl	Σr
Multiple of				Coefficient of the sine of the argument	Coefficient of the cosine of the argument
D	*M*	*M'*	*F*		
0	0	1	0	6 288 774	-20 905 355
2	0	-1	0	1 274 027	-3 699 111
2	0	0	0	658 314	-2 955 968
0	0	2	0	213 618	-569 925
0	1	0	0	-185 116	48 888
0	0	0	2	-114 332	-3 149
2	0	-2	0	58 793	246 158
2	-1	-1	0	57 066	-152 138
2	0	1	0	53 322	-170 733
2	-1	0	0	45 758	-204 586
0	1	-1	0	-40 923	-129 620
1	0	0	0	-34 720	108 743
0	1	1	0	-30 383	104 755
2	0	0	-2	15 327	10 321
0	0	1	2	-12 528	
0	0	1	-2	10 980	79 661
4	0	-1	0	10 675	-34 782
0	0	3	0	10 034	-23 210
4	0	-2	0	8 548	-21 636
2	1	-1	0	-7 888	24 208
2	1	0	0	-6 766	30 824
1	0	-1	0	-5 163	-8 379
1	1	0	0	4 987	-16 675
2	-1	1	0	4 036	-12 831
2	0	2	0	3 994	-10 445
4	0	0	0	3 861	-11 650
2	0	-3	0	3 665	14 403
0	1	-2	0	-2 689	-7 003
2	0	-1	2	-2 602	
2	-1	-2	0	2 390	10 056
1	0	1	0	-2 348	6 322
2	-2	0	0	2 236	-9 884

TABLE 45.A (cont.)

Argument Multiple of *D*	 *M*	 *M'*	 *F*	Σl Coefficient of the sine of the argument	Σr Coefficient of the cosine of the argument
0	1	2	0	-2 120	5 751
0	2	0	0	-2 069	
2	-2	-1	0	2 048	-4 950
2	0	1	-2	-1 773	4 130
2	0	0	2	-1 595	
4	-1	-1	0	1 215	-3 958
0	0	2	2	-1 110	
3	0	-1	0	-892	3 258
2	1	1	0	-810	2 616
4	-1	-2	0	759	-1 897
0	2	-1	0	-713	-2 117
2	2	-1	0	-700	2 354
2	1	-2	0	691	
2	-1	0	-2	596	
4	0	1	0	549	-1 423
0	0	4	0	537	-1 117
4	-1	0	0	520	-1 571
1	0	-2	0	-487	-1 739
2	1	0	-2	-399	
0	0	2	-2	-381	-4 421
1	1	1	0	351	
3	0	-2	0	-340	
4	0	-3	0	330	
2	-1	2	0	327	
0	2	1	0	-323	1165
1	1	-1	0	299	
2	0	3	0	294	
2	0	-1	-2		8 752

TABLE 45.B

Periodic terms for the latitude of the Moon (Σb).
The unit is 0.000 001 degree.

Argument Multiple of *D*	*M*	*M'*	*F*	Σb Coefficient of the sine of the argument	*Argument* Multiple of *D*	*M*	*M'*	*F*	Σb Coefficient of the sine of the argument
0	0	0	1	5 128 122	0	0	1	-3	777
0	0	1	1	280 602	4	0	-2	1	671
0	0	1	-1	277 693	2	0	0	-3	607
2	0	0	-1	173 237	2	0	2	-1	596
2	0	-1	1	55 413	2	-1	1	-1	491
2	0	-1	-1	46 271	2	0	-2	1	-451
2	0	0	1	32 573	0	0	3	-1	439
0	0	2	1	17 198	2	0	2	1	422
2	0	1	-1	9 266	2	0	-3	-1	421
0	0	2	-1	8 822	2	1	-1	1	-366
2	-1	0	-1	8 216	2	1	0	1	-351
2	0	-2	-1	4 324	4	0	0	1	331
2	0	1	1	4 200	2	-1	1	1	315
2	1	0	-1	-3 359	2	-2	0	-1	302
2	-1	-1	1	2 463	0	0	1	3	-283
2	-1	0	1	2 211	2	1	1	-1	-229
2	-1	-1	-1	2 065	1	1	0	-1	223
0	1	-1	-1	-1 870	1	1	0	1	223
4	0	-1	-1	1 828	0	1	-2	-1	-220
0	1	0	1	-1 794	2	1	-1	-1	-220
0	0	0	3	-1 749	1	0	1	1	-185
0	1	-1	1	-1 565	2	-1	-2	-1	181
1	0	0	1	-1 491	0	1	2	1	-177
0	1	1	1	-1 475	4	0	-2	-1	176
0	1	1	-1	-1 410	4	-1	-1	-1	166
0	1	0	-1	-1 344	1	0	1	-1	-164
1	0	0	-1	-1 335	4	0	1	-1	132
0	0	3	1	1 107	1	0	-1	-1	-119
4	0	0	-1	1 021	4	-1	0	-1	115
4	0	-1	1	833	2	-2	0	1	107

Additive to Σl :

$$+3958 \sin A_1$$
$$+1962 \sin (L' - F)$$
$$+\ 318 \sin A_2$$

Additive to Σb :

$$-2235 \sin L'$$
$$+\ 382 \sin A_3$$
$$+\ 175 \sin (A_1 - F)$$
$$+\ 175 \sin (A_1 + F)$$
$$+\ 127 \sin (L' - M')$$
$$-\ 115 \sin (L' + M')$$

The coordinates of the Moon are then given by

$$\lambda = L' + \frac{\Sigma l}{1\,000\,000} \qquad \text{(in degrees)}$$

$$\beta = \frac{\Sigma b}{1\,000\,000} \qquad \text{(in degrees)}$$

$$\Delta = 385\,000.56 + \frac{\Sigma r}{1000} \qquad \text{(in kilometers)}$$

Dividing the sums by 10^6 or by 10^3 is needed because in Tables 45.A and 45.B the coefficients are given in units of 10^{-6} degree or of 10^{-3} kilometer.

Example 45.a — Calculate the geocentric longitude, latitude, distance, and equatorial parallax of the Moon on 1992 April 12, at 0^h TD.

We find successively :

JDE = 2448 724.5	A_1 = 109°.57
T = −0.077 221 081 451	A_2 = 123°.78
L' = 134°.290 186	A_3 = 229°.53
D = 113°.842 309	E = 1.000 194
M = 97°.643 514	Σl = −1 127 527 } with the ad-
M' = 5°.150 839	Σb = −3 229 127 } ditive terms
F = 219°.889 726	Σr = −16 590 875

From which we deduce

$$\lambda = 134°.290\,186 - 1°.127\,527 = 133°.162\,659$$

$$\beta = -3°.229\,127 = -3°13'45''$$

$$\Delta = 385\,000.56 - 16\,590.875 = 368\,409.7 \text{ km}$$

$$\pi = \text{arcsine}\,(6378.14/368\,409.7) = 0°.991\,990 = 0°59'31''.2$$

The apparent longitude of the Moon is obtained by adding to λ the nutation in longitude ($\Delta\psi$), which is equal to $+16''.595 = +0°.004\,610$

(Chapter 21). Consequently,

$$\text{apparent } \lambda = 133^\circ.162\,659 + 0^\circ.004\,610 = 133^\circ.167\,269 = 133^\circ 10' 02''$$

For the given instant, the true obliquity of the ecliptic is (Chapter 21)

$$\varepsilon = \varepsilon_0 + \Delta\varepsilon = 23^\circ 26' 26''.29 = 23^\circ.440\,636$$

The Moon's apparent right ascension and declination are then found by means of expressions (12.3) and (12.4) :

$$\alpha = 134^\circ.688\,473 = 8^h 58^m 45^s.2$$

$$\delta = +13^\circ.768\,366 = +13^\circ 46' 06''$$

The exact values, obtained by using the complete ELP-2000/82 theory, are :

$\lambda = 133^\circ 10' 00''$ $\alpha = 8^h 58^m 45^s.1$

$\beta = -3^\circ 13' 45''$ $\delta = +13^\circ 46' 06''$

$\Delta = 368\,405.6$ km $\pi = 0^\circ 59' 31''.2$

Lunar node and lunar perigee

According to Chapront [2], the longitude of the (mean) ascending node Ω and that of the (mean) perigee π of the lunar orbit, in degrees, are given by

$$\Omega = 125.044\,5550 - 1934.136\,1849\,T + 0.002\,0762\,T^2 + T^3/467\,410 - T^4/60\,616\,000 \qquad (45.7)$$

$$\pi = 83^\circ.353\,2430 + 4069.013\,7111\,T - 0.010\,3238\,T^2 - T^3/80053 + T^4/18\,999\,000$$

where T has the same meaning as before. These longitudes are tropical, that is, they are measured from the mean equinox of the date.

From the formula for Ω we can find the instants when the (mean) ascending or descending node of the lunar orbit coincides with the vernal equinox, that is, when Ω is equal to 0° or to 180°, respectively. During the period 1910-2110, this occurs at the following dates :

$\Omega = 0°$	$\Omega = 180°$
1913 May 27	1922 Sep 16
1932 Jan 6	1941 Apr 27
1950 Aug 17	1959 Dec 7
1969 Mar 29	1978 Jul 19
1987 Nov 8	1997 Feb 27
2006 Jun 19	2015 Oct 10
2025 Jan 29	2034 May 21
2043 Sep 10	2052 Dec 30
2062 Apr 22	2071 Aug 12
2080 Dec 1	2090 Mar 23
2099 Jul 13	2108 Nov 3

References

1. M. Chapront-Touzé and J. Chapront, 'The lunar ephemeris ELP 2000', *Astronomy and Astrophysics*, Vol. 124, pages 50-62 (1983). — This article gives a description of that new lunar theory and discusses its accuracy. It does not give, however, the list of the many periodic terms. "ELP" means *Éphémérides Lunaires Parisiennes*, although this work is not an ephemeris (a list of calculated positions) but rather an analytic theory (a series of periodic terms).

2. M. Chapront-Touzé and J. Chapront, *Astronomy and Astrophysics*, Vol. 190, page 346 (1988).

Chapter 46

Illuminated Fraction of the Moon's Disk

The illuminated fraction k of the disk of the Moon depends on the selenocentric elongation of the Earth from the Sun, called the *phase angle* (i). Selenocentric means "as seen from the center of the Moon". The formula is

$$k = \frac{1 + \cos i}{2} \tag{46.1}$$

and this is the value of both the ratio of the illuminated *area* of the disk to the total area, and the ratio of the illuminated *length* of the diameter perpendicular to the line of cusps to the complete diameter (see the Figure).

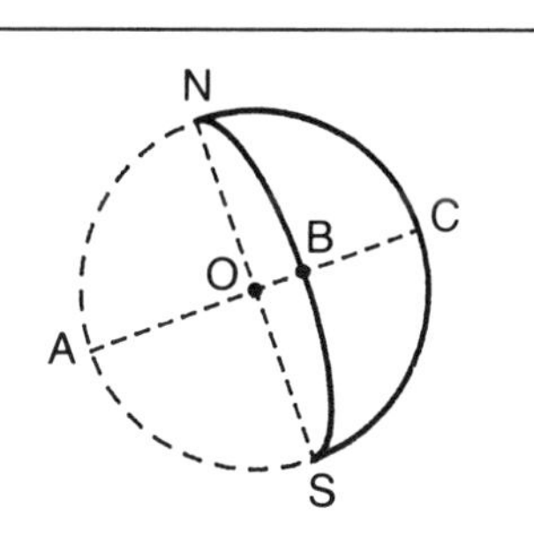

NCS = illuminated limb
N = northern cusp
S = southern cusp
C = midpoint of the illuminated limb
NOS = line of cusps
NBS = terminator (an ellipse)

Illuminated fraction k = ratio of lengths BC:AC = ratio of the areas NBSC:NASC

The phase angle i of the Moon, for a geocentric observer, can be found as follows. First, find the geocentric elongation ψ of the Moon from the Sun by means of one of the relations

$$\cos \psi = \sin \delta_\odot \sin \delta + \cos \delta_\odot \cos \delta \cos(\alpha_\odot - \alpha) \tag{46.2}$$

$$\cos \psi = \cos \beta \cos(\lambda - \lambda_\odot)$$

where $\alpha_\odot$, $\delta_\odot$, $\lambda_\odot$ and α, δ, λ are the geocentric right ascensions, declinations and longitudes of the

Sun and the Moon, respectively, and β the geocentric latitude of the Moon. Then we have

$$\tan i = \frac{R \sin \psi}{\Delta - R \cos \psi} \tag{46.3}$$

where R is the distance Earth-Sun, and Δ the distance Earth-Moon, both in the same units, for instance in kilometers. The angles ψ and i are always between 0 and 180 degrees. Once i is known, the illuminated fraction k can be obtained by means of formula (46.1).

Of course, for the calculation of k it is not needed to calculate the geocentric positions of the Moon and the Sun with high precision. An accuracy of, say, 1' will be sufficient.

If no high accuracy is required, it will suffice to put $\cos i = -\cos \psi$. The resulting error in k will never exceed 0.0014.

Lower accuracy, though still a good result, is obtained by neglecting the Moon's latitude and by calculating an approximate value of i as follows :

$$\begin{aligned} i = 180^\circ - D & - 6^\circ.289 \sin M' \\ & + 2^\circ.100 \sin M \\ & - 1^\circ.274 \sin (2D - M') \\ & - 0^\circ.658 \sin 2D \\ & - 0^\circ.214 \sin 2M' \\ & - 0^\circ.110 \sin D \end{aligned} \tag{46.4}$$

where the angles D, M and M' can be found by means of formulae (45.2) to (45.4). In this case, the geocentric positions of the Sun and the Moon are not needed.

Position Angle of the Moon's bright limb

The position angle of the Moon's bright limb is the position angle χ of the *midpoint* of the illuminated limb of the Moon (C in the Figure on page 315), reckoned eastward from the North Point of the disk (*not* from the axis of rotation of the lunar globe). It can be obtained from

$$\tan \chi = \frac{\cos \delta_0 \sin (\alpha_0 - \alpha)}{\sin \delta_0 \cos \delta - \cos \delta_0 \sin \delta \cos (\alpha_0 - \alpha)} \tag{46.5}$$

where α_0, δ_0, α and δ have the same meaning as before.

The angle χ is in the vicinity of 270° near First Quarter, near 90° after Full Moon. The angle χ is found in the correct quadrant by applying the ATN2 function to the numerator and the denominator of the fraction in formula (46.5) — see 'the correct quadrant' in Chapter 1.

If χ is the position angle of the (midpoint of the) bright limb, then the position angles of the cusps are $\chi - 90^\circ$ and $\chi + 90^\circ$.

The angle χ has the advantage that it unambiguously defines the illuminated limb of the Moon.

It should be noted that the angle χ is *not* measured from the direction of the observer's zenith. The *zenith* angle of the bright limb is $\chi - q$, where q is the parallactic angle (see Chapter 13).

Finally, it should be mentioned that formula (46.5) is valid in the case of a *planet* too.

Example 46.a — The Moon on 1992 April 12, at 0^h TD.

From Example 45.a we have, for that instant,

$$\alpha = 134°.6885$$
$$\delta = +13°.7684$$
$$\Delta = 368\,408 \text{ km}$$

The apparent position and the distance of the Sun at the same instant are

$$\alpha_0 = 1^h 22^m 37^s.9 = 20°.6579$$
$$\delta_0 = +8°41'47'' = +8°.6964$$
$$R = 1.002\,4977 \text{ AU} = 149\,971\,520 \text{ km}$$

The first formula (46.2) then gives $\cos\psi = -0.354\,991$, whence $\psi = 110°.7929$. Then

$$\tan i = +2.615\,403 \quad \text{by formula (46.3)}$$
$$i = 69°.0756$$

and, by formula (46.1), $k = 0.6786$, which should be rounded to 0.68.

If we use the approximate relation $\cos i = -\cos\psi$, we find $k = 0.6775$, which again rounds to 0.68.

Let us now use the approximate formula (46.4). In Example 45.a, we have found for the given instant

$$D = 113°.8423$$
$$M = 97°.6435$$
$$M' = 5°.1508$$

Then formula (46.4) gives $i = 68°.88$, whence, by (46.1), $k = 0.6802$, which again rounds to 0.68.

Finally, formula (46.5) gives

$$\tan\chi = \frac{-0.90283}{+0.24266} \quad \text{whence} \quad \chi = 285°.0$$

Chapter 47

Phases of the Moon

By definition, the times of New Moon, First Quarter, Full Moon and Last Quarter are the times at which the excess of the apparent longitude of the Moon over the apparent longitude of the Sun is 0°, 90°, 180°, and 270°, respectively.

Hence, to calculate the instants of these lunar phases, it is necessary to calculate the apparent longitudes of the Moon and the Sun separately. (However, the effect of the nutation may be neglected here, since the nutation in longitude $\Delta\psi$ will not affect the *difference* between the longitudes of Moon and Sun.)

However, if no high accuracy is required, the instants of the lunar phases can be calculated by the method described in this Chapter. The expressions are based on Chapront's ELP-2000/82 theory for the Moon (with improved expressions for the arguments M, M', etc., as mentioned in Chapter 45), and on Bretagnon's and Francou's VSOP87 theory for the Sun. The resulting times will be expressed in Julian Ephemeris Days (JDE), hence in Dynamical Time.

The times of the *mean* phases of the Moon, already affected by the Sun's aberration and by the Moon's light-time, are given by

$$\begin{aligned} \text{JDE} = 2451\,550.09765 &+ 29.530\,588\,853\,k \\ &+ 0.000\,1337\,T^2 \\ &- 0.000\,000\,150\,T^3 \\ &+ 0.000\,000\,000\,73\,T^4 \end{aligned} \qquad (47.1)$$

where an integer value of k gives a New Moon, an integer increased

by 0.25 gives a First Quarter,
by 0.50 gives a Full Moon,
by 0.75 gives a Last Quarter.

Any other value for k will give meaningless results !
The value $k = 0$ corresponds to the New Moon of 2000 January 6. Negative values of k give lunar phases before the year 2000.

For example,

+479.00 and -2793.00 correspond to a New Moon,
+479.25 and -2792.75 correspond to a First Quarter,
+479.50 and -2792.50 correspond to a Full Moon,
+479.75 and -2792.25 correspond to a Last Quarter.

An approximate value of k is given by

$$k \cong (\text{year} - 2000) \times 12.3685 \tag{47.2}$$

where the 'year' should be taken with decimals, for example 1987.25 for the end of March 1987 (because this is 0.25 year since the beginning of the year 1987). The sign ≅ means "is approximately equal to".

Finally, in formula (47.1) T is the time in Julian centuries since the epoch 2000.0; it is obtained with a sufficient accuracy from

$$T = \frac{k}{1236.85} \tag{47.3}$$

and hence is negative before the epoch 2000.0.

Calculate E by means of formula (45.6), and then the following angles, which are expressed in degrees and may be reduced to the interval 0-360 degrees and, if necessary, to radians before going further on.

Sun's mean anomaly at time JDE :

$$\begin{aligned} M = 2.5534 &+ 29.105\,356\,69\,k \\ &- 0.000\,0218\,T^2 \\ &- 0.000\,000\,11\,T^3 \end{aligned} \tag{47.4}$$

Moon's mean anomaly :

$$\begin{aligned} M' = 201.5643 &+ 385.816\,935\,28\,k \\ &+ 0.010\,7438\,T^2 \\ &+ 0.000\,012\,39\,T^3 \\ &- 0.000\,000\,058\,T^4 \end{aligned} \tag{47.5}$$

Moon's argument of latitude :

$$\begin{aligned} F = 160.7108 &+ 390.670\,502\,74\,k \\ &- 0.001\,6341\,T^2 \\ &- 0.000\,002\,27\,T^3 \\ &+ 0.000\,000\,011\,T^4 \end{aligned} \tag{47.6}$$

Longitude of the ascending node of the lunar orbit :

$$\begin{aligned} \Omega = 124.7746 &- 1.563\,755\,80\,k \\ &+ 0.002\,0691\,T^2 \\ &+ 0.000\,002\,15\,T^3 \end{aligned} \tag{47.7}$$

Planetary arguments :

$$
\begin{aligned}
A_1 &= 299.77 + 0.107\,408\,k - 0.009\,173\,T^2 \\
A_2 &= 251.88 + 0.016\,321\,k \\
A_3 &= 251.83 + 26.651\,886\,k \\
A_4 &= 349.42 + 36.412\,478\,k \\
A_5 &= 84.66 + 18.206\,239\,k \\
A_6 &= 141.74 + 53.303\,771\,k \\
A_7 &= 207.14 + 2.453\,732\,k \\
A_8 &= 154.84 + 7.306\,860\,k \\
A_9 &= 34.52 + 27.261\,239\,k \\
A_{10} &= 207.19 + 0.121\,824\,k \\
A_{11} &= 291.34 + 1.844\,379\,k \\
A_{12} &= 161.72 + 24.198\,154\,k \\
A_{13} &= 239.56 + 25.513\,099\,k \\
A_{14} &= 331.55 + 3.592\,518\,k
\end{aligned}
$$

To obtain the time of the *true* (apparent) phase, add the following corrections (in days) to the JDE obtained above.

New Moon	Full Moon	× sin
−0.40720	−0.40614	M'
+0.17241 × E	+0.17302 × E	M
+0.01608	+0.01614	$2M'$
+0.01039	+0.01043	$2F$
+0.00739 × E	+0.00734 × E	$M' - M$
−0.00514 × E	−0.00515 × E	$M' + M$
+0.00208 × E^2	+0.00209 × E^2	$2M$
−0.00111	−0.00111	$M' - 2F$
−0.00057	−0.00057	$M' + 2F$
+0.00056 × E	+0.00056 × E	$2M' + M$
−0.00042	−0.00042	$3M'$
+0.00042 × E	+0.00042 × E	$M + 2F$
+0.00038 × E	+0.00038 × E	$M - 2F$
−0.00024 × E	−0.00024 × E	$2M' - M$
−0.00017	−0.00017	Ω
−0.00007	−0.00007	$M' + 2M$
+0.00004	+0.00004	$2M' - 2F$
+0.00004	+0.00004	$3M$
+0.00003	+0.00003	$M' + M - 2F$
+0.00003	+0.00003	$2M' + 2F$
−0.00003	−0.00003	$M' + M + 2F$
+0.00003	+0.00003	$M' - M + 2F$
−0.00002	−0.00002	$M' - M - 2F$
−0.00002	−0.00002	$3M' + M$
+0.00002	+0.00002	$4M'$

First and Last Quarters

−0.62801		× sin M'
+0.17172	× E	M
−0.01183	× E	$M' + M$
+0.00862		$2M'$
+0.00804		$2F$
+0.00454	× E	$M' - M$
+0.00204	× E^2	$2M$
−0.00180		$M' - 2F$
−0.00070		$M' + 2F$
−0.00040		$3M'$
−0.00034	× E	$2M' - M$
+0.00032	× E	$M + 2F$
+0.00032	× E	$M - 2F$
−0.00028	× E^2	$M' + 2M$
+0.00027	× E	$2M' + M$
−0.00017		Ω
−0.00005		$M' - M - 2F$
+0.00004		$2M' + 2F$
−0.00004		$M' + M + 2F$
+0.00004		$M' - 2M$
+0.00003		$M' + M - 2F$
+0.00003		$3M$
+0.00002		$2M' - 2F$
+0.00002		$M' - M + 2F$
−0.00002		$3M' + M$

Calculate, for the Quarter phases only,

$$W = 0.00306 - 0.00038\,E \cos M + 0.00026 \cos M' - 0.00002 \cos(M' - M) + 0.00002 \cos(M' + M) + 0.00002 \cos 2F$$

Additional corrections :

for First Quarter : $+W$
for Last Quarter : $-W$

Additional corrections for all phases :

+0.000325 × sin	A_1	+0.000056 × sin	A_8
165	A_2	047	A_9
164	A_3	042	A_{10}
126	A_4	040	A_{11}
110	A_5	037	A_{12}
062	A_6	035	A_{13}
060	A_7	023	A_{14}

Example 47.a — Calculate the instant of the New Moon which took place in February 1977.

Mid-February 1977 corresponds to 1977.13, so we find by (47.2)

$$k \cong (1977.13 - 2000) \times 12.3685 = -282.87$$

whence $k = -283$, since k should be an integer for the New Moon phase. Then, by formula (47.3), $T = -0.22881$, and then formula (47.1) gives

$$\text{JDE} = 2443\,192.94101$$

With $k = -283$ and $T = -0.22881$, we further find

$$\begin{aligned} E &= 1.000\,5753 \\ M &= -8\,234^\circ.2625 = 45^\circ.7375 \\ M' &= -108\,984^\circ.6278 = 95^\circ.3722 \\ F &= -110\,399^\circ.0416 = 120^\circ.9584 \\ \Omega &= 567^\circ.3176 = 207^\circ.3176 \end{aligned}$$

The sum of the first group of periodic terms (for New Moon) is -0.28916, that of the 14 additional corrections is -0.00068. Consequently, the time of the true New Moon is

$$\text{JDE} = 2443\,192.94101 - 0.28916 - 0.00068 = 2443\,192.65117,$$

which corresponds to 1977 February 18.15117 TD
= 1977 February 18, at $3^h37^m41^s$ TD.

The correct value, calculated by means of the ELP-2000/82 theory, is $3^h37^m40^s$ TD.

In February 1977, the quantity ΔT = TD - UT was equal to 48 seconds. Hence, the New Moon of 1977 February 18 occurred at 3^h37^m Universal Time. See also Example 9.a, page 74.

Example 47.b — Calculate the time of the first Last Quarter of A.D. 2044.

For 'year' = 2044, formula (47.2) gives $k \cong +544.21$, so we shall use the value $k = +544.75$.

Then, by formula (47.1), JDE = 2467 636.88595.

Sum of the first group of periodic terms (for Last Quarter) = -0.39153.

Additional correction for Last Quarter = $-W$ = -0.00251.

Sum of additional 14 corrections = -0.00007.

Consequently, the time of the Last Quarter is

$$2467\,636.88595 - 0.39153 - 0.00251 - 0.00007 = 2467\,636.49184$$

which corresponds to 2044 January 21, at $23^h48^m15^s$ TD.

For the period 1980 to mid-2020, we compared the results of the method described in this Chapter with the accurate times obtained with the ELP-2000/82 and VSOP 87 theories.

	Mean error	*Maximum error*
New Moon :	3.6 sec.	16.4 sec.
First Quarter :	3.8	15.3
Full Moon :	3.8	17.4
Last Quarter :	3.8	13.0

Mean error of all phases = 3.72 seconds

If an error of a few minutes is not important, one may, of course, drop the smallest periodic terms and the fourteen additional terms.

The *mean* time interval between two consecutive New Moons is 29.530 589 days, or 29 days 12 hours 44 minutes 03 seconds. This is the length of the synodic period of the Moon. However, mainly by reason of the perturbing action of the Sun, the actual time interval between consecutive New Moons, or *lunation*, varies greatly. See Table 47.A, taken from [1].

TABLE 47.A

The shortest and the longest lunations, 1900 to 2100

From the New Moon of	*to that of*	*Duration of the lunation*
1903 June 25	1903 July 24	29 days 06 hours 35 min.
2035 June 6	2035 July 5	29 – 06 – 39 –
2053 June 16	2053 July 15	29 – 06 – 35 –
2071 June 27	2071 July 27	29 – 06 – 36 –
1955 Dec. 14	1956 Jan. 13	29 days 19 hours 54 min.
1973 Dec. 24	1974 Jan. 23	29 – 19 – 55 –

Reference

1. J. Meeus, 'Les durées extrêmes de la lunaison', *l'Astronomie* (Société Astronomique de France), Vol. 102, pages 288 - 289 (July-August 1988).

Chapter 48

Perigee and Apogee of the Moon

In this Chapter a method is given for the calculation of approximate times when the distance between the Earth and the Moon is a minimum (perigee) or a maximum (apogee). The resulting times will be expressed in Julian Ephemeris Days (JDE), hence in the uniform time scale of Dynamical Time. Our expressions are based on Chapront's lunar theory ELP-2000/82, with improved expressions for the arguments D, M, etc., as mentioned in Chapter 45.

First, calculate the time of the *mean* perigee or apogee by the formula

$$\begin{aligned} \text{JDE} = 2451\,534.6698 &+ 27.554\,549\,88\, k \\ &- 0.000\,6886\, T^2 \\ &- 0.000\,001\,098\, T^3 \\ &+ 0.000\,000\,0052\, T^4 \end{aligned} \qquad (48.1)$$

where an integer value of k gives a perigee, and an integer increased by 0.5 an apogee. *Important :* any other value for k will give meaningless results !

The value $k = 0$ corresponds to the perigee of 1999 December 22.

So, for example,

$k = +318$ and $k = -25$ will give a perigee,
$k = +429.5$ and $k = -1209.5$ will give an apogee,
$k = +224.87$ is an incorrect value.

An approximate value of k is given by

$$k \cong (\text{year} - 1999.97) \times 13.2555 \qquad (48.2)$$

where the 'year' should be taken with decimals, for instance 2041.33 represents the end of April of the year 2041.

Finally, in formula (48.1) T is the time in Julian centuries since the epoch 2000.0. It is obtained with a sufficient accuracy from

$$T = \frac{k}{1325.55} \qquad (48.3)$$

Calculate the following angles; they are expressed in degrees and may be reduced to the interval 0 - 360 degrees and, if necessary, to radians before calculating further.

Moon's mean elongation at time JDE :

$$\begin{aligned} D = 171.9179 &+ 335.910\,6046\,k \\ &- 0.010\,0250\,T^2 \\ &- 0.000\,011\,56\,T^3 \\ &+ 0.000\,000\,055\,T^4 \end{aligned}$$

Sun's mean anomaly :

$$\begin{aligned} M = 347.3477 &+ 27.157\,7721\,k \\ &- 0.000\,8323\,T^2 \\ &- 0.000\,0010\,T^3 \end{aligned}$$

Moon's argument of latitude :

$$\begin{aligned} F = 316.6109 &+ 364.528\,7911\,k \\ &- 0.012\,5131\,T^2 \\ &- 0.000\,0148\,T^3 \end{aligned}$$

To the JDE given by (48.1), add the sum of the periodic terms of Table 48.A, taking either those for perigee or for apogee, according to the case.

The Moon's equatorial horizontal parallax is obtained by making the sum of the terms given in Table 48.B.

From Tables 48.A and 48.B it appears that :

- for the periodic terms for the instant, the sine of the argument should be taken, while for the value of the corresponding parallax the cosine must be used;
- up to a given value of the coefficient, there are more periodic terms for the perigee than for the apogee;
- the successive coefficients in the same "$2D$" series (for example the terms in $2D-M$, $4D-M$, $6D-M$, etc.) have alternate signs for the perigee, while for the apogee all have the same sign;
- the coefficient of the largest periodic term (the term with argument $2D$) is much larger in the case of the perigee than for the apogee. As a consequence, the largest possible difference between the time of a *mean* and a *true* passage is 45 hours for the perigee, but only 13 hours for the apogee. Also, the Moon's perigee distance varies in a larger interval (approximately between 356370 and 370350 kilometers) than does the apogee distance (404050 to 406720 km).

Example 48.a — The Moon's apogee of October 1988.

Because the beginning of October corresponds to 0.75 year since the beginning of the calendar year, we put the value year = 1988.75 in formula (48.2). This gives $k \cong -148.73$. We therefore take the value $k = -148.5$ (apogee !).

Formulae (48.3) and (48.1) then give

$$T = -0.112\,029 \qquad \text{JDE} = 2447\,442.8191$$

Then we find

$$D = -49\,710^\circ.8070 = 329^\circ.1930$$
$$M = -3\,685^\circ.5815 = 274^\circ.4185$$
$$F = -53\,815^\circ.9147 = 184^\circ.0853$$

Sum of the terms in Table 48.A (apogee) = −0.4654 day
Sum of the terms in Table 48.B (apogee) = 3240.679

Hence, the time of the apogee is

$$\text{JDE} = 2447\,442.8191 - 0.4654 = 2447\,442.3537$$

which corresponds to 1988 October 7, at 20^h29^m TD. The corresponding value of the Moon's equatorial horizontal parallax is $3240''.679$, or $0^\circ 54' 00''.679$.

The exact values are 20^h30^m TD and $0^\circ 54' 00''.671$.

TABLE 48.A

Periodic terms for the time, in days

For the perigee

Argument of sine	Coefficient	Argument of sine	Coefficient
$2D$	−1.6769	$2D - 2M$	−0.0027
$4D$	+0.4589	$4D - 2M$	+0.0024
$6D$	−0.1856	$6D - 2M$	−0.0021
$8D$	+0.0883	$22D$	−0.0021
$2D - M$	$-0.0773 + 0.00019\,T$	$18D - M$	−0.0021
M	$+0.0502 - 0.00013\,T$	$6D + M$	+0.0019
$10D$	−0.0460	$11D$	−0.0018
$4D - M$	$+0.0422 - 0.00011\,T$	$8D + M$	−0.0014
$6D - M$	−0.0256	$4D - 2F$	−0.0014
$12D$	+0.0253	$6D + 2F$	−0.0014
D	+0.0237	$3D + M$	+0.0014
$8D - M$	+0.0162	$5D + M$	−0.0014
$14D$	−0.0145	$13D$	+0.0013
$2F$	+0.0129	$20D - M$	+0.0013
$3D$	−0.0112	$3D + 2M$	+0.0011
$10D - M$	−0.0104	$4D + 2F - 2M$	−0.0011
$16D$	+0.0086	$D + 2M$	−0.0010
$12D - M$	+0.0069	$22D - M$	−0.0009
$5D$	+0.0066	$4F$	−0.0008
$2D + 2F$	−0.0053	$6D - 2F$	+0.0008
$18D$	−0.0052	$2D - 2F + M$	+0.0008
$14D - M$	−0.0046	$2M$	+0.0007
$7D$	−0.0041	$2F - M$	+0.0007
$2D + M$	+0.0040	$2D + 4F$	+0.0007
$20D$	+0.0032	$2F - 2M$	−0.0006
$D + M$	−0.0032	$2D - 2F + 2M$	−0.0006
$16D - M$	+0.0031	$24D$	+0.0006
$4D + M$	−0.0029	$4D - 4F$	+0.0005
$9D$	+0.0027	$2D + 2M$	+0.0005
$4D + 2F$	+0.0027	$D - M$	−0.0004

TABLE 48.A (cont.)

For the apogee

Argument of sine	*Coefficient*	*Argument of sine*	*Coefficient*
$2D$	+0.4392	$8D - M$	+0.0011
$4D$	+0.0684	$4D - 2M$	+0.0010
M	+0.0456 - 0.00011 T	$10D$	+0.0009
$2D - M$	+0.0426 - 0.00011 T	$3D + M$	+0.0007
$2F$	+0.0212	$2M$	+0.0006
D	-0.0189	$2D + M$	+0.0005
$6D$	+0.0144	$2D + 2M$	+0.0005
$4D - M$	+0.0113	$6D + 2F$	+0.0004
$2D + 2F$	+0.0047	$6D - 2M$	+0.0004
$D + M$	+0.0036	$10D - M$	+0.0004
$8D$	+0.0035	$5D$	-0.0004
$6D - M$	+0.0034	$4D - 2F$	-0.0004
$2D - 2F$	-0.0034	$2F + M$	+0.0003
$2D - 2M$	+0.0022	$12D$	+0.0003
$3D$	-0.0017	$2D + 2F - M$	+0.0003
$4D + 2F$	+0.0013	$D - M$	-0.0003

TABLE 48.B

Terms for the parallax, in arcseconds

For the perigee

3629".215	
+63.224	× cos 2D
−6.990	4D
+2.834 −0.0071 T	2D − M
+1.927	6D
−1.263	D
−0.702	8D
+0.696 −0.0017 T	M
−0.690	2F
−0.629 +0.0016 T	4D − M
−0.392	2D − 2F
+0.297	10D
+0.260	6D − M
+0.201	3D
−0.161	2D + M
+0.157	D + M
−0.138	12D
−0.127	8D − M
+0.104	2D + 2F
+0.104	2D − 2M
−0.079	5D
+0.068	14D
+0".067	× cos 10D − M
+0.054	4D + M
−0.038	12D − M
−0.038	4D − 2M
+0.037	7D
−0.037	4D + 2F
−0.035	16D
−0.030	3D + M
+0.029	D − M
−0.025	6D + M
+0.023	2M
+0.023	14D − M
−0.023	2D + 2M
+0.022	6D − 2M
−0.021	2D − 2F − M
−0.020	9D
+0.019	18D
+0.017	6D + 2F
+0.014	2F − M
−0.014	16D − M
+0.013	4D − 2F
+0.012	8D + M
+0.011	11D
+0.010	5D + M
−0.010	20D

For the apogee

3245".251	
−9.147	× cos 2D
−0.841	D
+0.697	2F
−0.656 +0.0016 T	M
+0.355	4D
+0.159	2D − M
+0.127	D + M
+0.065	4D − M
+0".052	× cos 6D
+0.043	2D + M
+0.031	2D + 2F
−0.023	2D − 2F
+0.022	2D − 2M
+0.019	2D + 2M
−0.016	2M
+0.014	6D − M
+0.010	8D

Using the method described in this Chapter, 600 perigee and 600 apogee passages of the Moon were calculated, namely from June 1977 to August 2022. The results were compared with accurate values obtained with the ELP-2000/82 theory. The largest errors are

for the time :	31 minutes for the perigee, 3 minutes for the apogee;
for the parallax :	0".124 for the perigee, 0".051 for the apogee.

The latter errors correspond to distance errors of 12 and 6 kilometers, respectively. The distribution of the errors of the 600 calculated times is as follows :

Number of errors less than	*Perigee*	*Apogee*
1 minute	151	478
2 minutes	264	589
3 minutes	385	599
4 minutes	460	
5 minutes	492	
10 minutes	572	

The *mean* time interval between two consecutive passages of the Moon through perigee is 27.55455 days, or 27 days 13 hours 19 minutes; this is the length of the anomalistic period of the Moon. However, mainly by reason of the perturbing action of the Sun, the actual time interval between consecutive perigees varies greatly, between the extremes 24 days 16 hours and 28 days 13 hours. Examples :

perigee on 1997 December 9 at $16^h.9$ perigee on 1998 January 3 at $8^h.5$	diff. = 24 days 16 hours
perigee on 1990 December 2 at $10^h.8$ perigee on 1990 December 30 at $23^h.8$	diff. = 28 days 13 hours

The time interval between two consecutive *apogees*, however, varies between narrower limits, namely between 26.98 and 27.90 days (26 days 23½ hours and 27 days 21½ hours).

Extreme perigee and apogee distances of the Moon

Between the years 1500 and 2500, fourteen times the Moon approaches the Earth to less than 356 425 kilometers, and the same number of times the distance grows to larger than 406 710 km. These cases are mentioned in Table 48.C. The dates are UT dates.

For the calculation, use has been made of Chapront's lunar theory ELP-2000/82, except that we neglected all periodic terms with a coefficient less than 0.0005 km (50 centimeters).

It appears that, during the time interval of ten centuries considered here, the extreme distances between the centers of Earth and Moon are

356 371 km on 2257 January 1
406 720 km on 2266 January 7

The smallest perigee distance of the 20th century was that of 1912 January 4, as was already found earlier by Roger W. Sinnott, Associate Editor of *Sky and Telescope* [1].

Further, we see that these extreme perigees and apogees all occur during the winter months of the northern hemisphere, the period of the year when the Earth is closest to the Sun. It is evident that the variable Earth-Sun distance somewhat affects the Earth-Moon distance.

TABLE 48.C : Extreme perigees and apogees, A.D. 1500 to 2500

perigee < 356 425 km		apogee > 406 710 km	
1548 Dec 15	356 407 km	1921 Jan 9	406 710 km
1566 Dec 26	356 399	1984 Mar 2	406 712
1771 Jan 30	356 422	2107 Jan 23	406 716
1893 Dec 23	356 396	2125 Feb 3	406 720
1912 Jan 4	356 375	2143 Feb 14	406 713
1930 Jan 15	356 397	2247 Dec 27	406 715
2052 Dec 6	356 421	2266 Jan 7	406 720
2116 Jan 29	356 403	2284 Jan 18	406 714
2134 Feb 9	356 416	2388 Nov 29	406 715
2238 Dec 22	356 406	2406 Dec 11	406 718
2257 Jan 1	356 371	2424 Dec 21	406 712
2275 Jan 12	356 378	2452 Jan 21	406 710
2461 Jan 26	356 408	2470 Feb 1	406 714
2479 Feb 7	356 404	2488 Feb 12	406 711

References

1. Roger W. Sinnott, letter of 1981 March 4 to Jean Meeus.

2. J. Meeus, 'Extreme Perigees and Apogees of the Moon', *Sky and Telescope*, Vol. 62, pages 110-111 (August 1981).

Chapter 49

Passages of the Moon through the Nodes

When the center of the Moon passes through the ascending or through the descending node of its orbit, its geocentric latitude is zero. Approximate times of the passages through the nodes can be obtained as follows. The results will be expressed as a Julian Ephemeris Day, JDE, hence in Dynamical Time.

For a passage through the *ascending* node, take k = an integer. For a passage at the *descending* node, take for k an integer increased by 0.5. *Important :* any other value for k will give meaningless results!

Successive values of k will provide successive passages of the Moon through the nodes, the value k = zero corresponding to the passage at the ascending node of 2000 January 21. Negative values of k yield passages before this date.

For instance, $k = +223.0$ and -147.0 correspond to an ascending node, $+223.5$ and -146.5 to a descending node, while $+144.76$ is not a valid value for k.

An approximate value of k is given by

$$k \cong (\text{year} - 2000.05) \times 13.4223 \qquad (49.1)$$

where 'year' may be taken with decimals. Then calculate

$$T = \frac{k}{1342.23}$$

and the following angles in *degrees* :

$$D = 183.6380 + 331.737\,356\,91\,k + 0.001\,5057\,T^2 + 0.000\,002\,09\,T^3 - 0.000\,000\,010\,T^4$$

$$M = 17.4006 + 26.820\,372\,50\,k + 0.000\,0999\,T^2 + 0.000\,000\,06\,T^3$$

$$M' = 38.3776 + 355.527\,473\,22\,k + 0.012\,3577\,T^2 + 0.000\,014\,628\,T^3 - 0.000\,000\,069\,T^4$$

$$\Omega = 123.9767 - 1.440\,989\,49\,k + 0.002\,0625\,T^2 + 0.000\,002\,14\,T^3 - 0.000\,000\,016\,T^4$$

$$V = 299.75 + 132.85\,T - 0.009\,173\,T^2$$

$$P = \Omega + 272.75 - 2.3\,T$$

The time of the passage through the node is then given by the following expression, where the terms involving M (the Sun's mean anomaly) should be multiplied by the quantity E given by formula (45.6). These terms are indicated by an asterisk.

$$
\begin{array}{rl}
\mathrm{JDE} = 2451\,565.1619 & + \; 27.212\,220\,817\,k \\
& + \; 0.000\,2572\,T^2 \\
& + \; 0.000\,000\,021\,T^3 \\
& - \; 0.000\,000\,000\,088\,T^4 \\
& - \; 0.4721 \sin M' \\
& - \; 0.1649 \sin 2D \\
& - \; 0.0868 \sin (2D - M') \\
& + \; 0.0084 \sin (2D + M') \\
* & - \; 0.0083 \sin (2D - M) \\
* & - \; 0.0039 \sin (2D - M - M') \\
& + \; 0.0034 \sin 2M' \\
& - \; 0.0031 \sin (2D - 2M') \\
* & + \; 0.0030 \sin (2D + M) \\
* & + \; 0.0028 \sin (M - M') \\
* & + \; 0.0026 \sin M \\
& + \; 0.0025 \sin 4D \\
& + \; 0.0024 \sin D \\
* & + \; 0.0022 \sin (M + M') \\
& + \; 0.0017 \sin \Omega \\
& + \; 0.0014 \sin (4D - M') \\
* & + \; 0.0005 \sin (2D + M - M') \\
* & + \; 0.0004 \sin (2D - M + M') \\
* & - \; 0.0003 \sin (2D - 2M) \\
* & + \; 0.0003 \sin (4D - M) \\
& + \; 0.0003 \sin V \\
& + \; 0.0003 \sin P
\end{array}
$$

Example 49.a — Calculate the instant of the passage of the Moon through the ascending node in May 1987.

Since mid-May corresponds to 0.37 year since the beginning of the year, we put year = 1987.37 in formula (49.1), which gives the approximate value −170.19 for k. For a passage through the ascending node, k should be an integer, so we take $k = -170$. Then we find

$$
\begin{aligned}
T &= -0.126\,655 \\
D &= -56211^\circ.71265 = 308^\circ.28735 \\
M &= -4542^\circ.06272 = 137^\circ.93728 \\
M' &= -60401^\circ.29265 = 78^\circ.70735 \\
\Omega &= 368^\circ.9449 = 8^\circ.9449 \\
V &= 282^\circ.92 \\
P &= 641^\circ.99 = 281^\circ.99 \\
E &= 1.000\,319
\end{aligned}
$$

The final result is JDE = 2446 938.76803, which corresponds to 1987 May 23.26803 = 1987 May 23, at $6^h26^m.0$ TD.

The correct value is May 23, at $6^h25^m.6$ TD.

The table below gives an idea of the accuracy of the results obtained by means of the algorithm given in this Chapter, as compared with the times obtained by an accurate calculation.

Years (A.D.)	*Node*	*Number of instants*	*Greatest error in seconds*	*Number of errors < 60 sec.*	*Number of errors > 120 sec.*
1980 to 2020	ascending	551	142	487	3
1980 to 2020	descending	551	132	469	2
0 to 40	ascending	551	144	444	5
0 to 40	descending	551	135	478	2

Chapter 50

Maximum Declinations of the Moon

The plane of the orbit of the Moon forms with the plane of the ecliptic an angle of 5°. Therefore, in the sky the Moon is moving *approximately* along the ecliptic, and during each revolution (27 days) it reaches its greatest northern declination (in Taurus, in Gemini, or in northern Orion), and two weeks later its greatest southern declination (in Sagittarius or in Ophiuchus).

Since the lunar orbit forms with the ecliptic an angle of 5°, and the ecliptic an angle of 23° with the celestial equator, the extreme declinations of the Moon are between 18° and 28° (North or South), approximately. When, as in 1987, the ascending node of the lunar orbit is in the vicinity of the vernal equinox (see page 314), the Moon reaches high northern and southern declinations, approximately $+28\frac{1}{2}$ and $-28\frac{1}{2}$ degrees. This situation is repeated at intervals of 18.6 years, the revolution period of the lunar nodes.

In this Chapter a method is given for the calculation of approximate times of the maximum declinations of the Moon, and the values of these maximum declinations. These data are *geocentric* and they refer to the center of the Moon's disk.

Let k be an integer, negative before the beginning of the year 2000. Successive values of k will give successive maximum northern or southern declinations of the Moon. The value $k = 0$ corresponds to January 2000. Important : a non-integer value of k will give meaningless results !

An approximate value of k is given by

$$k \cong (\text{year} - 2000.03) \times 13.3686 \qquad (50.1)$$

where the 'year' can be taken with decimals. Then calculate

$$T = \frac{k}{1336.86}$$

TABLE 50.A

Periodic terms (days) for the time of the Moon's maximum declination

Coefficient for declination north	*Coefficient for* declination south		*Coefficient for* declination north	*Coefficient for* declination south	
d	d		d	d	
+0.8975	-0.8975	cos F	+0.0030	+0.0030	sin $(2D + M')$
-0.4726	-0.4726	sin M'	-0.0029	+0.0029	cos $(M' + 2F)$
-0.1030	-0.1030	sin $2F$	-0.0029	-0.0029	sin $(2D - M)$ *
-0.0976	-0.0976	sin $(2D - M')$	-0.0027	-0.0027	sin $(M' + F)$
-0.0462	+0.0541	cos $(M' - F)$	+0.0024	+0.0024	sin $(M - M')$ *
-0.0461	+0.0516	cos $(M' + F)$	-0.0021	-0.0021	sin $(M' - 3F)$
-0.0438	-0.0438	sin $2D$	+0.0019	-0.0019	sin $(2M' + F)$
+0.0162	+0.0112	sin M *	+0.0018	-0.0006	cos $(2D - 2M' - F)$
-0.0157	+0.0157	cos $3F$	+0.0018	-0.0018	sin $3F$
+0.0145	+0.0023	sin $(M' + 2F)$	+0.0017	-0.0017	cos $(M' + 3F)$
+0.0136	-0.0136	cos $(2D - F)$	+0.0017	+0.0017	cos $2M'$
-0.0095	+0.0110	cos $(2D - M' - F)$	-0.0014	+0.0014	cos $(2D - M')$
-0.0091	+0.0091	cos $(2D - M' + F)$	+0.0013	-0.0013	cos $(2D + M' + F)$
-0.0089	+0.0089	cos $(2D + F)$	+0.0013	-0.0013	cos M'
+0.0075	+0.0075	sin $2M'$	+0.0012	+0.0012	sin $(3M' + F)$
-0.0068	-0.0030	sin $(M' - 2F)$	+0.0011	+0.0011	sin $(2D - M' + F)$
+0.0061	-0.0061	cos $(2M' - F)$	-0.0011	+0.0011	cos $(2D - 2M')$
-0.0047	-0.0047	sin $(M' + 3F)$	+0.0010	+0.0010	cos $(D + F)$
-0.0043	-0.0043	sin $(2D - M - M')$ *	+0.0010	+0.0010	sin $(M + M')$ *
-0.0040	+0.0040	cos $(M' - 2F)$	-0.0009	-0.0009	sin $(2D - 2F)$
-0.0037	-0.0037	sin $(2D - 2M')$	+0.0007	-0.0007	cos $(2M' + F)$
+0.0031	-0.0031	sin F	-0.0007	-0.0007	cos $(3M' + F)$

and the following angles, in *degrees*; the quantities between square *brackets* should be used for *southern* declinations.

$D = 152.2029 + 333.0705546\,k - 0.0004025\,T^2 + 0.00000011\,T^3$
[345.6676]

$M = 14.8591 + 26.9281592\,k - 0.0000544\,T^2 - 0.00000010\,T^3$
[1.3951]

$M' = 4.6881 + 356.9562795\,k + 0.0103126\,T^2 + 0.00001251\,T^3$
[186.2100]

$F = 325.8867 + 1.4467806\,k - 0.0020708\,T^2 - 0.00000215\,T^3$
[145.1633]

TABLE 50.B

Periodic terms (deg.) for the value of the Moon's maximum declination

Coefficient for			*Coefficient for*		
declination north	*declination south*		*declination north*	*declination south*	
°	°		°	°	
+5.1093	-5.1093	sin F	+0.0038	-0.0038	cos $(2M' - F)$
+0.2658	+0.2658	cos $2F$	-0.0034	+0.0034	cos $(M' - 2F)$
+0.1448	-0.1448	sin $(2D - F)$	-0.0029	-0.0029	sin $2M'$
-0.0322	+0.0322	sin $3F$	+0.0029	+0.0029	sin $(3M' + F)$
+0.0133	+0.0133	cos $(2D - 2F)$	-0.0028	+0.0028	cos $(2D + M - F)$ *
+0.0125	+0.0125	cos $2D$	-0.0028	-0.0028	cos $(M' - F)$
-0.0124	-0.0015	sin $(M' - F)$	-0.0023	+0.0023	cos $3F$
-0.0101	+0.0101	sin $(M' + 2F)$	-0.0021	+0.0021	sin $(2D + F)$
+0.0097	-0.0097	cos F	+0.0019	+0.0019	cos $(M' + 3F)$
-0.0087	+0.0087	sin $(2D + M - F)$ *	+0.0018	+0.0018	cos $(D + F)$
+0.0074	+0.0074	sin $(M' + 3F)$	+0.0017	-0.0017	sin $(2M' - F)$
+0.0067	+0.0067	sin $(D + F)$	+0.0015	+0.0015	cos $(3M' + F)$
+0.0063	-0.0063	sin $(M' - 2F)$	+0.0014	+0.0014	cos $(2D+2M'+F)$
+0.0060	-0.0060	sin $(2D - M - F)$ *	-0.0012	+0.0012	sin $(2D-2M'-F)$
-0.0057	+0.0057	sin $(2D - M' - F)$	-0.0012	-0.0012	cos $2M'$
-0.0056	-0.0056	cos $(M' + F)$	-0.0010	+0.0010	cos M'
+0.0052	-0.0052	cos $(M' + 2F)$	-0.0010	-0.0010	sin $2F$
+0.0041	-0.0041	cos $(2M' + F)$	+0.0006	+0.0037	sin $(M' + F)$
-0.0040	-0.0040	cos $(M' - 3F)$			

The time of greatest northern or southern declination is then

$$\begin{aligned} \text{JDE} = 2451\,562.5897 &+ 27.321\,582\,241\,k + 0.000\,100\,695\,T^2 \\ [2451\,548.9289] & \qquad - 0.000\,000\,141\,T^3 \\ & \qquad + \text{periodic terms of Table 50.A} \end{aligned}$$

In Table 50.A, the terms involving M, the Sun's mean anomaly, should be multiplied by the quantity E given by formula (45.6). These terms are indicated by an asterisk.

The value of the greatest declination, in degrees, is

$$\delta = 23.6961 - 0.013\,004\,T + \text{periodic terms of Table 50.B.}$$

In Table 50.B, again, the terms indicated by an asterisk should be multiplied by E. It should be noted that the *absolute* value of the maximum declination is obtained; in the case of a greatest southern declination, this declination thus is *not* affected by the minus sign.

Example 50.a — Greatest northern declination of the Moon in December 1988.

Inserting the value year = 1988.95 in formula (50.1), we obtain $k \cong -148.12$, so we take $k = -148$. We then find

$T = -0.110\,707$ $\qquad$ $M' = -52\,824°.8411 = 95°.1589$

$D = -49\,142°.2392 = 177°.7608$ $\qquad$ $F = 111°.7631$

$M = -3970°.5085 = 349°.4915$ $\qquad$ $E = 1.000\,278$

We obtain JDE = 2447 518.3347, which corresponds to 1988 December 22.8347 = 1988 December 22, at 20^h02^m TD. The correct value is December 22, at 20^h01^m TD.

For the value of that maximum northern declination, we obtain 28°.1562 = +28°09'22". The correct value is +28°09'13".

Example 50.b — If we calculate the maximum southern declination for $k = +659$, we obtain JDE = 2469 553.0834, which corresponds to 2049 April 21, at 14^h TD, and $\delta = 22°.1384$, so the greatest southern declination is −22°08'.

Example 50.c — To find the Moon's greatest northern declination of mid-March of the year −4, we have 'year' = 0.20 year after the beginning of the year −4; so 'year' = − 4 + 0.20 = −3.80, and *not* −4.20 !

This gives for k the approximate value −26788.40, whence k = −26 788 (an integer!).

We then obtain JDE = 1719 672.1337, which corresponds to March 16 at 15^h TD of the year −4;
greatest northern declination = 28°.9739 = +28°58'.

Using the method described in this Chapter, 600 maximum northern and 600 maximum southern declinations were calculated, namely from 1977 August to 2022 June. The maximum errors are 10 minutes for the time, and 26" for the value of the maximum declination. For 69% of the cases, the calculated time is less than 3 minutes in error, and in 74% of the cases the calculated declination is less than 10" in error.

The coefficients of the periodic terms in Tables 50.A and 50.B have been calculated using for the obliquity of the ecliptic its value for the epoch 2000.0. As a consequence, the error resulting from using these terms will increase with time, but the maximum possible error will not exceed half an hour between the years −1000 and +5000.

Chapter 51

Ephemeris for Physical Observations of the Moon

Optical librations

The mean period of rotation of the Moon is equal to the mean sidereal period of revolution around the Earth, and the mean plane of the lunar equator intersects the ecliptic at a constant inclination, I, in the line of nodes of the lunar orbit, with the descending node of the equator at the ascending node of the orbit.

On the average, therefore, the same hemisphere of the Moon is always turned towards the Earth. However, because of the apparent oscillations known as *optical librations*, which are due to variations in the geometric position of the Earth relative to the lunar surface during the course of the orbital motion of the Moon, about 59 percent of the surface can be observed altogether.

The mean center of the Moon's apparent disk is the origin of the system of selenographic coordinates on the surface of the Moon. Selenographic longitudes are measured from the lunar meridian that passes through the mean center of the apparent disk, positive in the direction towards *Mare Crisium*, that is, towards the west on the geocentric celestial sphere. Selenographic latitudes are measured from the lunar equator, positive towards the north, that is, they are positive in the hemisphere containing *Mare Serenitatis*.

The displacement, at any time, of the mean center of the disk from the apparent center, represents the amount of libration, and is measured by the selenographic coordinates of the apparent center of the disk at that time.

The selenographic longitude and latitude of the Earth, as given in the almanacs, are the geocentric selenographic coordinates of the apparent central point of the disk; at this point on the surface of the Moon, the Earth is in the zenith. When the *libration in longi-*

tude, that is the selenographic longitude of the Earth, is positive, the mean central point of the disk is displaced eastwards on the celestial sphere, exposing to view a region on the west limb. When the *libration in latitude*, or selenographic latitude of the Earth, is positive, the mean central point of the disk is displaced towards the south, and a region on the north limb is exposed to view.

The optical librations in longitude (l') and in latitude (b') can be obtained as follows. Let

I = the inclination of the mean lunar equator to the ecliptic, namely 1°32'32".7 = 1°.54242. This is the value adopted by the International Astronomical Union;

λ = apparent geocentric longitude of the Moon;

β = apparent geocentric latitude of the Moon;

$\Delta\psi$ = nutation in longitude (see Chapter 21);

F = argument of latitude of the Moon, obtained from (45.5);

Ω = mean longitude of the ascending node of the lunar orbit, obtained from formula (45.7).

Then we have

$$\left.\begin{aligned} W &= \lambda - \Delta\psi - \Omega \\ \tan A &= \frac{\sin W \cos\beta \cos I - \sin\beta \sin I}{\cos W \cos\beta} \\ l' &= A - F \\ \sin b' &= -\sin W \cos\beta \sin I - \sin\beta \cos I \end{aligned}\right\} \qquad (51.1)$$

In the calculation of λ, the effect of the nutation is supposed be included, so $\lambda - \Delta\psi$ represents in fact the 'apparent longitude of the Moon without the effect of the nutation'.

Physical librations

There is an actual rotational motion of the Moon about its mean rotation; this is called the physical libration. The physical libration is much smaller than the optical libration, and can never be larger than 0.04 degree in both longitude and latitude.

The physical librations in longitude (l'') and in latitude (b'') can be calculated as follows, and the total librations are the sums of the optical and physical librations :

$$l = l' + l'', \qquad b = b' + b''.$$

Calculate the quantities ρ, σ and τ (in degrees) by means of the following expressions due to D.H. Eckhardt [1], where the angles D, M and M' are obtained by means of expressions (45.2) to (45.4); find E by means of (45.6), and the angles K_1 and K_2 (in degrees) from

$$K_1 = 119.75 + 131.849\,T$$
$$K_2 = 72.56 + 20.186\,T$$

where, as elsewhere in this book, T is the time measured in Julian centuries of 36525 days from the Epoch J2000.0 = JDE 2451 545.0.

$$\begin{aligned}
\rho = & -0.02752 \cos M' \\
& -0.02245 \sin F \\
& +0.00684 \cos (M' - 2F) \\
& -0.00293 \cos 2F \\
& -0.00085 \cos (2F - 2D) \\
& -0.00054 \cos (M' - 2D) \\
& -0.00020 \sin (M' + F) \\
& -0.00020 \cos (M' + 2F) \\
& -0.00020 \cos (M' - F) \\
& +0.00014 \cos (M' + 2F - 2D)
\end{aligned}$$

$$\begin{aligned}
\sigma = & -0.02816 \sin M' \\
& +0.02244 \cos F \\
& -0.00682 \sin (M' - 2F) \\
& -0.00279 \sin 2F \\
& -0.00083 \sin (2F - 2D) \\
& +0.00069 \sin (M' - 2D) \\
& +0.00040 \cos (M' + F) \\
& -0.00025 \sin 2M' \\
& -0.00023 \sin (M' + 2F) \\
& +0.00020 \cos (M' - F) \\
& +0.00019 \sin (M' - F) \\
& +0.00013 \sin (M' + 2F - 2D) \\
& -0.00010 \cos (M' - 3F)
\end{aligned}$$

$$\begin{aligned}
\tau = & +0.02520\, E \sin M \\
& +0.00473 \sin (2M' - 2F) \\
& -0.00467 \sin M' \\
& +0.00396 \sin K_1 \\
& +0.00276 \sin (2M' - 2D) \\
& +0.00196 \sin \Omega \\
& -0.00183 \cos (M' - F) \\
& +0.00115 \sin (M' - 2D) \\
& -0.00096 \sin (M' - D) \\
& +0.00046 \sin (2F - 2D) \\
& -0.00039 \sin (M' - F) \\
& -0.00032 \sin (M' - M - D) \\
& +0.00027 \sin (2M' - M - 2D) \\
& +0.00023 \sin K_2 \\
& -0.00014 \sin 2D \\
& +0.00014 \cos (2M' - 2F) \\
& -0.00012 \sin (M' - 2F) \\
& -0.00012 \sin 2M' \\
& +0.00011 \sin (2M' - 2M - 2D)
\end{aligned}$$

Then we have

$$\begin{aligned}
l'' &= -\tau + (\rho \cos A + \sigma \sin A) \tan b' \\
b'' &= \sigma \cos A - \rho \sin A
\end{aligned} \qquad (51.2)$$

Position Angle of Axis

The position angle of the Moon's axis of rotation, P, is defined as for the planets (see Chapters 41 and 42). It can be calculated as follows; the effect of the physical libration is taken into account.

I, Ω, $\Delta\psi$, ρ, σ and b have the same meaning as before, and let α be the apparent geocentric right ascension of the Moon, and ε the true obliquity of the ecliptic. Then

$$V = \Omega + \Delta\psi + \frac{\sigma}{\sin I}$$

$$X = \sin(I+\rho)\sin V$$

$$Y = \sin(I+\rho)\cos V\cos\varepsilon - \cos(I+\rho)\sin\varepsilon$$

$$\tan\omega = X/Y$$

$$\sin P = \frac{\sqrt{X^2+Y^2}\ \cos(\alpha-\omega)}{\cos b}$$

The angle ω can be obtained in the correct quadrant by using the 'second' arctangent function: ω = ATN2 (X, Y). If this function is not available, divide X by Y, take the usual arctangent of the result, then add 180° if $Y < 0$.

The angle P is to be taken in the first or in the fourth quadrant.

Example 51.a — The Moon on 1992 April 12, at 0^h TD.

For this instant we have (see Example 45.a):

D =	113°.842 309		
M =	97°.643 514	λ =	133°.167 269
M' =	5°.150 839	β =	−3°.229 127
F =	219°.889 726	$\lambda - \Delta\psi$ =	133°.162 659
$\Delta\psi$ =	+0°.004 610	ε =	23°.440 636
E =	1.000 194	α =	134°.688 473

Then we obtain :

Ω =	274°.400 655	l'' =	−0°.025
W =	218°.762 004	b'' =	+0°.006
A =	218°.683 937	l =	−1°.23
l' =	−1°.206	b =	+4°.20
b' =	+4°.194	V =	273°.820 506
K_1 =	109°.57	$I+\rho$ =	1°.532 00
K_2 =	71°.00	X =	−0.026 676
ρ =	−0.01 042	Y =	−0.396 022
σ =	−0.01 574	ω =	183°.8536
τ =	+0.02 673	P =	15°.08

Topocentric librations

For precise reductions of observations, the geocentric values of the librations and position angle of the axis should be reduced to the values at the place of the observer on the surface of the Earth. For the librations, the differences may reach 1°, and have important effects on the limb-contour.

The topocentric librations in longitude and latitude, and the topocentric position angle of the axis, may be calculated either by direct calculation or by differential corrections of the geocentric values.

a. Direct calculation. — The formulae given before are used, but the geocentric coordinates λ, β, α of the Moon are replaced by the topocentric ones. For this purpose, the topocentric right ascension and declination of the Moon are obtained by means of formulae (39.2) and (39.3); then they are transformed to the ecliptical coordinates λ and β by the usual conversion formulae (12.1) and (12.2) to obtain the topocentric longitude and latitude.

b. Differential corrections. — Let ϕ be the observer's latitude, δ the geocentric declination of the Moon, H the local hour angle of the Moon (calculated from the local sidereal time and the *geocentric* right ascension), and π the geocentric horizontal parallax of the Moon. Then calculate

$$\tan Q = \frac{\cos\phi \sin H}{\cos\delta \sin\phi - \sin\delta \cos\phi \cos H}$$

$$\cos z = \sin\delta \sin\phi + \cos\delta \cos\phi \cos H$$

$$\pi' = \pi\,(\sin z + 0.0084 \sin 2z)$$

Then the corrections to the geocentric librations (l, b) and to the position angle (P) are

$$\Delta l = \frac{-\pi' \sin(Q - P)}{\cos b}$$

$$\Delta b = +\pi' \cos(Q - P)$$

$$\Delta P = +\Delta l \sin(b + \Delta b) - \pi' \sin Q \tan\delta$$

These formulae were given in Reference [2].

The selenographic position of the Sun

The selenographic coordinates of the Sun determine the regions of the lunar surface that are illuminated.

The selenographic longitude l_0 and latitude b_0 of the subsolar point on the lunar surface – the point where the Sun is at the zenith – are obtained by replacing, in the formulae (51.1) for the selenographic coordinates of the Earth, the geocentric ecliptical coordinates λ, β of the Moon by the *heliocentric* ecliptical coordinates λ_H, β_H of the Moon. With sufficient accuracy we have

$$\lambda_H = \lambda_0 + 180° + \frac{\Delta}{R} \times 57°.296 \cos\beta \sin(\lambda_0 - \lambda)$$

$$\beta_H = \frac{\Delta}{R}\beta$$

where λ_0 is the apparent geocentric longitude of the Sun. The fraction Δ/R is the ratio of the distance Earth-Moon to the distance Earth-Sun; hence, Δ and R should be expressed in the same units, for instance kilometers. If, instead, R is expressed in astronomical units, and π is the equatorial horizontal parallax of the Moon expressed in *seconds* of arc ("), the fraction Δ/R is equal to

$$\frac{8.794}{\pi R}$$

Hence, to find l_0 and b_0, first calculate λ_H and β_H. Then use expressions (51.1), replacing λ by λ_H, and β by β_H; this will give l'_0 and b'_0. The quantities ρ, σ and τ are found by the unchanged expressions, and finally l''_0 and b''_0 by (51.2), using b'_0 instead of b'. Then

$$l_0 = l'_0 + l''_0 \qquad \text{and} \qquad b_0 = b'_0 + b''_0$$

Subtracting l_0 from 90° or 450° gives the selenographic *colongitude* of the Sun (c_0), which is tabulated in some ephemerides.

The quantities l_0 (or c_0) and b_0 determine the exact position of the terminator on the surface of the Moon. The subsolar point at l_0, b_0 is the pole of the great circle on the lunar surface that bounds the illuminated hemisphere. The morning terminator, where the Sun is rising on the Moon, is at selenographic longitude $l_0 - 90°$, or $360° - c_0$. The evening terminator is at longitude $l_0 + 90°$, or $180° - c_0$. When $c_0 = 0°$, the Sun is rising at selenographic longitude 0°; this occurs near First Quarter. At Full Moon, Last Quarter, and New Moon, respectively, c_0 is approximately 90°, 180°, and 270°, and the morning terminator is approximately at selenographic longitudes 270°, 180°, and 90°.

It should be noted that, while l_0 is *decreasing* with time, the colongitude c_0 is *increasing*. Their mean daily motion is equal to that of the Moon's mean elongation D, namely 12.190 749 degrees.

At a point on the lunar surface at selenographic longitude η and latitude θ, sunrise occurs approximately when $c_0 = 360° - \eta$, noon when $c_0 = 90° - \eta$, and sunset when $c_0 = 180° - \eta$. The exact altitude h of the Sun above the lunar horizon at any time may be calculated from

$$\sin h = \sin b_0 \sin \theta + \cos b_0 \cos \theta \sin (c_0 + \eta)$$

Example 51.b — The Moon on 1992 April 12, at 0^h TD.

For this instant, we have (from accurate calculations using the VSOP87 and ELP-2000/82 theories):

$$\lambda_0 = 22°.33978$$
$$\Delta = 368\,406 \text{ kilometers}$$
$$R = 1.002\,497\,69 \text{ AU} = 149\,971\,500 \text{ km}$$

The other relevant quantities have been found in Example 51.a. We then find

$\lambda_H = 202°.208\,438$	$l''_0 = -0°.026$
$\beta_H = -0°.007\,932$	$b''_0 = -0°.015$
$W = 287°.803\,173$	$l_0 = 67°.89$
$A = 287°.809\,284$	$b_0 = +1°.46$
$l'_0 = 67°.920$	$c_0 = 22°.11$
$b'_0 = +1°.476$	

References

1. D.H. Eckhardt, 'Theory of the Libration of the Moon', *Moon and Planets*, Vol. 25, page 3 (1981).

2. *Explanatory Supplement to the Astronomical Ephemeris* (London, 1961), page 324.

Chapter 52

Eclipses

Without too much calculation, it is possible to obtain with good accuracy the principal characteristics of an eclipse of the Sun or the Moon. For a solar eclipse, the situation is complicated by the fact that the phases of the event are different for different observers at the Earth's surface, while in the case of a lunar eclipse all observers see the same phase at the same instant.

For this reason, we will not consider here the calculation of the local circumstances of a solar eclipse. The interested reader may calculate these circumstances from the Besselian Elements published yearly in the *Astronomical Ephemeris* (renamed *Astronomical Almanac* in 1981). Besselian elements for all solar eclipses for the years -2003 to +2526 can be found in the *Canon* by Mucke and Meeus [1]. For modern times, accurate Besselian elements have been published by Meeus [2]. Besides the elements, these two works give the formulae needed for their use, together with numerical examples.

Espenak published a *Canon* [3] giving data about the paths of total and annular solar eclipses from 1986 to 2035, with beautiful world maps for all eclipses in that period. That work does not contain Besselian elements, however, so it does not provide the possibility to calculate extra data, such as local circumstances for places outside the path of total or annular phase.

Let us also mention the work by Stephenson and Houlden [4], which contains data and charts for the total and annular eclipses visible in East Asia from 1500 B.C. to A.D. 1900.

General data

First, calculate the instant (JDE) of the *mean* New or Full Moon, by means of formulae (47.1) to (47.3). Remember that k must be an integer for a New Moon (solar eclipse), and an integer increased by 0.5 for a Full Moon (lunar eclipse).

Then, calculate the values of the angles M, M', F and Ω for that instant, by means of expressions (47.4) to (47.7), and the value of E by formula (45.6).

The value of F will give a first information about the occurrence of a solar or lunar eclipse. If F differs from the nearest multiple of 180° by less than 13°.9, then there is certainly an eclipse; if the difference is larger than 21°.0, there is no eclipse; between these two values, the eclipse is uncertain at this stage and the case must be examined further. Use can be made of the following rule: there is no eclipse if $|\sin F| > 0.36$.

Note that, after one lunation, F increases by 30°.6705.

If F is near 0° or 360°, the eclipse occurs near the Moon's ascending node. If F is near 180°, the eclipse takes places near the descending node of the Moon's orbit.

Calculate

$$F_1 = F - 0°.02665 \sin \Omega$$

$$A_1 = 299°.77 + 0°.107\,408\, k - 0.009\,173\, T^2$$

Then, to obtain the *time of maximum eclipse* (for the Earth generally in the case of a solar eclipse), the following corrections (in days) should be added to the time of mean conjunction given by expression (47.1).

$$\begin{array}{lll}
-0.4075 & \times \sin & M' \\
+0.1721 \times E & & M \\
+0.0161 & & 2M' \\
-0.0097 & & 2F_1 \\
+0.0073 \times E & & M' - M \\
-0.0050 \times E & & M' + M \\
-0.0023 & & M' - 2F_1 \\
+0.0021 \times E & & 2M \\
+0.0012 & & M' + 2F_1 \\
+0.0006 \times E & & 2M' + M \\
-0.0004 & & 3M' \\
-0.0003 \times E & & M + 2F_1 \\
+0.0003 & & A_1 \\
-0.0002 \times E & & M - 2F_1 \\
-0.0002 \times E & & 2M' - M \\
-0.0002 & & \Omega
\end{array} \qquad (52.1)$$

(For the first two terms: for *lunar* eclipses, change the constants to −0.4065 and +0.1727)

This algorithm should not be used, of course, if high accuracy is needed. For the 221 solar eclipses of the years A.D. 1951 to 2050, the method gives a mean error of 0.36 minute, and a greatest error of 1.1 minute in the times of maximum eclipse.

Then calculate

$$P = +0.2070 \times E \times \sin M + 0.0024 \times E \sin 2M - 0.0392 \sin M' + 0.0116 \sin 2M' - 0.0073 \times E \sin (M' + M) + 0.0067 \times E \sin (M' - M) + 0.0118 \sin 2F_1$$

$$Q = +5.2207 - 0.0048 \times E \times \cos M + 0.0020 \times E \cos 2M - 0.3299 \cos M' - 0.0060 \times E \cos (M' + M) + 0.0041 \times E \cos (M' - M)$$

$$W = |\cos F_1|$$

$$\gamma = (P \cos F_1 + Q \sin F_1) \times (1 - 0.0048\, W)$$

$$u = 0.0059 + 0.0046\, E \cos M - 0.0182 \cos M' + 0.0004 \cos 2M' - 0.0005 \cos (M + M')$$

Solar eclipses

In the case of a solar eclipse, γ represents the least distance from the axis of the Moon's shadow to the center of the Earth, in units of the equatorial radius of the Earth. The quantity γ is positive or negative, depending upon the axis of the shadow passing north or south of the Earth's center. When γ is between +0.9972 and -0.9972, the solar eclipse is central: there exists a line of central eclipse on the Earth's surface.

The quantity u denotes the radius of the Moon's *umbral* cone in the fundamental plane, again in units of the Earth's equatorial radius. (The fundamental plane is the plane through the center of the Earth and perpendicular to the axis of the Moon's shadow). The radius of the *penumbral* cone in the fundamental plane is

$$u + 0.5461$$

If $|\gamma|$ is between 0.9972 and $1.5433 + u$, the eclipse is not central. In most cases, it is then a partial eclipse. However, when $|\gamma|$ is between 0.9972 and 1.0260, a part of the umbral cone may touch the surface of the Earth (within the polar regions), while the axis of the cone does *not* touch the Earth. These *non-central* total or annular eclipses occur when $0.9972 < |\gamma| < 0.9972 + |u|$. Between the years 1950 and 2100, there are seven eclipses of this type :

1950 March 18	annular, not central
1957 April 30	annular, not central
1957 October 23	total, not central
1967 November 2	total, not central
2014 April 29	annular, not central
2043 April 9	Total, not central
2043 October 3	annular, not central

If $|\gamma| > 1.5433 + u$, no eclipse is visible from the Earth's surface.

In the case of a *central* eclipse, the type of the eclipse can be determined by the following rules :

if $u < 0$, the eclipse is total;

if $u > +0.0047$, the eclipse is annular;

if u is between 0 and +0.0047, the eclipse is either annular or annular-total.

In this latter case, the ambiguity is removed as follows. Calculate

$$\omega = 0.00464 \sqrt{1 - \gamma^2} > 0$$

Then, if $u < \omega$, the eclipse is annular-total; otherwise it is an annular one.

In the case of a *partial* solar eclipse, the greatest magnitude is attained at the point of the surface of the Earth which comes closest to the axis of shadow. The magnitude of the eclipse at that point is

$$\frac{1.5433 + u - |\gamma|}{0.5461 + 2u} \tag{52.2}$$

Lunar eclipses

In the case of a lunar eclipse, γ represents the least distance from the center of the Moon to the axis of the Earth's shadow, in units of the Earth's equatorial radius. The quantity γ is positive or negative depending upon the Moon's center passing north or south of the axis of shadow. The radii at the distance of the Moon, again in equatorial Earth radii, are :

for the penumbra : $\rho = 1.2848 + u$

for the umbra : $\sigma = 0.7403 - u$

while the magnitude of the eclipse may be found as follows :

for penumbral eclipses : $$\frac{1.5573 + u - |\gamma|}{0.5450} \tag{52.3}$$

for umbral eclipses : $$\frac{1.0128 - u - |\gamma|}{0.5450} \tag{52.4}$$

If the magnitude is negative, this indicates that there is no eclipse.

The *semidurations* of the partial and total phases in the *umbra* can be found as follows. Calculate

$$P = 1.0128 - u$$
$$T = 0.4678 - u$$
$$n = 0.5458 + 0.0400 \cos M'$$

Then the semidurations in *minutes* are :

partial phase : $\frac{60}{n}\sqrt{P^2 - \gamma^2}$ total phase : $\frac{60}{n}\sqrt{T^2 - \gamma^2}$

For the semiduration of the partial phase in the *penumbra*, find $H = 1.5573 + u$, and then the semiduration in minutes is

$$\frac{60}{n}\sqrt{H^2 - \gamma^2}$$

It must be noted that the contacts of the Moon with the penumbra cannot be observed, and that most penumbral eclipses (in which the Moon enters only the penumbra of the Earth) cannot be discerned visually. Only at eclipses occurring deep in the penumbra can a weak shading on the Moon's northern or southern limb be seen.

In the formulae given above, the increase of the theoretical radii of the shadow cones by the Earth's atmosphere is taken into account. However, instead of the traditional rule consisting of increasing by 1/50 the theoretical radii, we have preferred the method used since 1951 by the *Connaissance des Temps* — see for instance Reference [5]. As compared with the results of the 'French rule', the magnitude of a lunar eclipse calculated by using the traditional rule is too large by about 0.005 for umbral eclipses, by about 0.026 for penumbral eclipses.

To obtain the results according to the traditional rule (1/50), the following changes should be made to the constants in the expressions given above :

replace 1.2848 by 1.2985
0.7403 by 0.7432
1.5573 by 1.5710
1.0128 by 1.0157
0.4678 by 0.4707

For the predictions of lunar eclipses, such as those published in the various almanacs, it is customary to assume the penumbra and the umbra to be exactly circular, and to use a mean radius for the Earth. In fact, the shadow will differ somewhat from a circular cone as the Earth is not a true sphere. By simple geometrical considerations, it is found that the Earth's shadow, at the Moon's distance, must be *more* flattened than the terrestrial globe, the mean value for the flattening of the umbra being 1/214 [6]. The true flattening of the umbra is perhaps even still larger. Soulsby [7] finds a mean oblateness of 1/102 from observations made at 18 lunar eclipses in the period 1974 - 1989.

Example 52.a — Solar eclipse of 1993 May 21.

Since May 21 is the 141th day of the year, the given date corresponds to 1993.38. Formula (47.2) then gives

$$k \cong -81.88, \quad \text{whence} \quad k = -82.$$

Then, by means of formulae (47.3) and (47.1),

$$\text{JDE} = 2449\,128.5894$$

We find further

$$M = 135°.9142$$
$$M' = 244°.5757$$
$$F = 165°.7296$$
$$\Omega = 253°.0026$$
$$F_1 = 165°.7551$$

Because $180° - F$ is between 13°.9 and 21°.0, the eclipse is uncertain. We further find :

$$P = +0.1842$$
$$Q = +5.3589$$
$$\gamma = +1.1348$$
$$u = +0.0097$$

Because $|\gamma|$ is between 0.9972 and $1.5433 + u$, the eclipse is a partial one. Using formula (52.2), we find that the maximum magnitude is

$$\frac{1.5433 + 0.0097 - 1.1348}{0.5461 + 0.0194} = 0.740$$

Because F is near 180°, the eclipse occurs near the Moon's descending node. Because γ is positive, the eclipse is visible in the northern hemisphere of the Earth.

To obtain the time of maximum eclipse, we add to JDE the terms given by formula (52.1). This gives

$$\text{JDE} = 2449\,128.5894 + 0.5085 = 2449\,129.0979$$

which corresponds to 1993 May 21, at $14^h21^m.0$ TD.

The correct values, resulting from an accurate calculation [2], are $14^h20^m14^s$ TD, $\gamma = +1.1370$, and a maximum magnitude of 0.735.

Notes about the accuracy

The algorithms given in this Chapter are not intended to obtain highly accurate results. Still, for lunar eclipses the results will generally be precise enough for historical research, or when high accuracy is not needed. On the other hand, as has been said at the beginning of this Chapter, accurate data for modern solar eclipses can be obtained by using our *Elements of Solar Eclipses* [2].

The formula given for γ does not yield rigorously exact results. This is quite evident, if we consider the fact that only twelve periodic terms are used to calculate the quantities P and Q, while hundreds of terms are needed to obtain accurate positions of the Sun and the Moon. Even formulae (52.2), (52.3) and (52.4), and the expressions for the quantities P, T, n and H, are not rigorously exact.

For the 221 solar eclipses of the period 1951 - 2050, the mean error of the values of γ as calculated by using the algorithm of this Chapter is 0.00065, while the maximum error is 0.0024, which corresponds to 15 kilometers. Considering the simplicity of our formulae, this accuracy is quite satisfactory.

From what precedes, it results that *in limiting cases* the type of an eclipse will still be unknown. In such a case, an accurate calculation is needed to settle the question.

Further, in a *search procedure* for eclipses, a small safety margin should be considered in order to be sure that no eclipse will be overlooked. For instance, while the correct condition for a central eclipse is indeed $|\gamma| < 0.9972$ (*), a limiting value of 1.000 or even 1.005 should be used in order to find *all* possible central eclipses when use is made of the value of γ obtained with the method described in this Chapter.

Here are some examples.

For the solar eclipse of 1935 January 5 ($k = -804$), our method gives $\gamma = -1.5395$ and $u = -0.00464$, whence $|\gamma| > u + 1.5433 = 1.5387$, so we might think there was no eclipse on that date. Formula (52.2) yields the value -0.002 (*negative !*) for the maximum magnitude. The correct value of γ was -1.5383, however, so there was a very small partial eclipse on 1935 January 5, with a maximum magnitude of 0.001.

For the annular solar eclipse of 1957 April 30 ($k = -528$), our algorithm yields the value $\gamma = +0.9966$, so one might think this was a central eclipse. The exact value was $\gamma = +0.9990$, so it was actually a non-central annular eclipse.

For the lunar eclipse of 1890 November 26 ($k = -1349.5$), our algorithm gives a magnitude (in the umbra) of -0.007. In fact, it was a very small partial eclipse in the umbra.

(*) In fact, the 'constant' 0.9972 may vary between 0.9970 and 0.9974 from one eclipse to another.

Exercices

Find the first solar eclipse of the year 1979, and show that it was a total one visible from the northern hemisphere.

Was the solar eclipse of April 1977 a total or an annular one?

Show that there was no eclipse of the Sun in July 1947.

Show that there are four solar eclipses in the year 2000, and that all four are partial eclipses.

Show that there will be no lunar eclipse in January 2008.

Show that there were three total eclipses of the Moon in 1982.

Find the first lunar eclipse of the year 1234. (Answer: the partial lunar eclipse of 1234 March 17).

References

1. H. Mucke, J. Meeus, *Canon of Solar Eclipses, -2003 to +2526*; Astronomisches Büro (Wien, 1983).

2. J. Meeus, *Elements of Solar Eclipses, 1951 to 2200* (Willmann-Bell, ed.; 1989).

3. F. Espenak, *Fifty Year Canon of Solar Eclipses: 1986-2035*; NASA Reference Publication 1178 (Washington, 1987).

4. F.R. Stephenson, M.A. Houlden, *Atlas of Historical Eclipse Maps*; Cambridge University Press (1986).

5. A. Danjon, 'Les éclipses de Lune par la pénombre en 1951', *l'Astronomie*, Vol. 65, pages 51-53 (February 1951).

6. J. Meeus, 'Die Abplattung des Erdschattens bei Mondfinsternissen', *Die Sterne*, Vol. 45, pages 116-117 (1969).

7. B.W. Soulsby, *Journal of the British Astron. Assoc.*, Vol. 100, page 297 (December 1990).

Chapter 53

Semidiameters of the Sun, Moon and Planets

Sun and Planets

The semidiameters s of the Sun and planets are computed from

$$s = \frac{s_o}{\Delta}$$

where s_o is the body's semidiameter at unit distance (1 AU),
Δ is the body's distance to the Earth, in AU.

For the Sun, the value adopted in the calculations is [1]

$$s_o = 15'59''.63 = 959''.63.$$

For the planets, the following values of s_o have been used for many years [2] :

Mercury	3″.34	Saturn :		
Venus	8.41	equatorial	83″.33	
		polar	74.57	
Mars	4.68	Uranus	34.28	(A)
Jupiter :		Neptune	36.56	
equatorial	98.47			
polar	91.91			

Later, the following values have been adopted [3] :

Mercury	3″.36	Saturn :			
Venus	8.34	equatorial	82″.73		
		polar	73.82		
Mars	4.68				(B)
		Uranus	35.02		
Jupiter :					
equatorial	98.44	Neptune	33.50		
polar	92.06	Pluto	2.07		

Note that, according to the latter values, Neptune is smaller than Uranus.

For Venus, the value 8″.34 refers to the planet's crust, not to the top cloud level as seen from the Earth. For this reason, we prefer to use the older value 8″.41 for Venus when calculating astronomical phenomena such as transits and occultations.

In the case of Saturn, let a and b be the equatorial and the polar semidiameters at unit distance. Then, while the apparent equatorial semidiameter s_e is given by $s_e = a/\Delta$, the apparent *polar* semidiameter should be calculated from

$$s_p = s_e \sqrt{1 - k \cos^2 B}$$

where $k = 1 - \left(\frac{b}{a}\right)^2$, and B is the Saturnicentric latitude of the Earth (see Chapter 44).

If the older values (A) are chosen, namely $a = 83''.33$ and $b = 74''.57$, then $k = 0.199\,197$. If one adopts the values from (B), then $k = 0.203\,800$.

Strictly speaking, this procedure should also be used in the case of Jupiter. But for this planet the angle B (called D_E in Chapter 42) can never exceed 4°, so it will generally be sufficient to put $s_p = b/\Delta$ here.

Moon

Let Δ be the distance between the centers of Earth and Moon in kilometers, π the equatorial horizontal parallax of the Moon, s its geocentric semidiameter, and k the ratio of the Moon's mean radius to the equatorial radius of the Earth. In the *Astronomical Ephemeris* for the years 1963 to 1968, the value $k = 0.272\,481$ was used in eclipse calculations, and we have used this value ever since.

Then we have rigorously

$$\sin \pi = \frac{6378.14}{\Delta} \qquad \text{and} \qquad \sin s = k \sin \pi$$

but in most cases it will be sufficient to use the formula

$$s \text{ (in arcseconds)} = \frac{358\,473\,400}{\Delta}$$

which gives an error less than 0.0005 arcsecond as compared with the results obtained by the expressions given before.

Computed in this way, the Moon's semidiameter is geocentric, that is, it applies to a fictitious observer located at the center of the Earth. The observed, topocentric semidiameter s' will be slightly larger than the geocentric semidiameter; it is given by

$$\sin s' = \frac{\sin s}{q} = \frac{k}{q} \sin \pi$$

while the topocentric distance of the Moon (that is, the distance from the observer to the center of the Moon) is $\Delta' = q\,\Delta$, q being given by formula (39.7).

Alternatively, the topocentric semidiameter s' of the Moon can be obtained, with an accuracy which is sufficient for many purposes, by multiplying the geocentric value s by

$$1 + \sin h \sin \pi$$

where h is the altitude of the Moon above the observer's horizon.

The increase in the Moon's semidiameter, due to the fact that the observer is not geocentric, is zero when the Moon is on the horizon, and a maximum (between 14" and 18") when the Moon is at the zenith.

References

1. A. Auwers, *Astronomische Nachrichten*, Vol. 128, No. 3068, column 367 (1891).

2. See, for instance, the *Astronomical Ephemeris* for 1980, page 550.

3. *Astronomical Almanac* for 1984, page E43.

Chapter 54

Stellar Magnitudes

Adding stellar magnitudes

If two stars have magnitudes m_1 and m_2, respectively, their combined magnitude m can be calculated as follows :

$$x = 0.4\,(m_2 - m_1)$$

$$m = m_2 - 2.5\,\log\,(10^x + 1)$$

where the logarithm is to the base 10.

Example 54.a — The magnitudes of the components of Castor (α Gem) are 1.96 and 2.89. Calculate the combined magnitude.

One finds

$$x = 0.4\,(2.89 - 1.96) = 0.372$$

$$m = 2.89 - 2.5\,\log\,(10^{0.372} + 1) = 1.58$$

If more than two stars are involved, with magnitudes m_1, m_2, ... m_i, ..., the combined magnitude m can better be found from

$$m = -2.5\,\log \sum 10^{-0.4\,m_i}$$

where, again, the logarithm is to the base 10. The symbol Σ indicates that the sum must be made of all quantities

$$10^{-0.4\,m_i}$$

Example 54.b — The triple star β Mon has components of magnitudes 4.73, 5.22 and 5.60, respectively. Calculate the combined magnitude.

$$m = -2.5 \log \left(10^{(-0.4)(4.73)} + 10^{(-0.4)(5.22)} + 10^{(-0.4)(5.60)}\right)$$

$$= -2.5 \log (0.01282 + 0.00817 + 0.00575) = 3.93$$

Example 54.c — A star cluster consists of

4 stars	of (mean)	magnitude	5.0
14	–	–	6.0
23	–	–	7.0
38	–	–	8.0

Calculate the combined magnitude.

$$4 \times 10^{(-0.4)(5)} = 0.04000$$

$$14 \times 10^{(-0.4)(6)} = 0.05574$$

$$23 \times 10^{(-0.4)(7)} = 0.03645$$

$$38 \times 10^{(-0.4)(8)} = 0.02398$$

Sum $\Sigma = 0.15617$

Combined magnitude $= -2.5 \log 0.15617 = +2.02$

Brightness ratio

If two stars have magnitudes m_1 and m_2, respectively, the ratio I_1/I_2 of their apparent luminosities can be found from

$$x = 0.4\,(m_2 - m_1) \qquad \frac{I_1}{I_2} = 10^x$$

If the brightness ratio I_1/I_2 is given, the corresponding magnitude difference $\Delta m = m_2 - m_1$ can be calculated from

$$\Delta m = 2.5 \log \frac{I_1}{I_2}$$

Example 54.d — How many times is Vega (magnitude 0.14) brighter than Polaris (magnitude 2.12) ?

$$x = 0.4\,(2.12 - 0.14) = 0.792$$

$$10^x = 6.19$$

Hence, Vega is 6.19 times as bright as the Pole Star.

Example 54.e — A star is 500 times as bright as another one.

The corresponding magnitude difference is

$$\Delta m = 2.5 \log 500 = 6.75$$

Distance and Absolute Magnitude

If π is a star's parallax expressed in seconds of a degree ("), this star's distance to us is equal to

$$\frac{1}{\pi} \text{ parsecs} \qquad \text{or} \qquad \frac{3.2616}{\pi} \text{ light-years}$$

If π is a star's parallax expressed in seconds of a degree ("), and m is the apparent magnitude of this star, its absolute magnitude M can be calculated from

$$M = m + 5 + 5 \log \pi$$

where, again, the logarithm is to the base 10.

If d is the star's distance in parsecs, we have

$$M = m + 5 - 5 \log d$$

Unlike the parallaxes within the solar system (see Chapter 39), the parallax considered here is, of course, the stellar, annual parallax resulting from the orbital motion of the Earth around the Sun; so it is *not* the parallax related to the dimensions of the Earth's *globe* !

The *parsec* is the unit of length equal to the distance at which the radius of the Earth's orbit (1 AU) subtends an angle of 1" (parallax = 1"). The name is a contraction of *par*allax and *sec*ond.

$$\begin{aligned} 1 \text{ parsec} &= 3.2616 \text{ light-years} \\ &= 206\,265 \text{ astronomical units} \\ &= 30.8568 \times 10^{12} \text{ kilometers} \end{aligned}$$

The *absolute magnitude* of a star is the apparent magnitude of this star if it were located at a distance of 10 parsecs.

Chapter 55

Binary Stars

The orbital elements of a binary star are the following ones :

P = the period of revolution expressed in mean solar years;

T = the time of perihelion passage, generally given as a year and decimals (for instance, 1945.62);

e = the eccentricity of the true orbit ;

a = the semimajor axis expressed in seconds of a degree ($''$);

i = the inclination of the plane of the true orbit to the plane at right angles to the line of sight. For direct motion in the apparent orbit, i ranges from 0° to 90° ; for retrograde motion, i is between 90 and 180 degrees. When i is 90°, the apparent orbit is a straight line passing through the primary star ;

Ω = the position angle of the ascending node ;

ω = the longitude of periastron; this is the angle in the plane of the true orbit measured from the ascending node to the periastron, taken always in the direction of motion.

When these orbital elements are known, the apparent position angle θ and the angular distance ρ can be calculated for any given time t, as follows.

$$n = \frac{360°}{P} \qquad\qquad M = n\,(t - T)$$

where t is expressed as a year and decimals (just as T); n is the mean annual motion of the companion, expressed in degrees and decimals, and is always positive. M is the companion's mean anomaly for the given time t.

Then solve Kepler's equation

$$E = M + e \sin E$$

by one of the methods described in Chapter 29, and then calculate the radius vector r and the true anomaly v from

$$r = a\,(1 - e \cos E)$$

$$\tan \frac{v}{2} = \sqrt{\frac{1+e}{1-e}}\,\tan \frac{E}{2}$$

Then find $(\theta - \Omega)$ from

$$\tan(\theta - \Omega) = \frac{\sin(v+\omega)\,\cos i}{\cos(v+\omega)} \tag{55.1}$$

Of course, this formula can be written

$$\tan(\theta - \Omega) = \tan(v+\omega)\,\cos i$$

but in this case the correct quadrant for $(\theta - \Omega)$ is not determined. As in previous cases mentioned in this book, one may apply the ATN2 function, if available in the programming language, to the numerator and the denominator of the fraction in (55.1). This will place the angle $(\theta - \Omega)$ at once in the correct quadrant.

When $(\theta - \Omega)$ is found, add Ω to obtain θ. If necessary, reduce the result to the interval $0° - 360°$.

Remember that, by definition, position angle 0° means northward on the sky, 90° east, 180° south, and 270° west. Consequently, if $0° < \theta < 180°$, the companion is 'following' the primary star in the diurnal motion of the celestial sphere; if θ is between 180° and 360°, the companion is 'preceding' the primary star.

The angular separation ρ is found from

$$\rho = \frac{r \cos(v+\omega)}{\cos(\theta - \Omega)}$$

Example 55.a — According to E. Silbernagel (1929), the orbital elements for η Coronae Borealis are :

P = 41.623 years	i = 59°.025
T = 1934.008	Ω = 23°.717
e = 0.2763	ω = 219°.907
a = 0".907	

Calculate θ and ρ for the epoch 1980.0.

We find successively :

$$n = 8.64906$$
$$t - T = 1980.0 - 1934.008 = 45.992$$
$$M = 397°.788 = 37°.788$$
$$E = 49°.897$$
$$r = 0''.74557$$
$$v = 63°.416$$

$$\tan(\theta - \Omega) = \frac{-0.500\,813}{+0.230\,440}$$

$$\theta - \Omega = -65°.291$$
$$\theta = -41°.574 = 318°.4$$
$$\rho = 0''.411$$

As an exercise, calculate an ephemeris for γ Virginis, using the following elements [1] :

P = 168.68 years	i = 148°.0
T = 2005.13	Ω = 36°.9 (2000.0)
e = 0.885	ω = 256°.5
a = 3''.697	

Answer. — Here is an ephemeris with an interval of four years, starting at 1980. The position angle θ decreases with time, since i is between 90 and 180 degrees.

year = 1980.0	θ = 296°.65	ρ = 3''.78
1984.0	293.10	3.43
1988.0	288.70	3.04
1992.0	282.89	2.60
1996.0	274.41	2.08
2000.0	259.34	1.45
2004.0	208.67	0.59
2008.0	35.54	1.04
2012.0	12.72	1.87

Least separation (0''.36) occurs at epoch 2005.21.

The position angles θ refer to the mean equator of 2000.0, that is, for the same epoch as the angle Ω.

Eccentricity of the apparent orbit

The apparent orbit of a binary star is an ellipse whose eccentricity e' is generally different from the eccentricity e of the true orbit. It may be interesting to know e', although this apparent eccentricity has no astrophysical significance.

The following formulae have been derived by the author [2] :

$$A = (1 - e^2 \cos^2 \omega) \cos^2 i$$
$$B = e^2 \sin \omega \cos \omega \cos i$$
$$C = 1 - e^2 \sin^2 \omega$$
$$D = (A - C)^2 + 4B^2$$

$$e'^2 = \frac{2\sqrt{D}}{A + C + \sqrt{D}}$$

It should be noted that e' is independent of the orbital elements a and Ω, and that it can be smaller as well as larger than the true eccentricity e.

Example 55.b — Find the eccentricity of the apparent orbit of η Coronae Borealis. The orbital elements are given in Example 55.a.

We find

$A = 0.25298$
$B = 0.01934$
$C = 0.96858$
$D = 0.51358$

$e' = 0.860$

Hence, for this binary the apparent orbit is much more elongated than the true orbit.

References

1. W.D. Heintz, 'Orbits of 15 visual binaries', *Astronomy and Astrophysics, Supplement Series*, Vol. 82, pages 65–69 (1990).

2. J. Meeus, 'The eccentricity of the apparent orbit of a binary star', *Journal of the British Astron. Assoc.*, Vol. 89, pages 485–488 (August 1979).

Chapter 56

Calculation of a Planar Sundial

by R. Sagot and D. Savoie (*)

One wishes to draw a planar sundial of any given orientation and inclination, provided with a straight stylus of length a perpendicular to its surface. Hence, this stylus generally is *not* directed towards the celestial pole. This sundial has the following principal parameters :

– the latitude ϕ of the place;

– the gnomonic declination D, that is, the azimuth of the perpendicular to the sundial's plane, measured from the southern meridian towards the west, from 0 to 360 degrees. So, if $D = 0°$, the sundial is 'due south'; if $D = 270°$, it is 'due east'; etc.;

– the zenithal distance z of the direction defined by the straight stylus. If $z = 0°$, the sundial is horizontal; in this case, D is meaningless – but see the special case later in this Chapter.
If $z = 90°$, the sundial is vertical.

The coordinates x and y of the tip of the shadow of the straight stylus of length a are measured in an orthogonal coordinate system situated in the sundial's plane. The origin of this system coincides with the footprint of the stylus; the x-axis is horizontal, while the y-axis coincides with the line of greatest slope of the sundial. In all cases, x is measured positively towards the right, while y is positive upwards.

The Sun's hour angle H is measured from the upper meridian transit (true noon); it increases by 15 degrees per hour. For example,

---

(*) Robert SAGOT and Denis SAVOIE are former president and president, respectively, of the 'Commission des Cadrans Solaires' (Sundials Section) of the Société Astronomique de France.

$H = -45°$ corresponds to 9 hours a.m. (true solar time), $H = +15°$ to 1 hour p.m., etc.

In the following formulae, for each hour angle H the declination δ of the Sun will take the successive values (in degrees) -23.44, -20.15, -11.47, 0, +11.47, +20.15, and +23.44, which correspond to the dates when the longitude of the Sun is a multiple of 30°.

In the course of a day, the tip of the shadow of the stylus will describe on the sundial's plane a curve which is a conic (a circle, an ellipse, a parabola, or an hyperbola). However, if $\delta = 0°$ the curve is always a straight line.

Calculate

$$P = \sin\phi\ \cos z - \cos\phi\ \sin z \cos D$$

$$Q = \sin D \sin z \sin H + (\cos\phi\ \cos z + \sin\phi\ \sin z \cos D) \cos H + P \tan\delta$$

$$N_x = \cos D \sin H - \sin D\,(\sin\phi\ \cos H - \cos\phi\ \tan\delta)$$

$$N_y = \cos z \sin D \sin H - (\cos\phi\ \sin z - \sin\phi\ \cos z \cos D) \cos H - (\sin\phi\ \sin z + \cos\phi\ \cos z \cos D) \tan\delta$$

Then the coordinates x and y are given by

$$x = a\frac{N_x}{Q} \qquad\qquad y = a\frac{N_y}{Q}$$

For each hour angle, one obtains a series of points; by connecting these points, an hour line is created on the sundial. The point (if it exists) to which the hour lines converge, is called the *center* of the sundial; it is also the point of fixation of the *polar stylus*, which is parallel to the Earth's axis of rotation. Its coordinates $x_\circ$ and $y_\circ$ are given by

$$x_\circ = \frac{a}{P}\cos\phi\ \sin D, \qquad y_\circ = -\frac{a}{P}(\sin\phi\ \sin z + \cos\phi\ \cos z \cos D)$$

The length u of the polar stylus, from its point of fixation to the tip of the perpendicular stylus of length a, is

$$u = \frac{a}{|P|}$$

while the angle ψ which the polar stylus makes with the sundial's plane is given by

$$\sin\psi = |P|$$

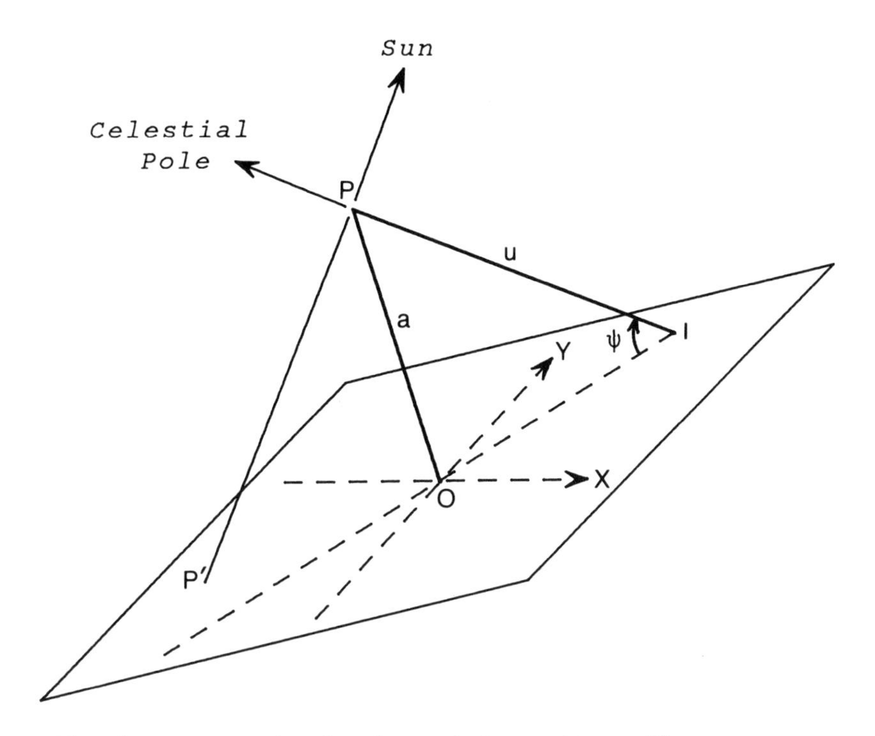

The plane represents the plane of the sundial. OP is the perpendicular stylus, of length a, while IP is the polar stylus, length u. P' is the shadow (x, y) of the tip of the stylus. The point I is called the center of the sundial, while O is the origin of the x-y system.

The formulae for the position of the polar stylus become meaningless when $P = 0$, that is when $\cos D \tan z = \tan \phi$. This means that the polar stylus is then parallel to the plane of the sundial.

It is proper to limit the drawing of the sundial to the useful lines. For example, a vertical sundial oriented 'due north' (D = 180°), at latitude +40°, can never show 11^h a.m. (true solar time). At the same latitude, a vertical sundial oriented 'due south' cannot indicate 19^h (= 7^h p.m.) near the June solstice.

In order to make sure that the sundial really works, two conditions should be fullfilled: the Sun must be above the horizon, and the plane of the sundial must be illuminated. Consequently, it is necessary, for each calculated point (x, y), to verify whether these two conditions are satisfied simultaneously.

In practice, for a given arc of declination, the calculation should start at the moment of the geometric rise of the Sun, or at the first integer hour following that rise, and stop at the moment of the geometric sunset. The Sun's hour angle H_0 at the time of sunrise or sunset is given by

$$\cos H_0 = -\tan\phi \tan\delta$$

with $H_0 < 0$ for sunrise, $H_0 > 0$ for sunset.

For each value of H, one should look at the sign of Q: if this quantity is negative, this means that the Sun does not illuminate the plane, and in that case one passes over to the next declination. Hence, only those values for which Q is positive must be retained.

It is possible that, on a given date, Q is at first positive, then becomes negative, and later is positive again.

Example 56.a — Consider an inclined sundial at latitude 40° N, with $D = 70°$, $z = 50°$, and $a = 1$. For $\delta = +23°.44$ (summer solstice), we have $H_0 = -111°.33$ (or 4^h35^m a.m., true solar time).

Beginning the calculations with $H = -105°$, we find $Q < 0$. This quantity is negative again for $H = -90°$, $-75°$, and $-60°$. Only from $H = -47°$ on, is the sundial illuminated, and it will remain illuminated till sunset. Hence, if a step of 15 degrees has been chosen, the values of x and y should be calculated for $H = -45°$ to $+105°$.

For $H = +30°$ and $\delta = +23°.44$, we find $x = -0.0390$, $y = -0.3615$.

For $H = -15°$ and $\delta = -11°.47$, we find $x = -2.0007$, $y = -1.1069$.

The coordinates of the center are $x_0 = +3.3880$, $y_0 = -3.1102$, and we have $\psi = 12°.2672$.

Example 56.b — Consider a vertical sundial at latitude $\phi = -35°$, with $D = 160°$, $z = 90°$, and $a = 1$.

For $\delta = 0°$ (equinox), we have $H_0 = -90°$ and $Q < 0$. Q becomes positive for $H = -57°$, so the calculations will be made for $H = -45°$ till sunset ($H_0 = +90°$).

For $H = +45°$ and $\delta = 0°$, we find $x = -0.8439$, $y = -0.9298$.

For $H = 0°$ and $\delta = +20°.15$, we find $x = +0.3640$, $y = -0.7410$.

The coordinates of the center are $x_0 = +0.3640$, $y_0 = +0.7451$, and we have $\psi = 50°.3315$.

Example 56.c — Inclined sundial at latitude 40° N, with $D = 160°$ and $z = 75°$.

For $\delta = +23°.44$, this sundial will be illuminated from sunrise (when $H = -111°$) until $H = -84°$. Then it will be illuminated again from $H = +2°$ until sunset ($H = +111°$). So, if a step of 15° has been chosen, the calculation will be made for $H = -105°$, $-90°$, and then for $+15°$ to $+105°$.

The formulae given above form the most general case which can occur in gnomonics. They allow the calculation of the classical hour lines of true solar time, but also the declination curves, the lines for mean time (when introducing the equation of time in the calculation of H), the lines for Universal Time or for zone time, azimuth and altitude lines, etc.

The formulae simplify greatly for some special cases, which we shall now examine briefly.

Special cases

(1) Equatorial sundial

The plane of this sundial is parallel to the plane of the equator and hence there are two sides: the northern side serves for the positive declinations (spring and summer), the southern side for the negative declinations of the Sun (autumn and winter). At a place of latitude ϕ, we have

for the northern side : $z = 90° - \phi$ and $D = 180°$

for the southern side : $z = 90° + \phi$ and $D = 0°$

The line of 12 hours ($H = 0°$) coincides with the line of greatest descending slope. Further,

$$Q = \pm \tan\delta \qquad x_0 = 0$$

$$x = -a\,\frac{\sin H}{\tan\delta} \qquad y_0 = 0$$

$$u = a$$

$$y = \mp a\,\frac{\cos H}{\tan\delta} \qquad \psi = 90°$$

where the upper sign is to be taken for the northern side, the lower sign for the southern side.

(2) Horizontal sundial

The sundial's plane is horizontal, so $z = 0°$. The angle D is not defined and the direction of the x-axis can be chosen at will. We shall consider the case $D = 0°$, where the x-axis is directed towards the east, the y-axis towards the north. The formulae simplify to

$$Q = \cos\phi\ \cos H + \sin\phi\ \tan\delta \qquad x_0 = 0$$

$$x = a\ \frac{\sin H}{Q} \qquad y_0 = -\frac{a}{\tan\phi}$$

$$y = a\ \frac{\sin\phi\ \cos H - \cos\phi\ \tan\delta}{Q} \qquad u = \frac{a}{|\sin\phi|}$$

$$\psi = |\phi|$$

(3) Vertical sundial

The plane of the sundial is vertical, so $z = 90°$. The x-axis is horizontal; the y-axis is directed towards the zenith. The formulae simplify to

$$Q = \sin D\ \sin H + \sin\phi\ \cos D\ \cos H - \cos\phi\ \cos D\ \tan\delta$$

$$x = a\ \frac{\cos D\ \sin H - \sin\phi\ \sin D\ \cos H + \cos\phi\ \sin D\ \tan\delta}{Q}$$

$$y = -a\ \frac{\cos\phi\ \cos H + \sin\phi\ \tan\delta}{Q}$$

$$x_0 = -a\ \tan D \qquad u = \frac{a}{|\cos\phi\ \cos D|}$$

$$y_0 = +a\ \tan\phi\ /\ \cos D$$

General Remarks

In the case of a sundial with a perpendicular stylus, as considered in this Chapter, it is the *extremity* of the umbra of that stylus which indicates the time, while in the case of a sundial with a polar stylus, it is the entire umbra which gives the time.

Because we give the coordinates x_0, y_0 of the center of the sundial, it is always possible to construct the polar stylus IP, if this is wanted: the polar stylus is the straight line connecting that center with the extremity of the perpendicular stylus. See the Figure on page 373.

The advantage of the system of axes $x-y$ used in this Chapter is that the perpendicular stylus does always exist; this is not always the case for the polar stylus.

Appendix I

Some Astronomical Terms

The following notes may be found helpful by those who are not familiar with the technical terms used in this book, but further guidance should be sought from textbooks on astronomy.

The **celestial equator** is the great circle that is the projection of the Earth's equator onto the celestial sphere. Its plane is perpendicular to the axis of rotation of the Earth.

The **celestial poles** are the poles of the celestial equator, or the intersections of the axis of rotation of the Earth with the celestial sphere.

The **ecliptic** is defined to be the plane of the (undisturbed) orbit of the Earth around the Sun.

The **equinox** or, better, the **vernal equinox**, which is the zero point of both right ascension and celestial longitude, is defined to be in the direction of the ascending node of the ecliptic on the equator. It is that intersection of equator and ecliptic where the ecliptic runs (eastwards) from negative to positive declinations. The other intersection, which is diametrically opposite, is the **autumnal equinox**.

The **equinoxes** are the instants when the apparent longitude of the Sun is 0° or 180°.

Solstices : both the points on the ecliptic 90 degrees away from the equinoxes, and the instants when the apparent longitude of the Sun is 90° or 270°.

Celestial longitude, or **ecliptical longitude**, often called simply **longitude**, is measured (from 0° to 360°) from the vernal equinox, positive to the east, along the ecliptic.

Celestial latitude, or **ecliptical latitude**, or simply **latitude**, is measured (from 0° to +90° or to -90°) from the ecliptic, positive to the north, negative to the south.

Right ascension is measured (from 0 to 24 hours, sometimes from 0° to 360°) from the vernal equinox, positive to the east, along the celestial equator.

Declination is measured (from 0° to ±90°) from the equator, positive to the north, negative to the south.

Owing to the effects of **precession** and **nutation**, the ecliptic and equator, and hence the equinoxes and the poles, are continuously in motion, and so the current celestial coordinates of a 'fixed' direction change continuously. The motion of the equator is primarily due to the action of the Sun and the Moon, while the (much slower) motion of the ecliptic is primarily due to the perturbing action of the planets.

Mean equator : the instantaneous celestial equator exclusive of the periodic perturbations of the nutation.

Mean equator and equinox, or simply **mean equinox** : an expression used to denote that the reference system takes into account the precession (secular effects) but not the nutation (periodic effects).

Coordinates : two (or three) numbers which define the position of a point on a surface (or in space). Examples: the longitude and latitude are the two geographical coordinates of a point on the surface of the Earth; the rectangular coordinates X, Y, Z of a point in three-dimensional space.

Heliocentric : referred to the center of the Sun, for instance a heliocentric orbit, or heliocentric coordinates.

Geocentric : referred to the center of the Earth, for instance a geocentric observer, or geocentric coordinates.

Topocentric : referred to the observer on the Earth's surface, for example the topocentric right ascension and declination of the Moon.

Aberration is the apparent displacement of the position of an object due to the finite speed of light. The **annual aberration** of a star is due to the orbital motion of the Earth around the Sun (or, more exactly, around the barycenter of the solar system).

Azimuth : the angular distance measured from the South, positive to the West, along the horizon, to the vertical circle through the point in question. Navigators and meteorologists measure the azimuth from the North, positive to the East.

Ascending node : that intersection of the orbital plane with the reference plane where the latitudinal coordinate is increasing (going north). The other intersection is the **descending node**.

Conjunction : that configuration of two celestial objects such that either their right ascensions or their celestial longitudes are equal.

Opposition : that configuration of two celestial objects such that their celestial longitudes differ by 180°. Most frequently used when one of the objects is the Sun.

Heliographic coordinate system : a coordinate system on the surface of the Sun.

Planetographic coordinate system : a coordinate system on the surface of a planet. In the case of Mars, the term **areographic** is generally used. For the Moon, the term is **selenographic**. Compare with **geographic** for the Earth.

Epoch : a particular fixed instant used as a reference point on a time scale, such as B1950.0 or J2000.0.

A **Julian century** is a time interval of 36525 days.

An **ephemeris day** is equal to 86400 seconds in the uniform time scale known as Dynamical Time.

The **sidereal time** is the measure of time defined by the motion of the vernal equinox in hour angle; it is the hour angle of that equinox (at a given place and for a given instant). The **true solar time** is the local hour angle of the Sun. The **mean solar time** is the hour angle of the mean Sun, and this is measured from mean noon. The **civil time** is the mean solar time increased by 12 hours, and thus is measured from mean midnight. [The expression 'mean time measured from midnight' is a *contradictio in terminis*, since the mean (solar) time by definition is measured from noon. Many people erroneously use the expression 'Greenwich Mean Time', when in fact Greenwich *Civil* Time is meant.]

Universal Time is the civil time on the meridian of Greenwich.

The **astronomical unit** (AU) is a unit of length used to measure distances in the solar system. It is often called 'mean distance of the Earth to the Sun'. But, rigorously, one AU is the radius of the circular orbit which a particle of negligible mass, and free of perturbations, would describe around the Sun with a period of $2\pi/k$ days, where k is the Gaussian gravitational constant

$$k = 0.017\,202\,098\,95.$$

As a consequence, the semimajor axis of the elliptical orbit of the Earth is not exactly 1 AU, but 1.000 001 018 AU.

Radius vector : the straight line connecting a body to the central body around which it revolves, or the distance between these bodies at a given instant. The radius vector of a planet or a comet is generally expressed in astronomical units.

Perihelion : the point of the orbit (of a planet, minor planet or comet) which is nearest to the Sun. For the corresponding point of the Moon's orbit with respect to the Earth, the term is **perigee**. For a satellite of Jupiter with respect to this planet, the traditional term is **perijove**. For a double star, one says **periastron**.

The **geometric** position of a planet is the 'true' position of that body at the given instant; that is no allowance is being made for the effects of aberration and light-time.

Astrometric position: see page 216.

Anomalies. – The **mean anomaly** (M) of a planet is the angular distance, as seen from the Sun, between the perihelion and the mean position of the planet. The angular distance measured from the perihelion to the true position of the planet is called the **true anomaly** (v). The **eccentric anomaly** is an auxiliary quantity needed to solve Kepler's equation and to obtain afterwards the true anomaly. The **equation of the center** is the difference between the true and the mean anomalies ($C = v - M$); it is the difference between the actual position of the body in the elliptic orbit and the position the body would have if its angular motion were uniform.

An **ephemeris** is a table of positions or other calculated data of a celestial body (Sun, Moon, planet, comet, etc.) for a series of (generally equidistant) instants. From the Greek εφημερος = *daily*.

Appendix II

Planets : Periodic Terms

In this Appendix, pages 382-422, the most important periodic terms from the French planetary theory VSOP87 are given. The successive columns contain the following data :

- the name of the planet;
- the label of the series;
- the current No. of the term in the series ;
- the quantities A, B, and C.

In each series, the terms are sorted by decreasing value of A.

For example :

Planet	Series	No.	A	B	C
VENUS	R0	1	72 334 821	0	0
		2	489 824	4.021 518	10 213.285 546
		3	1 658	4.902 1	20 426.571 1
		4	1 632	2.845 5	7 860.419 4
		5	1 378	1.128 5	11 790.629 1
		6	498	2.587	9 683.595
		7	374	1.423	3 930.210
		8	264	5.529	9 437.763
		9	237	2.551	15 720.839
		10	222	2.013	19 367.189
		11	126	2.728	1 577.344
		12	119	3.020	10 404.734
VENUS	R1	1	34 551	0.891 99	10 213.285 55
		2	234	1.772	20 426.571
		3	234	3.142	0

For more explanation about the use of these terms, see Chapter 31.

MERCURY	L0	1	440 250 710	0	0
		2	40 989 415	1.483 020 34	26 087.903 141 57
		3	5 046 294	4.477 854 9	52 175.806 283 1
		4	855 347	1.165 203	78 263.709 425
		5	165 590	4.119 692	104 351.612 566
		6	34 562	0.779 31	130 439.515 71
		7	7 583	3.713 5	156 527.418 8
		8	3 560	1.512 0	1 109.378 6
		9	1 803	4.103 3	5 661.332 0
		10	1 726	0.358 3	182 615.322 0
		11	1 590	2.995 1	25 028.521 2
		12	1 365	4.599 2	27 197.281 7
		13	1 017	0.880 3	31 749.235 2
		14	714	1.541	24 978.525
		15	644	5.303	21 535.950
		16	451	6.050	51 116.424
		17	404	3.282	208 703.225
		18	352	5.242	20 426.571
		19	345	2.792	15 874.618
		20	343	5.765	955.600
		21	339	5.863	25 558.212
		22	325	1.337	53 285.185
		23	273	2.495	529.691
		24	264	3.917	57 837.138
		25	260	0.987	4 551.953
		26	239	0.113	1 059.382
		27	235	0.267	11 322.664
		28	217	0.660	13 521.751
		29	209	2.092	47 623.853
		30	183	2.629	27 043.503
		31	182	2.434	25 661.305
		32	176	4.536	51 066.428
		33	173	2.452	24 498.830
		34	142	3.360	37 410.567
		35	138	0.291	10 213.286
		36	125	3.721	39 609.655
		37	118	2.781	77 204.327
		38	106	4.206	19 804.827
MERCURY	L1	1	2608 814 706 223	0	0
		2	1 126 008	6.217 039 7	26 087.903 141 6
		3	303 471	3.055 655	52 175.806 283
		4	80 538	6.104 55	78 263.709 42
		5	21 245	2.835 32	104 351.612 57
		6	5 592	5.826 8	130 439.515 7
		7	1 472	2.518 5	156 527.418 8
		8	388	5.480	182 615.322
		9	352	3.052	1 109.379
		10	103	2.149	208 703.225
		11	94	6.12	27 197.28
		12	91	0.00	24 978.52
		13	52	5.62	5 661.33
		14	44	4.57	25 028.52
		15	28	3.04	51 066.43
		16	27	5.09	234 791.13
MERCURY	L2	1	53 050	0	0
		2	16 904	4.690 72	26 087.903 14
		3	7 397	1.347 4	52 175.806 3
		4	3 018	4.456 4	78 263.709 4
		5	1 107	1.262 3	104 351.612 6
		6	378	4.320	130 439.516
		7	123	1.069	156 527.419

MERCURY (cont.)	L2	8	39	4.08	182 615.32
		9	15	4.63	1 109.38
		10	12	0.79	208 703.23
MERCURY	L3	1	188	0.035	52 175.806
		2	142	3.125	26 087.903
		3	97	3.00	78 263.71
		4	44	6.02	104 351.61
		5	35	0	0
		6	18	2.78	130 439.52
		7	7	5.82	156 527.42
		8	3	2.57	182 615.32
MERCURY	L4	1	114	3.141 6	0
		2	3	2.03	26 087.90
		3	2	1.42	78 263.71
		4	2	4.50	52 175.81
		5	1	4.50	104 351.61
		6	1	1.27	130 439.52
MERCURY	L5	1	1	3.14	0
MERCURY	B0	1	11 737 529	1.983 574 99	26 087.903 141 57
		2	2 388 077	5.037 389 6	52 175.806 283 1
		3	1 222 840	3.141 592 7	0
		4	543 252	1.796 444	78 263.709 425
		5	129 779	4.832 325	104 351.612 566
		6	31 867	1.580 88	130 439.515 71
		7	7 963	4.609 7	156 527.418 8
		8	2 014	1.353 2	182 615.322 0
		9	514	4.378	208 703.225
		10	209	2.020	24 978.525
		11	208	4.918	27 197.282
		12	132	1.119	234 791.128
		13	121	1.813	53 285.185
		14	100	5.657	20 426.571
MERCURY	B1	1	429 151	3.501 698	26 087.903 142
		2	146 234	3.141 593	0
		3	22 675	0.015 15	52 175.806 28
		4	10 895	0.485 40	78 263.709 42
		5	6 353	3.429 4	104 351.612 6
		6	2 496	0.160 5	130 439.515 7
		7	860	3.185	156 527.419
		8	278	6.210	182 615.322
		9	86	2.95	208 703.23
		10	28	0.29	27 197.28
		11	26	5.98	234 791.13
MERCURY	B2	1	11 831	4.790 66	26 087.903 14
		2	1 914	0	0
		3	1 045	1.212 2	52 175.806 3
		4	266	4.434	78 263.709
		5	170	1.623	104 351.613
		6	96	4.80	130 439.52
		7	45	1.61	156 527.42
		8	18	4.67	182 615.32
		9	7	1.43	208 703.23
MERCURY	B3	1	235	0.354	26 087.903
		2	161	0	0
		3	19	4.36	52 175.81
		4	6	2.51	78 263.71

MERCURY (cont.)	B3	5	5	6.14	104 351.61
		6	3	3.12	130 439.52
		7	2	6.27	156 527.42
MERCURY	B4	1	4	1.75	26 087.90
		2	1	3.14	0
MERCURY	R0	1	39 528 272	0	0
		2	7 834 132	6.192 337 2	26 087.903 141 6
		3	795 526	2.959 897	52 175.806 283
		4	121 282	6.010 642	78 263.709 425
		5	21 922	2.778 20	104 351.612 57
		6	4 354	5.828 9	130 439.515 7
		7	918	2.597	156 527.419
		8	290	1.424	25 028.521
		9	260	3.028	27 197.282
		10	202	5.647	182 615.322
		11	201	5.592	31 749.235
		12	142	6.253	24 978.525
		13	100	3.734	21 535.950
MERCURY	R1	1	217 348	4.656 172	26 087.903 142
		2	44 142	1.423 86	52 175.806 28
		3	10 094	4.474 66	78 263.709 42
		4	2 433	1.242 3	104 351.612 6
		5	1 624	0	0
		6	604	4.293	130 439.516
		7	153	1.061	156 527.419
		8	39	4.11	182 615.32
MERCURY	R2	1	3 118	3.082 3	26 087.903 1
		2	1 245	6.151 8	52 175.806 3
		3	425	2.926	78 263.709
		4	136	5.980	104 351.613
		5	42	2.75	130 439.52
		6	22	3.14	0
		7	13	5.80	156 527.42
MERCURY	R3	1	33	1.68	26 087.90
		2	24	4.63	52 175.81
		3	12	1.39	78 263.71
		4	5	4.44	104 351.61
		5	2	1.21	130 439.52
VENUS	L0	1	317 614 667	0	0
		2	1 353 968	5.593 133 2	10 213.285 546 2
		3	89 892	5.306 50	20 426.571 09
		4	5 477	4.416 3	7 860.419 4
		5	3 456	2.699 6	11 790.629 1
		6	2 372	2.993 8	3 930.209 7
		7	1 664	4.250 2	1 577.343 5
		8	1 438	4.157 5	9 683.594 6
		9	1 317	5.186 7	26.298 3
		10	1 201	6.153 6	30 639.856 6
		11	769	0.816	9 437.763
		12	761	1.950	529.691
		13	708	1.065	775.523
		14	585	3.998	191.448
		15	500	4.123	15 720.839
		16	429	3.586	19 367.189
		17	327	5.677	5 507.553
		18	326	4.591	10 404.734

VENUS (cont.)	L0	19	232	3.163	9 153.904
		20	180	4.653	1 109.379
		21	155	5.570	19 651.048
		22	128	4.226	20.775
		23	128	0.962	5 661.332
		24	106	1.537	801.821
VENUS	L1	1	1021 352 943 053	0	0
		2	95 708	2.464 24	10 213.285 55
		3	14 445	0.516 25	20 426.571 09
		4	213	1.795	30 639.857
		5	174	2.655	26.298
		6	152	6.106	1 577.344
		7	82	5.70	191.45
		8	70	2.68	9 437.76
		9	52	3.60	775.52
		10	38	1.03	529.69
		11	30	1.25	5 507.55
		12	25	6.11	10 404.73
VENUS	L2	1	54 127	0	0
		2	3 891	0.345 1	10 213.285 5
		3	1 338	2.020 1	20 426.571 1
		4	24	2.05	26.30
		5	19	3.54	30 639.86
		6	10	3.97	775.52
		7	7	1.52	1 577.34
		8	6	1.00	191.45
VENUS	L3	1	136	4.804	10 213.286
		2	78	3.67	20 426.57
		3	26	0	0
VENUS	L4	1	114	3.141 6	0
		2	3	5.21	20 426.57
		3	2	2.51	10 213.29
VENUS	L5	1	1	3.14	0
VENUS	B0	1	5 923 638	0.267 027 8	10 213.285 546 2
		2	40 108	1.147 37	20 426.571 09
		3	32 815	3.141 59	0
		4	1 011	1.089 5	30 639.856 6
		5	149	6.254	18 073.705
		6	138	0.860	1 577.344
		7	130	3.672	9 437.763
		8	120	3.705	2 352.866
		9	108	4.539	22 003.915
VENUS	B1	1	513 348	1.803 643	10 213.285 546
		2	4 380	3.386 2	20 426.571 1
		3	199	0	0
		4	197	2.530	30 639.857
VENUS	B2	1	22 378	3.385 09	10 213.285 55
		2	282	0	0
		3	173	5.256	20 426.571
		4	27	3.87	30 639.86
VENUS	B3	1	647	4.992	10 213.286
		2	20	3.14	0
		3	6	0.77	20 426.57
		4	3	5.44	30 639.86

VENUS	B4	1	14	0.32	10 213.29
VENUS	R0	1	72 334 821	0	0
		2	489 824	4.021 518	10 213.285 546
		3	1 658	4.902 1	20 426.571 1
		4	1 632	2.845 5	7 860.419 4
		5	1 378	1.128 5	11 790.629 1
		6	498	2.587	9 683.595
		7	374	1.423	3 930.210
		8	264	5.529	9 437.763
		9	237	2.551	15 720.839
		10	222	2.013	19 367.189
		11	126	2.728	1 577.344
		12	119	3.020	10 404.734
VENUS	R1	1	34 551	0.891 99	10 213.285 55
		2	234	1.772	20 426.571
		3	234	3.142	0
VENUS	R2	1	1 407	5.063 7	10 213.285 5
		2	16	5.47	20 426.57
		3	13	0	0
VENUS	R3	1	50	3.22	10 213.29
VENUS	R4	1	1	0.92	10 213.29

EARTH	L0	1	175 347 046	0	0
		2	3 341 656	4.669 256 8	6 283.075 850 0
		3	34 894	4.626 10	12 566.151 70
		4	3 497	2.744 1	5 753.384 9
		5	3 418	2.828 9	3.523 1
		6	3 136	3.627 7	77 713.771 5
		7	2 676	4.418 1	7 860.419 4
		8	2 343	6.135 2	3 930.209 7
		9	1 324	0.742 5	11 506.769 8
		10	1 273	2.037 1	529.691 0
		11	1 199	1.109 6	1 577.343 5
		12	990	5.233	5 884.927
		13	902	2.045	26.298
		14	857	3.508	398.149
		15	780	1.179	5 223.694
		16	753	2.533	5 507.553
		17	505	4.583	18 849.228
		18	492	4.205	775.523
		19	357	2.920	0.067
		20	317	5.849	11 790.629
		21	284	1.899	796.298
		22	271	0.315	10 977.079
		23	243	0.345	5 486.778
		24	206	4.806	2 544.314
		25	205	1.869	5 573.143
		26	202	2.458	6 069.777
		27	156	0.833	213.299
		28	132	3.411	2 942.463
		29	126	1.083	20.775
		30	115	0.645	0.980
		31	103	0.636	4 694.003
		32	102	0.976	15 720.839
		33	102	4.267	7.114
		34	99	6.21	2 146.17
		35	98	0.68	155.42
		36	86	5.98	161 000.69

EARTH (cont.)	L0	37	85	1.30	6 275.96
		38	85	3.67	71 430.70
		39	80	1.81	17 260.15
		40	79	3.04	12 036.46
		41	75	1.76	5 088.63
		42	74	3.50	3 154.69
		43	74	4.68	801.82
		44	70	0.83	9 437.76
		45	62	3.98	8 827.39
		46	61	1.82	7 084.90
		47	57	2.78	6 286.60
		48	56	4.39	14 143.50
		49	56	3.47	6 279.55
		50	52	0.19	12 139.55
		51	52	1.33	1 748.02
		52	51	0.28	5 856.48
		53	49	0.49	1 194.45
		54	41	5.37	8 429.24
		55	41	2.40	19 651.05
		56	39	6.17	10 447.39
		57	37	6.04	10 213.29
		58	37	2.57	1 059.38
		59	36	1.71	2 352.87
		60	36	1.78	6 812.77
		61	33	0.59	17 789.85
		62	30	0.44	83 996.85
		63	30	2.74	1 349.87
		64	25	3.16	4 690.48
EARTH	L1	1	628 331 966 747	0	0
		2	206 059	2.678 235	6 283.075 850
		3	4 303	2.635 1	12 566.151 7
		4	425	1.590	3.523
		5	119	5.796	26.298
		6	109	2.966	1 577.344
		7	93	2.59	18 849.23
		8	72	1.14	529.69
		9	68	1.87	398.15
		10	67	4.41	5 507.55
		11	59	2.89	5 223.69
		12	56	2.17	155.42
		13	45	0.40	796.30
		14	36	0.47	775.52
		15	29	2.65	7.11
		16	21	5.34	0.98
		17	19	1.85	5 486.78
		18	19	4.97	213.30
		19	17	2.99	6 275.96
		20	16	0.03	2 544.31
		21	16	1.43	2 146.17
		22	15	1.21	10 977.08
		23	12	2.83	1 748.02
		24	12	3.26	5 088.63
		25	12	5.27	1 194.45
		26	12	2.08	4 694.00
		27	11	0.77	553.57
		28	10	1.30	6 286.60
		29	10	4.24	1 349.87
		30	9	2.70	242.73
		31	9	5.64	951.72
		32	8	5.30	2 352.87
		33	6	2.65	9 437.76
		34	6	4.67	4 690.48

EARTH	L2	1	52 919	0	0
		2	8 720	1.072 1	6 283.075 8
		3	309	0.867	12 566.152
		4	27	0.05	3.52
		5	16	5.19	26.30
		6	16	3.68	155.42
		7	10	0.76	18 849.23
		8	9	2.06	77 713.77
		9	7	0.83	775.52
		10	5	4.66	1 577.34
		11	4	1.03	7.11
		12	4	3.44	5 573.14
		13	3	5.14	796.30
		14	3	6.05	5 507.55
		15	3	1.19	242.73
		16	3	6.12	529.69
		17	3	0.31	398.15
		18	3	2.28	553.57
		19	2	4.38	5 223.69
		20	2	3.75	0.98
EARTH	L3	1	289	5.844	6 283.076
		2	35	0	0
		3	17	5.49	12 566.15
		4	3	5.20	155.42
		5	1	4.72	3.52
		6	1	5.30	18 849.23
		7	1	5.97	242.73
EARTH	L4	1	114	3.142	0
		2	8	4.13	6 283.08
		3	1	3.84	12 566.15
EARTH	L5	1	1	3.14	0
EARTH	B0	1	280	3.199	84 334.662
		2	102	5.422	5 507.553
		3	80	3.88	5 223.69
		4	44	3.70	2 352.87
		5	32	4.00	1 577.34
EARTH	B1	1	9	3.90	5 507.55
		2	6	1.73	5 223.69
EARTH	R0	1	100 013 989	0	0
		2	1 670 700	3.098 463 5	6 283.075 850 0
		3	13 956	3.055 25	12 566.151 70
		4	3 084	5.198 5	77 713.771 5
		5	1 628	1.173 9	5 753.384 9
		6	1 576	2.846 9	7 860.419 4
		7	925	5.453	11 506.770
		8	542	4.564	3 930.210
		9	472	3.661	5 884.927
		10	346	0.964	5 507.553
		11	329	5.900	5 223.694
		12	307	0.299	5 573.143
		13	243	4.273	11 790.629
		14	212	5.847	1 577.344
		15	186	5.022	10 977.079
		16	175	3.012	18 849.228
		17	110	5.055	5 486.778
		18	98	0.89	6 069.78
		19	86	5.69	15 720.84

EARTH (cont.)	R0	20	86	1.27	161 000.69
		21	65	0.27	17 260.15
		22	63	0.92	529.69
		23	57	2.01	83 996.85
		24	56	5.24	71 430.70
		25	49	3.25	2 544.31
		26	47	2.58	775.52
		27	45	5.54	9 437.76
		28	43	6.01	6 275.96
		29	39	5.36	4 694.00
		30	38	2.39	8 827.39
		31	37	0.83	19 651.05
		32	37	4.90	12 139.55
		33	36	1.67	12 036.46
		34	35	1.84	2 942.46
		35	33	0.24	7 084.90
		36	32	0.18	5 088.63
		37	32	1.78	398.15
		38	28	1.21	6 286.60
		39	28	1.90	6 279.55
		40	26	4.59	10 447.39
EARTH	R1	1	103 019	1.107 490	6 283.075 850
		2	1 721	1.064 4	12 566.151 7
		3	702	3.142	0
		4	32	1.02	18 849.23
		5	31	2.84	5 507.55
		6	25	1.32	5 223.69
		7	18	1.42	1 577.34
		8	10	5.91	10 977.08
		9	9	1.42	6 275.96
		10	9	0.27	5 486.78
EARTH	R2	1	4 359	5.784 6	6 283.075 8
		2	124	5.579	12 566.152
		3	12	3.14	0
		4	9	3.63	77 713.77
		5	6	1.87	5 573.14
		6	3	5.47	18 849.23
EARTH	R3	1	145	4.273	6 283.076
		2	7	3.92	12 566.15
EARTH	R4	1	4	2.56	6 283.08
MARS	L0	1	620 347 712	0	0
		2	18 656 368	5.050 371 00	3 340.612 426 70
		3	1 108 217	5.400 998 4	6 681.224 853 4
		4	91 798	5.754 79	10 021.837 28
		5	27 745	5.970 50	3.523 12
		6	12 316	0.849 56	2 810.921 46
		7	10 610	2.939 59	2 281.230 50
		8	8 927	4.157 0	0.017 3
		9	8 716	6.110 1	13 362.449 7
		10	7 775	3.339 7	5 621.842 9
		11	6 798	0.364 6	398.149 0
		12	4 161	0.228 1	2 942.463 4
		13	3 575	1.661 9	2 544.314 4
		14	3 075	0.857 0	191.448 3
		15	2 938	6.078 9	0.067 3
		16	2 628	0.648 1	3 337.089 3

MARS (cont.)	L0	17	2580	0.0300	3344.1355
		18	2389	5.0390	796.2980
		19	1799	0.6563	529.6910
		20	1546	2.9158	1751.5395
		21	1528	1.1498	6151.5339
		22	1286	3.0680	2146.1654
		23	1264	3.6228	5092.1520
		24	1025	3.6933	8962.4553
		25	892	0.183	16703.062
		26	859	2.401	2914.014
		27	833	4.495	3340.630
		28	833	2.464	3340.595
		29	749	3.822	155.420
		30	724	0.675	3738.761
		31	713	3.663	1059.382
		32	655	0.489	3127.313
		33	636	2.922	8432.764
		34	553	4.475	1748.016
		35	550	3.810	0.980
		36	472	3.625	1194.447
		37	426	0.554	6283.076
		38	415	0.497	213.299
		39	312	0.999	6677.702
		40	307	0.381	6684.748
		41	302	4.486	3532.061
		42	299	2.783	6254.627
		43	293	4.221	20.775
		44	284	5.769	3149.164
		45	281	5.882	1349.867
		46	274	0.542	3340.545
		47	274	0.134	3340.680
		48	239	5.372	4136.910
		49	236	5.755	3333.499
		50	231	1.282	3870.303
		51	221	3.505	382.897
		52	204	2.821	1221.849
		53	193	3.357	3.590
		54	189	1.491	9492.146
		55	179	1.006	951.718
		56	174	2.414	553.569
		57	172	0.439	5486.778
		58	160	3.949	4562.461
		59	144	1.419	135.065
		60	140	3.326	2700.715
		61	138	4.301	7.114
		62	131	4.045	12303.068
		63	128	2.208	1592.596
		64	128	1.807	5088.629
		65	117	3.128	7903.073
		66	113	3.701	1589.073
		67	110	1.052	242.729
		68	105	0.785	8827.390
		69	100	3.243	11773.377
MARS	L1	1	334085627474	0	0
		2	1458227	3.6042605	3340.6124267
		3	164901	3.926313	6681.224853
		4	19963	4.26594	10021.83728
		5	3452	4.7321	3.5231
		6	2485	4.6128	13362.4497
		7	842	4.459	2281.230
		8	538	5.016	398.149

MARS (cont.)	L1	9	521	4.994	3344.136
		10	433	2.561	191.448
		11	430	5.316	155.420
		12	382	3.539	796.298
		13	314	4.963	16703.062
		14	283	3.160	2544.314
		15	206	4.569	2146.165
		16	169	1.329	3337.089
		17	158	4.185	1751.540
		18	134	2.233	0.980
		19	134	5.974	1748.016
		20	118	6.024	6151.534
		21	117	2.213	1059.382
		22	114	2.129	1194.447
		23	114	5.428	3738.761
		24	91	1.10	1349.87
		25	85	3.91	553.57
		26	83	5.30	6684.75
		27	81	4.43	529.69
		28	80	2.25	8962.46
		29	73	2.50	951.72
		30	73	5.84	242.73
		31	71	3.86	2914.01
		32	68	5.02	382.90
		33	65	1.02	3340.60
		34	65	3.05	3340.63
		35	62	4.15	3149.16
		36	57	3.89	4136.91
		37	48	4.87	213.30
		38	48	1.18	3333.50
		39	47	1.31	3185.19
		40	41	0.71	1592.60
		41	40	2.73	7.11
		42	40	5.32	20043.67
		43	33	5.41	6283.08
		44	28	0.05	9492.15
		45	27	3.89	1221.85
		46	27	5.11	2700.72
MARS	L2	1	58016	2.04979	3340.61243
		2	54188	0	0
		3	13908	2.45742	6681.22485
		4	2465	2.8000	10021.8373
		5	398	3.141	13362.450
		6	222	3.194	3.523
		7	121	0.543	155.420
		8	62	3.49	16703.06
		9	54	3.54	3344.14
		10	34	6.00	2281.23
		11	32	4.14	191.45
		12	30	2.00	796.30
		13	23	4.33	242.73
		14	22	3.45	398.15
		15	20	5.42	553.57
		16	16	0.66	0.98
		17	16	6.11	2146.17
		18	16	1.22	1748.02
		19	15	6.10	3185.19
		20	14	4.02	951.72
		21	14	2.62	1349.87
		22	13	0.60	1194.45
		23	12	3.86	6684.75

MARS (cont.)	L2	24	11	4.72	2544.31
		25	10	0.25	382.90
		26	9	0.68	1059.38
		27	9	3.83	20043.67
		28	9	3.88	3738.76
		29	8	5.46	1751.54
		30	7	2.58	3149.16
		31	7	2.38	4136.91
		32	6	5.48	1592.60
		33	6	2.34	3097.88
MARS	L3	1	1482	0.4443	3340.6124
		2	662	0.885	6681.225
		3	188	1.288	10021.837
		4	41	1.65	13362.45
		5	26	0	0
		6	23	2.05	155.42
		7	10	1.58	3.52
		8	8	2.00	16703.06
		9	5	2.82	242.73
		10	4	2.02	3344.14
		11	3	4.59	3185.19
		12	3	0.65	553.57
MARS	L4	1	114	3.1416	0
		2	29	5.64	6681.22
		3	24	5.14	3340.61
		4	11	6.03	10021.84
		5	3	0.13	13362.45
		6	3	3.56	155.42
		7	1	0.49	16703.06
		8	1	1.32	242.73
MARS	L5	1	1	3.14	0
		2	1	4.04	6681.22
MARS	B0	1	3197135	3.7683204	3340.6124267
		2	298033	4.106170	6681.224853
		3	289105	0	0
		4	31366	4.44651	10021.83728
		5	3484	4.7881	13362.4497
		6	443	5.026	3344.136
		7	443	5.652	3337.089
		8	399	5.131	16703.062
		9	293	3.793	2281.230
		10	182	6.136	6151.534
		11	163	4.264	529.691
		12	160	2.232	1059.382
		13	149	2.165	5621.843
		14	143	1.182	3340.595
		15	143	3.213	3340.630
		16	139	2.418	8962.455
MARS	B1	1	350069	5.368478	3340.612427
		2	14116	3.14159	0
		3	9671	5.4788	6681.2249
		4	1472	3.2021	10021.8373
		5	426	3.408	13362.450
		6	102	0.776	3337.089
		7	79	3.72	16703.06
		8	33	3.46	5621.84
		9	26	2.48	2281.23

MARS	B2	1	16727	0.60221	3340.61243
		2	4987	3.1416	0
		3	302	3.559	6681.225
		4	26	1.90	13362.45
		5	21	0.92	10021.84
		6	12	2.24	3337.09
		7	8	2.25	16703.06
MARS	B3	1	607	1.981	3340.612
		2	43	0	0
		3	14	1.80	6681.22
		4	3	3.45	10021.84
MARS	B4	1	13	0	0
		2	11	3.46	3340.61
		3	1	0.50	6681.22
MARS	R0	1	153033488	0	0
		2	14184953	3.47971284	3340.61242670
		3	660776	3.817834	6681.224853
		4	46179	4.15595	10021.83728
		5	8110	5.5596	2810.9215
		6	7485	1.7724	5621.8429
		7	5523	1.3644	2281.2305
		8	3825	4.4941	13362.4497
		9	2484	4.9255	2942.4634
		10	2307	0.0908	2544.3144
		11	1999	5.3606	3337.0893
		12	1960	4.7425	3344.1355
		13	1167	2.1126	5092.1520
		14	1103	5.0091	398.1490
		15	992	5.839	6151.534
		16	899	4.408	529.691
		17	807	2.102	1059.382
		18	798	3.448	796.298
		19	741	1.499	2146.165
		20	726	1.245	8432.764
		21	692	2.134	8962.455
		22	633	0.894	3340.595
		23	633	2.924	3340.630
		24	630	1.287	1751.540
		25	574	0.829	2914.014
		26	526	5.383	3738.761
		27	473	5.199	3127.313
		28	348	4.832	16703.062
		29	284	2.907	3532.061
		30	280	5.257	6283.076
		31	276	1.218	6254.627
		32	275	2.908	1748.016
		33	270	3.764	5884.927
		34	239	2.037	1194.447
		35	234	5.105	5486.778
		36	228	3.255	6872.673
		37	223	4.199	3149.164
		38	219	5.583	191.448
		39	208	5.255	3340.545
		40	208	4.846	3340.680
		41	186	5.699	6677.702
		42	183	5.081	6684.748
		43	179	4.184	3333.499
		44	176	5.953	3870.303
		45	164	3.799	4136.910

MARS	R1	1	1107433	2.0325052	3340.6124267
		2	103176	2.370718	6681.224853
		3	12877	0	0
		4	10816	2.70888	10021.83728
		5	1195	3.0470	13362.4497
		6	439	2.888	2281.230
		7	396	3.423	3344.136
		8	183	1.584	2544.314
		9	136	3.385	16703.062
		10	128	6.043	3337.089
		11	128	0.630	1059.382
		12	127	1.954	796.298
		13	118	2.998	2146.165
		14	88	3.42	398.15
		15	83	3.86	3738.76
		16	76	4.45	6151.53
		17	72	2.76	529.69
		18	67	2.55	1751.54
		19	66	4.41	1748.02
		20	58	0.54	1194.45
		21	54	0.68	8962.46
		22	51	3.73	6684.75
		23	49	5.73	3340.60
		24	49	1.48	3340.63
		25	48	2.58	3149.16
		26	48	2.29	2914.01
		27	39	2.32	4136.91
MARS	R2	1	44242	0.47931	3340.61243
		2	8138	0.8700	6681.2249
		3	1275	1.2259	10021.8373
		4	187	1.573	13362.450
		5	52	3.14	0
		6	41	1.97	3344.14
		7	27	1.92	16703.06
		8	18	4.43	2281.23
		9	12	4.53	3185.19
		10	10	5.39	1059.38
		11	10	0.42	796.30
MARS	R3	1	1113	5.1499	3340.6124
		2	424	5.613	6681.225
		3	100	5.997	10021.837
		4	20	0.08	13362.45
		5	5	3.14	0
		6	3	0.43	16703.06
MARS	R4	1	20	3.58	3340.61
		2	16	4.05	6681.22
		3	6	4.46	10021.84
		4	2	4.84	13362.45
JUPITER	L0	1	59954691	0	0
		2	9695899	5.0619179	529.6909651
		3	573610	1.444062	7.113547
		4	306389	5.417347	1059.381930
		5	97178	4.14265	632.78374
		6	72903	3.64043	522.57742
		7	64264	3.41145	103.09277
		8	39806	2.29377	419.48464
		9	38858	1.27232	316.39187
		10	27965	1.78455	536.80451

JUPITER (cont.)	L0	11	13 590	5.774 81	1 589.072 90
		12	8 769	3.630 0	949.175 6
		13	8 246	3.582 3	206.185 5
		14	7 368	5.081 0	735.876 5
		15	6 263	0.025 0	213.299 1
		16	6 114	4.513 2	1 162.474 7
		17	5 305	4.186 3	1 052.268 4
		18	5 305	1.306 7	14.227 1
		19	4 905	1.320 8	110.206 3
		20	4 647	4.699 6	3.932 2
		21	3 045	4.316 8	426.598 2
		22	2 610	1.566 7	846.082 8
		23	2 028	1.063 8	3.181 4
		24	1 921	0.971 7	639.897 3
		25	1 765	2.141 5	1 066.495 5
		26	1 723	3.880 4	1 265.567 5
		27	1 633	3.582 0	515.463 9
		28	1 432	4.296 8	625.670 2
		29	973	4.098	95.979
		30	884	2.437	412.371
		31	733	6.085	838.969
		32	731	3.806	1 581.959
		33	709	1.293	742.990
		34	692	6.134	2 118.764
		35	614	4.109	1 478.867
		36	582	4.540	309.278
		37	495	3.756	323.505
		38	441	2.958	454.909
		39	417	1.036	2.448
		40	390	4.897	1 692.166
		41	376	4.703	1 368.660
		42	341	5.715	533.623
		43	330	4.740	0.048
		44	262	1.877	0.963
		45	261	0.820	380.128
		46	257	3.724	199.072
		47	244	5.220	728.763
		48	235	1.227	909.819
		49	220	1.651	543.918
		50	207	1.855	525.759
		51	202	1.807	1 375.774
		52	197	5.293	1 155.361
		53	175	3.730	942.062
		54	175	3.226	1 898.351
		55	175	5.910	956.289
		56	158	4.365	1 795.258
		57	151	3.906	74.782
		58	149	4.377	1 685.052
		59	141	3.136	491.558
		60	138	1.318	1 169.588
		61	131	4.169	1 045.155
		62	117	2.500	1 596.186
		63	117	3.389	0.521
		64	106	4.554	526.510
JUPITER	L1	1	52 993 480 757	0	0
		2	489 741	4.220 667	529.690 965
		3	228 919	6.026 475	7.113 547
		4	27 655	4.572 66	1 059.381 93
		5	20 721	5.459 39	522.577 42
		6	12 106	0.169 86	536.804 51
		7	6 068	4.424 2	103.092 8
		8	5 434	3.984 8	419.484 6

JUPITER (cont.)	L1	9	4 238	5.890 1	14.227 1
		10	2 212	5.267 7	206.185 5
		11	1 746	4.926 7	1 589.072 9
		12	1 296	5.551 3	3.181 4
		13	1 173	5.856 5	1 052.268 4
		14	1 163	0.514 5	3.932 2
		15	1 099	5.307 0	515.463 9
		16	1 007	0.464 8	735.876 5
		17	1 004	3.150 4	426.598 2
		18	848	5.758	110.206
		19	827	4.803	213.299
		20	816	0.586	1 066.495
		21	725	5.518	639.897
		22	568	5.989	625.670
		23	474	4.132	412.371
		24	413	5.737	95.979
		25	345	4.242	632.784
		26	336	3.732	1 162.475
		27	234	4.035	949.176
		28	234	6.243	309.278
		29	199	1.505	838.969
		30	195	2.219	323.505
		31	187	6.086	742.990
		32	184	6.280	543.918
		33	171	5.417	199.072
		34	131	0.626	728.763
		35	115	0.680	846.083
		36	115	5.286	2 118.764
		37	108	4.493	956.289
		38	80	5.82	1 045.15
		39	72	5.34	942.06
		40	70	5.97	532.87
		41	67	5.73	21.34
		42	66	0.13	526.51
		43	65	6.09	1 581.96
		44	59	0.59	1 155.36
		45	58	0.99	1 596.19
		46	57	5.97	1 169.59
		47	57	1.41	533.62
		48	55	5.43	10.29
		49	52	5.73	117.32
		50	52	0.23	1 368.66
		51	50	6.08	525.76
		52	47	3.63	1 478.87
		53	47	0.51	1 265.57
		54	40	4.16	1 692.17
		55	34	0.10	302.16
		56	33	5.04	220.41
		57	32	5.37	508.35
		58	29	5.42	1 272.68
		59	29	3.36	4.67
		60	29	0.76	88.87
		61	25	1.61	831.86
JUPITER	L2	1	47 234	4.321 48	7.113 55
		2	38 966	0	0
		3	30 629	2.930 21	529.690 97
		4	3 189	1.055 0	522.577 4
		5	2 729	4.845 5	536.804 5
		6	2 723	3.414 1	1 059.381 9
		7	1 721	4.187 3	14.227 1
		8	383	5.768	419.485
		9	378	0.760	515.464

JUPITER (cont.)	L2	10	367	6.055	103.093
		11	337	3.786	3.181
		12	308	0.694	206.186
		13	218	3.814	1 589.073
		14	199	5.340	1 066.495
		15	197	2.484	3.932
		16	156	1.406	1 052.268
		17	146	3.814	639.897
		18	142	1.634	426.598
		19	130	5.837	412.371
		20	117	1.414	625.670
		21	97	4.03	110.21
		22	91	1.11	95.98
		23	87	2.52	632.78
		24	79	4.64	543.92
		25	72	2.22	735.88
		26	58	0.83	199.07
		27	57	3.12	213.30
		28	49	1.67	309.28
		29	40	4.02	21.34
		30	40	0.62	323.51
		31	36	2.33	728.76
		32	29	3.61	10.29
		33	28	3.24	838.97
		34	26	4.50	742.99
		35	26	2.51	1 162.47
		36	25	1.22	1 045.15
		37	24	3.01	956.29
		38	19	4.29	532.87
		39	18	0.81	508.35
		40	17	4.20	2 118.76
		41	17	1.83	526.51
		42	15	5.81	1 596.19
		43	15	0.68	942.06
		44	15	4.00	117.32
		45	14	5.95	316.39
		46	14	1.80	302.16
		47	13	2.52	88.87
		48	13	4.37	1 169.59
		49	11	4.44	525.76
		50	10	1.72	1 581.96
		51	9	2.18	1 155.36
		52	9	3.29	220.41
		53	9	3.32	831.86
		54	8	5.76	846.08
		55	8	2.71	533.62
		56	7	2.18	1 265.57
		57	6	0.50	949.18
JUPITER	L3	1	6 502	2.598 6	7.113 5
		2	1 357	1.346 4	529.691 0
		3	471	2.475	14.227
		4	417	3.245	536.805
		5	353	2.974	522.577
		6	155	2.076	1 059.382
		7	87	2.51	515.46
		8	44	0	0
		9	34	3.83	1 066.50
		10	28	2.45	206.19
		11	24	1.28	412.37
		12	23	2.98	543.92
		13	20	2.10	639.90
		14	20	1.40	419.48

JUPITER (cont.)	L3	15	19	1.59	103.09
		16	17	2.30	21.34
		17	17	2.60	1 589.07
		18	16	3.15	625.67
		19	16	3.36	1 052.27
		20	13	2.76	95.98
		21	13	2.54	199.07
		22	13	6.27	426.60
		23	9	1.76	10.29
		24	9	2.27	110.21
		25	7	3.43	309.28
		26	7	4.04	728.76
		27	6	2.52	508.35
		28	5	2.91	1 045.15
		29	5	5.25	323.51
		30	4	4.30	88.87
		31	4	3.52	302.16
		32	4	4.09	735.88
		33	3	1.43	956.29
		34	3	4.36	1 596.19
		35	3	1.25	213.30
		36	3	5.02	838.97
		37	3	2.24	117.32
		38	2	2.90	742.99
		39	2	2.36	942.06
JUPITER	L4	1	669	0.853	7.114
		2	114	3.142	0
		3	100	0.743	14.227
		4	50	1.65	536.80
		5	44	5.82	529.69
		6	32	4.86	522.58
		7	15	4.29	515.46
		8	9	0.71	1 059.38
		9	5	1.30	543.92
		10	4	2.32	1 066.50
		11	4	0.48	21.34
		12	3	3.00	412.37
		13	2	0.40	639.90
		14	2	4.26	199.07
		15	2	4.91	625.67
		16	2	4.26	206.19
		17	1	5.26	1 052.27
		18	1	4.72	95.98
		19	1	1.29	1 589.07
JUPITER	L5	1	50	5.26	7.11
		2	16	5.25	14.23
		3	4	0.01	536.80
		4	2	1.10	522.58
		5	1	3.14	0
JUPITER	B0	1	2 268 616	3.558 526 1	529.690 965 1
		2	110 090	0	0
		3	109 972	3.908 093	1 059.381 930
		4	8 101	3.605 1	522.577 4
		5	6 438	0.306 3	536.804 5
		6	6 044	4.258 8	1 589.072 9
		7	1 107	2.985 3	1 162.474 7
		8	944	1.675	426.598
		9	942	2.936	1 052.268
		10	894	1.754	7.114
		11	836	5.179	103.093

JUPITER (cont.)	B0	12	767	2.155	632.784
		13	684	3.678	213.299
		14	629	0.643	1 066.495
		15	559	0.014	846.083
		16	532	2.703	110.206
		17	464	1.173	949.176
		18	431	2.608	419.485
		19	351	4.611	2 118.764
		20	132	4.778	742.990
		21	123	3.350	1 692.166
		22	116	1.387	323.505
		23	115	5.049	316.392
		24	104	3.701	515.464
		25	103	2.319	1 478.867
		26	102	3.153	1 581.959
JUPITER	B1	1	177 352	5.701 665	529.690 965
		2	3 230	5.779 4	1 059.381 9
		3	3 081	5.474 6	522.577 4
		4	2 212	4.734 8	536.804 5
		5	1 694	3.141 6	0
		6	346	4.746	1 052.268
		7	234	5.189	1 066.495
		8	196	6.186	7.114
		9	150	3.927	1 589.073
		10	114	3.439	632.784
		11	97	2.91	949.18
		12	82	5.08	1 162.47
		13	77	2.51	103.09
		14	77	0.61	419.48
		15	74	5.50	515.46
		16	61	5.45	213.30
		17	50	3.95	735.88
		18	46	0.54	110.21
		19	45	1.90	846.08
		20	37	4.70	543.92
		21	36	6.11	316.39
		22	32	4.92	1 581.96
JUPITER	B2	1	8 094	1.463 2	529.691 0
		2	813	3.141 6	0
		3	742	0.957	522.577
		4	399	2.899	536.805
		5	342	1.447	1 059.382
		6	74	0.41	1 052.27
		7	46	3.48	1 066.50
		8	30	1.93	1 589.07
		9	29	0.99	515.46
		10	23	4.27	7.11
		11	14	2.92	543.92
		12	12	5.22	632.78
		13	11	4.88	949.18
		14	6	6.21	1 045.15
JUPITER	B3	1	252	3.381	529.691
		2	122	2.733	522.577
		3	49	1.04	536.80
		4	11	2.31	1 052.27
		5	8	2.77	515.46
		6	7	4.25	1 059.38
		7	6	1.78	1 066.50
		8	4	1.13	543.92
		9	3	3.14	0

JUPITER	B4	1	15	4.53	522.58
		2	5	4.47	529.69
		3	4	5.44	536.80
		4	3	0	0
		5	2	4.52	515.46
		6	1	4.20	1 052.27
JUPITER	B5	1	1	0.09	522.58
JUPITER	R0	1	520 887 429	0	0
		2	25 209 327	3.491 086 40	529.690 965 09
		3	610 600	3.841 154	1 059.381 930
		4	282 029	2.574 199	632.783 739
		5	187 647	2.075 904	522.577 418
		6	86 793	0.710 01	419.484 64
		7	72 063	0.214 66	536.804 51
		8	65 517	5.979 96	316.391 87
		9	30 135	2.161 32	949.175 61
		10	29 135	1.677 59	103.092 77
		11	23 947	0.274 58	7.113 55
		12	23 453	3.540 23	735.876 51
		13	22 284	4.193 63	1 589.072 90
		14	13 033	2.960 43	1 162.474 70
		15	12 749	2.715 50	1 052.268 38
		16	9 703	1.906 7	206.185 5
		17	9 161	4.413 5	213.299 1
		18	7 895	2.479 1	426.598 2
		19	7 058	2.181 8	1 265.567 5
		20	6 138	6.264 2	846.082 8
		21	5 477	5.657 3	639.897 3
		22	4 170	2.016 1	515.463 9
		23	4 137	2.722 2	625.670 2
		24	3 503	0.565 3	1 066.495 5
		25	2 617	2.009 9	1 581.959 3
		26	2 500	4.551 8	838.969 3
		27	2 128	6.127 5	742.990 1
		28	1 912	0.856 2	412.371 1
		29	1 611	3.088 7	1 368.660 3
		30	1 479	2.680 3	1 478.866 6
		31	1 231	1.890 4	323.505 4
		32	1 217	1.801 7	110.206 3
		33	1 015	1.386 7	454.909 4
		34	999	2.872	309.278
		35	961	4.549	2 118.764
		36	886	4.148	533.623
		37	821	1.593	1 898.351
		38	812	5.941	909.819
		39	777	3.677	728.763
		40	727	3.988	1 155.361
		41	655	2.791	1 685.052
		42	654	3.382	1 692.166
		43	621	4.823	956.289
		44	615	2.276	942.062
		45	562	0.081	543.918
		46	542	0.284	525.759
JUPITER	R1	1	1 271 802	2.649 375 1	529.690 965 1
		2	61 662	3.000 76	1 059.381 93
		3	53 444	3.897 18	522.577 42
		4	41 390	0	0
		5	31 185	4.882 77	536.804 51
		6	11 847	2.413 30	419.484 64
		7	9 166	4.759 8	7.113 5

JUPITER (cont.)	R1	8	3404	3.3469	1589.0729
		9	3203	5.2108	735.8765
		10	3176	2.7930	103.0928
		11	2806	3.7422	515.4639
		12	2677	4.3305	1052.2684
		13	2600	3.6344	206.1855
		14	2412	1.4695	426.5982
		15	2101	3.9276	639.8973
		16	1646	5.3095	1066.4955
		17	1641	4.4163	625.6702
		18	1050	3.1611	213.2991
		19	1025	2.5543	412.3711
		20	806	2.678	632.784
		21	741	2.171	1162.475
		22	677	6.250	838.969
		23	567	4.577	742.990
		24	485	2.469	949.176
		25	469	4.710	543.918
		26	445	0.403	323.505
		27	416	5.368	728.763
		28	402	4.605	309.278
		29	347	4.681	14.227
		30	338	3.168	956.289
		31	261	5.343	846.083
		32	247	3.923	942.062
		33	220	4.842	1368.660
		34	203	5.600	1155.361
		35	200	4.439	1045.155
		36	197	3.706	2118.764
		37	196	3.759	199.072
		38	184	4.265	95.979
		39	180	4.402	532.872
		40	170	4.846	526.510
		41	146	6.130	533.623
		42	133	1.322	110.206
		43	132	4.512	525.759
JUPITER	R2	1	79645	1.35866	529.69097
		2	8252	5.7777	522.5774
		3	7030	3.2748	536.8045
		4	5314	1.8384	1059.3819
		5	1861	2.9768	7.1135
		6	964	5.480	515.464
		7	836	4.199	419.485
		8	498	3.142	0
		9	427	2.228	639.897
		10	406	3.783	1066.495
		11	377	2.242	1589.073
		12	363	5.368	206.186
		13	342	6.099	1052.268
		14	339	6.127	625.670
		15	333	0.003	426.598
		16	280	4.262	412.371
		17	257	0.963	632.784
		18	230	0.705	735.877
		19	201	3.069	543.918
		20	200	4.429	103.093
		21	139	2.932	14.227
		22	114	0.787	728.763
		23	95	1.70	838.97
		24	86	5.14	323.51
		25	83	0.06	309.28
		26	80	2.98	742.99

JUPITER (cont.)	R2	27	75	1.60	956.29
		28	70	1.51	213.30
		29	67	5.47	199.07
		30	62	6.10	1 045.15
		31	56	0.96	1 162.47
		32	52	5.58	942.06
		33	50	2.72	532.87
		34	45	5.52	508.35
		35	44	0.27	526.51
		36	40	5.95	95.98
JUPITER	R3	1	3 519	6.058 0	529.691 0
		2	1 073	1.673 2	536.804 5
		3	916	1.413	522.577
		4	342	0.523	1 059.382
		5	255	1.196	7.114
		6	222	0.952	515.464
		7	90	3.14	0
		8	69	2.27	1 066.50
		9	58	1.41	543.92
		10	58	0.53	639.90
		11	51	5.98	412.37
		12	47	1.58	625.67
		13	43	6.12	419.48
		14	37	1.18	14.23
		15	34	1.67	1 052.27
		16	34	0.85	206.19
		17	31	1.04	1 589.07
		18	30	4.63	426.60
		19	21	2.50	728.76
		20	15	0.89	199.07
		21	14	0.96	508.35
		22	13	1.50	1 045.15
		23	12	2.61	735.88
		24	12	3.56	323.51
		25	11	1.79	309.28
		26	11	6.28	956.29
		27	10	6.26	103.09
		28	9	3.45	838.97
JUPITER	R4	1	129	0.084	536.805
		2	113	4.249	529.691
		3	83	3.30	522.58
		4	38	2.73	515.46
		5	27	5.69	7.11
		6	18	5.40	1 059.38
		7	13	6.02	543.92
		8	9	0.77	1 066.50
		9	8	5.68	14.23
		10	7	1.43	412.37
		11	6	5.12	639.90
		12	5	3.34	625.67
		13	3	3.40	1 052.27
		14	3	4.16	728.76
		15	3	2.90	426.60
JUPITER	R5	1	11	4.75	536.80
		2	4	5.92	522.58
		3	2	5.57	515.46
		4	2	4.30	543.92
		5	2	3.69	7.11
		6	2	4.13	1 059.38
		7	2	5.49	1 066.50

SATURN	L0	1	87 401 354	0	0
		2	11 107 660	3.962 050 90	213.299 095 44
		3	1 414 151	4.585 815 2	7.113 547 0
		4	398 379	0.521 120	206.185 548
		5	350 769	3.303 299	426.598 191
		6	206 816	0.246 584	103.092 774
		7	79 271	3.840 07	220.412 64
		8	23 990	4.669 77	110.206 32
		9	16 574	0.437 19	419.484 64
		10	15 820	0.938 09	632.783 74
		11	15 054	2.716 70	639.897 29
		12	14 907	5.769 03	316.391 87
		13	14 610	1.565 19	3.932 15
		14	13 160	4.448 91	14.227 09
		15	13 005	5.981 19	11.045 70
		16	10 725	3.129 40	202.253 40
		17	6 126	1.763 3	277.035 0
		18	5 863	0.236 6	529.691 0
		19	5 228	4.207 8	3.181 4
		20	5 020	3.177 9	433.711 7
		21	4 593	0.619 8	199.072 0
		22	4 006	2.244 8	63.735 9
		23	3 874	3.222 8	138.517 5
		24	3 269	0.774 9	949.175 6
		25	2 954	0.982 8	95.979 2
		26	2 461	2.031 6	735.876 5
		27	1 758	3.265 8	522.577 4
		28	1 640	5.505 0	846.082 8
		29	1 581	4.372 7	309.278 3
		30	1 391	4.023 3	323.505 4
		31	1 124	2.837 3	415.552 5
		32	1 087	4.183 4	2.447 7
		33	1 017	3.717 0	227.526 2
		34	957	0.507	1 265.567
		35	853	3.421	175.166
		36	849	3.191	209.367
		37	789	5.007	0.963
		38	749	2.144	853.196
		39	744	5.253	224.345
		40	687	1.747	1 052.268
		41	654	1.599	0.048
		42	634	2.299	412.371
		43	625	0.970	210.118
		44	580	3.093	74.782
		45	546	2.127	350.332
		46	543	1.518	9.561
		47	530	4.449	117.320
		48	478	2.965	137.033
		49	474	5.475	742.990
		50	452	1.044	490.334
		51	449	1.290	127.472
		52	372	2.278	217.231
		53	355	3.013	838.969
		54	347	1.539	340.771
		55	343	0.246	0.521
		56	330	0.247	1 581.959
		57	322	0.961	203.738
		58	322	2.572	647.011
		59	309	3.495	216.480
		60	287	2.370	351.817
		61	278	0.400	211.815
		62	249	1.470	1 368.660
		63	227	4.910	12.530

SATURN (cont.)	L0	64	220	4.204	200.769
		65	209	1.345	625.670
		66	208	0.483	1162.475
		67	208	1.283	39.357
		68	204	6.011	265.989
		69	185	3.503	149.563
		70	184	0.973	4.193
		71	182	5.491	2.921
		72	174	1.863	0.751
		73	165	0.440	5.417
		74	149	5.736	52.690
		75	148	1.535	5.629
		76	146	6.231	195.140
		77	140	4.295	21.341
		78	131	4.068	10.295
		79	125	6.277	1898.351
		80	122	1.976	4.666
		81	118	5.341	554.070
		82	117	2.679	1155.361
		83	114	5.594	1059.382
		84	112	1.105	191.208
		85	110	0.166	1.484
		86	109	3.438	536.805
		87	107	4.012	956.289
		88	104	2.192	88.866
		89	103	1.197	1685.052
		90	101	4.965	269.921
SATURN	L1	1	21354295596	0	0
		2	1296855	1.8282054	213.2990954
		3	564348	2.885001	7.113547
		4	107679	2.277699	206.185548
		5	98323	1.08070	426.59819
		6	40255	2.04128	220.41264
		7	19942	1.27955	103.09277
		8	10512	2.74880	14.22709
		9	6939	0.4049	639.8973
		10	4803	2.4419	419.4846
		11	4056	2.9217	110.2063
		12	3769	3.6497	3.9322
		13	3385	2.4169	3.1814
		14	3302	1.2626	433.7117
		15	3071	2.3274	199.0720
		16	1953	3.5639	11.0457
		17	1249	2.6280	95.9792
		18	922	1.961	227.526
		19	706	4.417	529.691
		20	650	6.174	202.253
		21	628	6.111	309.278
		22	487	6.040	853.196
		23	479	4.988	522.577
		24	468	4.617	63.736
		25	417	2.117	323.505
		26	408	1.299	209.367
		27	352	2.317	632.784
		28	344	3.959	412.371
		29	340	3.634	316.392
		30	336	3.772	735.877
		31	332	2.861	210.118
		32	289	2.733	117.320
		33	281	5.744	2.448
		34	266	0.543	647.011

SATURN (cont.)	L1	35	230	1.644	216.480
		36	192	2.965	224.345
		37	173	4.077	846.083
		38	167	2.597	21.341
		39	136	2.286	10.295
		40	131	3.441	742.990
		41	128	4.095	217.231
		42	109	6.161	415.552
		43	98	4.73	838.97
		44	94	3.48	1 052.27
		45	92	3.95	88.87
		46	87	1.22	440.83
		47	83	3.11	625.67
		48	78	6.24	302.16
		49	67	0.29	4.67
		50	66	5.65	9.56
		51	62	4.29	127.47
		52	62	1.83	195.14
		53	58	2.48	191.96
		54	57	5.02	137.03
		55	55	0.28	74.78
		56	54	5.13	490.33
		57	51	1.46	536.80
		58	47	1.18	149.56
		59	47	5.15	515.46
		60	46	2.23	956.29
		61	44	2.71	5.42
		62	40	0.41	269.92
		63	40	3.89	728.76
		64	38	0.65	422.67
		65	38	2.53	12.53
		66	37	3.78	2.92
		67	35	6.08	5.63
		68	34	3.21	1 368.66
		69	33	4.64	277.03
		70	33	5.43	1 066.50
		71	33	0.30	351.82
		72	32	4.39	1 155.36
		73	31	2.43	52.69
		74	30	2.84	203.00
		75	30	6.19	284.15
		76	30	3.39	1 059.38
		77	29	2.03	330.62
		78	28	2.74	265.99
		79	26	4.51	340.77

SATURN	L2	1	116 441	1.179 879	7.113 547
		2	91 921	0.074 25	213.299 10
		3	90 592	0	0
		4	15 277	4.064 92	206.185 55
		5	10 631	0.257 78	220.412 64
		6	10 605	5.409 64	426.598 19
		7	4 265	1.046 0	14.227 1
		8	1 216	2.918 6	103.092 8
		9	1 165	4.609 4	639.897 3
		10	1 082	5.691 3	433.711 7
		11	1 045	4.042 1	199.072 0
		12	1 020	0.633 7	3.181 4
		13	634	4.388	419.485
		14	549	5.573	3.932
		15	457	1.268	110.206
		16	425	0.209	227.526

SATURN (cont.)	L2	17	274	4.288	95.979
		18	162	1.381	11.046
		19	129	1.566	309.278
		20	117	3.881	853.196
		21	105	4.900	647.011
		22	101	0.893	21.341
		23	96	2.91	316.39
		24	95	5.63	412.37
		25	85	5.73	209.37
		26	83	6.05	216.48
		27	82	1.02	117.32
		28	75	4.76	210.12
		29	67	0.46	522.58
		30	66	0.48	10.29
		31	64	0.35	323.51
		32	61	4.88	632.78
		33	53	2.75	529.69
		34	46	5.69	440.83
		35	45	1.67	202.25
		36	42	5.71	88.87
		37	32	0.07	63.74
		38	32	1.67	302.16
		39	31	4.16	191.96
		40	27	0.83	224.34
		41	25	5.66	735.88
		42	20	5.94	217.23
		43	18	4.90	625.67
		44	17	1.63	742.99
		45	16	0.58	515.46
		46	14	0.21	838.97
		47	14	3.76	195.14
		48	12	4.72	203.00
		49	12	0.13	234.64
		50	12	3.12	846.08
		51	11	5.92	536.80
		52	11	5.60	728.76
		53	11	3.20	1 066.50
		54	10	4.99	422.67
		55	10	0.26	330.62
		56	10	4.15	860.31
		57	9	0.46	956.29
		58	8	2.14	269.92
		59	8	5.25	429.78
		60	8	4.03	9.56
		61	7	5.40	1 052.27
		62	6	4.46	284.15
		63	6	5.93	405.26
SATURN	L3	1	16 039	5.739 45	7.113 55
		2	4 250	4.585 4	213.299 1
		3	1 907	4.760 8	220.412 6
		4	1 466	5.913 3	206.185 5
		5	1 162	5.619 7	14.227 1
		6	1 067	3.608 2	426.598 2
		7	239	3.861	433.712
		8	237	5.768	199.072
		9	166	5.116	3.181
		10	151	2.736	639.897
		11	131	4.743	227.526
		12	63	0.23	419.48
		13	62	4.74	103.09
		14	40	5.47	21.34

SATURN (cont.)	L3	15	40	5.96	95.98
		16	39	5.83	110.21
		17	28	3.01	647.01
		18	25	0.99	3.93
		19	19	1.92	853.20
		20	18	4.97	10.29
		21	18	1.03	412.37
		22	18	4.20	216.48
		23	18	3.32	309.28
		24	16	3.90	440.83
		25	16	5.62	117.32
		26	13	1.18	88.87
		27	11	5.58	11.05
		28	11	5.93	191.96
		29	10	3.95	209.37
		30	9	3.39	302.16
		31	8	4.88	323.51
		32	7	0.38	632.78
		33	6	2.25	522.58
		34	6	1.06	210.12
		35	5	4.64	234.64
		36	4	3.14	0
		37	4	2.31	515.46
		38	3	2.20	860.31
		39	3	0.59	529.69
		40	3	4.93	224.34
		41	3	0.42	625.67
		42	2	4.77	330.62
		43	2	3.35	429.78
		44	2	3.20	202.25
		45	2	1.19	1 066.50
		46	2	1.35	405.26
		47	2	4.16	223.59
		48	2	3.07	654.12
SATURN	L4	1	1 662	3.998 3	7.113 5
		2	257	2.984	220.413
		3	236	3.902	14.227
		4	149	2.741	213.299
		5	114	3.142	0
		6	110	1.515	206.186
		7	68	1.72	426.60
		8	40	2.05	433.71
		9	38	1.24	199.07
		10	31	3.01	227.53
		11	15	0.83	639.90
		12	9	3.71	21.34
		13	6	2.42	419.48
		14	6	1.16	647.01
		15	4	1.45	95.98
		16	4	2.12	440.83
		17	3	4.09	110.21
		18	3	2.77	412.37
		19	3	3.01	88.87
		20	3	0.00	853.20
		21	3	0.39	103.09
		22	2	3.78	117.32
		23	2	2.83	234.64
		24	2	5.08	309.28
		25	2	2.24	216.48
		26	2	5.19	302.16
		27	1	1.55	191.96

SATURN	L5	1	124	2.259	7.114
		2	34	2.16	14.23
		3	28	1.20	220.41
		4	6	1.22	227.53
		5	5	0.24	433.71
		6	4	6.23	426.60
		7	3	2.97	199.07
		8	3	4.29	206.19
		9	2	6.25	213.30
		10	1	5.28	639.90
		11	1	0.24	440.83
		12	1	3.14	0
SATURN	B0	1	4330678	3.6028443	213.2990954
		2	240348	2.852385	426.598191
		3	84746	0	0
		4	34116	0.57297	206.18555
		5	30863	3.48442	220.41264
		6	14734	2.11847	639.89729
		7	9917	5.7900	419.4846
		8	6994	4.7360	7.1135
		9	4808	5.4331	316.3919
		10	4788	4.9651	110.2063
		11	3432	2.7326	433.7117
		12	1506	6.0130	103.0928
		13	1060	5.6310	529.6910
		14	969	5.204	632.784
		15	942	1.396	853.196
		16	708	3.803	323.505
		17	552	5.131	202.253
		18	400	3.359	227.526
		19	319	3.626	209.367
		20	316	1.997	647.011
		21	314	0.465	217.231
		22	284	4.886	224.345
		23	236	2.139	11.046
		24	215	5.950	846.083
		25	209	2.120	415.552
		26	207	0.730	199.072
		27	179	2.954	63.736
		28	141	0.644	490.334
		29	139	4.595	14.227
		30	139	1.998	735.877
		31	135	5.245	742.990
		32	122	3.115	522.577
		33	116	3.109	216.480
		34	114	0.963	210.118
SATURN	B1	1	397555	5.332900	213.299095
		2	49479	3.14159	0
		3	18572	6.09919	426.59819
		4	14801	2.30586	206.18555
		5	9644	1.6967	220.4126
		6	3757	1.2543	419.4846
		7	2717	5.9117	639.8973
		8	1455	0.8516	433.7117
		9	1291	2.9177	7.1135
		10	853	0.436	316.392
		11	298	0.919	632.784
		12	292	5.316	853.196
		13	284	1.619	227.526
		14	275	3.889	103.093
		15	172	0.052	647.011

SATURN (cont.)	B1	16	166	2.444	199.072
		17	158	5.209	110.206
		18	128	1.207	529.691
		19	110	2.457	217.231
		20	82	2.76	210.12
		21	81	2.86	14.23
		22	69	1.66	202.25
		23	65	1.26	216.48
		24	61	1.25	209.37
		25	59	1.82	323.51
		26	46	0.82	440.83
		27	36	1.82	224.34
		28	34	2.84	117.32
		29	33	1.31	412.37
		30	32	1.19	846.08
		31	27	4.65	1 066.50
		32	27	4.44	11.05
SATURN	B2	1	20 630	0.504 82	213.299 10
		2	3 720	3.998 3	206.185 5
		3	1 627	6.181 9	220.412 6
		4	1 346	0	0
		5	706	3.039	419.485
		6	365	5.099	426.598
		7	330	5.279	433.712
		8	219	3.828	639.897
		9	139	1.043	7.114
		10	104	6.157	227.526
		11	93	1.98	316.39
		12	71	4.15	199.07
		13	52	2.88	632.78
		14	49	4.43	647.01
		15	41	3.16	853.20
		16	29	4.53	210.12
		17	24	1.12	14.23
		18	21	4.35	217.23
		19	20	5.31	440.83
		20	18	0.85	110.21
		21	17	5.68	216.48
		22	16	4.26	103.09
		23	14	3.00	412.37
		24	12	2.53	529.69
		25	8	3.32	202.25
		26	7	5.56	209.37
		27	7	0.29	323.51
		28	6	1.16	117.32
		29	6	3.61	860.31
SATURN	B3	1	666	1.990	213.299
		2	632	5.698	206.186
		3	398	0	0
		4	188	4.338	220.413
		5	92	4.84	419.48
		6	52	3.42	433.71
		7	42	2.38	426.60
		8	26	4.40	227.53
		9	21	5.85	199.07
		10	18	1.99	639.90
		11	11	5.37	7.11
		12	10	2.55	647.01
		13	7	3.46	316.39
		14	6	4.80	632.78
		15	6	0.02	210.12

SATURN (cont.)	B3	16	6	3.52	440.83
		17	5	5.64	14.23
		18	5	1.22	853.20
		19	4	4.71	412.37
		20	3	0.63	103.09
		21	2	3.72	216.48
SATURN	B4	1	80	1.12	206.19
		2	32	3.12	213.30
		3	17	2.48	220.41
		4	12	3.14	0
		5	9	0.38	419.48
		6	6	1.56	433.71
		7	5	2.63	227.53
		8	5	1.28	199.07
		9	1	1.43	426.60
		10	1	0.67	647.01
		11	1	1.72	440.83
		12	1	6.18	639.90
SATURN	B5	1	8	2.82	206.19
		2	1	0.51	220.41
SATURN	R0	1	955 758 136	0	0
		2	52 921 382	2.392 262 20	213.299 095 44
		3	1 873 680	5.235 496 1	206.185 548 4
		4	1 464 664	1.647 630 5	426.598 190 9
		5	821 891	5.935 200	316.391 870
		6	547 507	5.015 326	103.092 774
		7	371 684	2.271 148	220.412 642
		8	361 778	3.139 043	7.113 547
		9	140 618	5.704 067	632.783 739
		10	108 975	3.293 136	110.206 321
		11	69 007	5.941 00	419.484 64
		12	61 053	0.940 38	639.897 29
		13	48 913	1.557 33	202.253 40
		14	34 144	0.195 19	277.034 99
		15	32 402	5.470 85	949.175 61
		16	20 937	0.463 49	735.876 51
		17	20 839	1.521 03	433.711 74
		18	20 747	5.332 56	199.072 00
		19	15 298	3.059 44	529.690 97
		20	14 296	2.604 34	323.505 42
		21	12 884	1.648 92	138.517 50
		22	11 993	5.980 51	846.082 83
		23	11 380	1.731 06	522.577 42
		24	9 796	5.204 8	1 265.567 5
		25	7 753	5.851 9	95.979 2
		26	6 771	3.004 3	14.227 1
		27	6 466	0.177 3	1 052.268 4
		28	5 850	1.455 2	415.552 5
		29	5 307	0.597 4	63.735 9
		30	4 696	2.149 2	227.526 2
		31	4 044	1.640 1	209.366 9
		32	3 688	0.780 2	412.371 1
		33	3 461	1.850 9	175.166 1
		34	3 420	4.945 5	1 581.959 3
		35	3 401	0.553 9	350.332 1
		36	3 376	3.695 3	224.344 8
		37	2 976	5.684 7	210.117 7
		38	2 885	1.387 6	838.969 3
		39	2 881	0.179 6	853.196 4
		40	2 508	3.538 5	742.990 1

SATURN (cont.)	R0	41	2 448	6.184 1	1 368.660 3
		42	2 406	2.965 6	117.319 9
		43	2 174	0.015 1	340.770 9
		44	2 024	5.054 1	11.045 7
SATURN	R1	1	6 182 981	0.258 435 2	213.299 095 4
		2	506 578	0.711 147	206.185 548
		3	341 394	5.796 358	426.598 191
		4	188 491	0.472 157	220.412 642
		5	186 262	3.141 593	0
		6	143 891	1.407 449	7.113 547
		7	49 621	6.017 44	103.092 77
		8	20 928	5.092 46	639.897 29
		9	19 953	1.175 60	419.484 64
		10	18 840	1.608 20	110.206 32
		11	13 877	0.758 86	199.072 00
		12	12 893	5.943 30	433.711 74
		13	5 397	1.288 5	14.227 1
		14	4 869	0.867 9	323.505 4
		15	4 247	0.393 0	227.526 2
		16	3 252	1.258 5	95.979 2
		17	3 081	3.436 6	522.577 4
		18	2 909	4.606 8	202.253 4
		19	2 856	2.167 3	735.876 5
		20	1 988	2.450 5	412.371 1
		21	1 941	6.023 9	209.366 9
		22	1 581	1.291 9	210.117 7
		23	1 340	4.308 0	853.196 4
		24	1 316	1.253 0	117.319 9
		25	1 203	1.866 5	316.391 9
		26	1 091	0.075 3	216.480 5
		27	966	0.480	632.784
		28	954	5.152	647.011
		29	898	0.983	529.691
		30	882	1.885	1 052.268
		31	874	1.402	224.345
		32	785	3.064	838.969
		33	740	1.382	625.670
		34	658	4.144	309.278
		35	650	1.725	742.990
		36	613	3.033	63.736
		37	599	2.549	217.231
		38	503	2.130	3.932
SATURN	R2	1	436 902	4.786 717	213.299 095
		2	71 923	2.500 70	206.185 55
		3	49 767	4.971 68	220.412 64
		4	43 221	3.869 40	426.598 19
		5	29 646	5.963 10	7.113 55
		6	4 721	2.475 3	199.072 0
		7	4 142	4.106 7	433.711 7
		8	3 789	3.097 7	639.897 3
		9	2 964	1.372 1	103.092 8
		10	2 556	2.850 7	419.484 6
		11	2 327	0	0
		12	2 208	6.275 9	110.206 3
		13	2 188	5.855 5	14.227 1
		14	1 957	4.924 5	227.526 2
		15	924	5.464	323.505
		16	706	2.971	95.979
		17	546	4.129	412.371
		18	431	5.178	522.577
		19	405	4.173	209.367

SATURN (cont.)	R2	20	391	4.481	216.480
		21	374	5.834	117.320
		22	361	3.277	647.011
		23	356	3.192	210.118
		24	326	2.269	853.196
		25	207	4.022	735.877
		26	204	0.088	202.253
		27	180	3.597	632.784
		28	178	4.097	440.825
		29	154	3.135	625.670
		30	148	0.136	302.165
		31	133	2.594	191.958
		32	132	5.933	309.278
SATURN	R3	1	20 315	3.021 87	213.299 10
		2	8 924	3.191 4	220.412 6
		3	6 909	4.351 7	206.185 5
		4	4 087	4.224 1	7.113 5
		5	3 879	2.010 6	426.598 2
		6	1 071	4.203 6	199.072 0
		7	907	2.283	433.712
		8	606	3.175	227.526
		9	597	4.135	14.227
		10	483	1.173	639.897
		11	393	0	0
		12	229	4.698	419.485
		13	188	4.590	110.206
		14	150	3.202	103.093
		15	121	3.768	323.505
		16	102	4.710	95.979
		17	101	5.819	412.371
		18	93	1.44	647.01
		19	84	2.63	216.48
		20	73	4.15	117.32
		21	62	2.31	440.83
		22	55	0.31	853.20
		23	50	2.39	209.37
		24	45	4.37	191.96
		25	41	0.69	522.58
		26	40	1.84	302.16
		27	38	5.94	88.87
		28	32	4.01	21.34
SATURN	R4	1	1 202	1.415 0	220.412 6
		2	708	1.162	213.299
		3	516	6.240	206.186
		4	427	2.469	7.114
		5	268	0.187	426.598
		6	170	5.959	199.072
		7	150	0.480	433.712
		8	145	1.442	227.526
		9	121	2.405	14.227
		10	47	5.57	639.90
		11	19	5.86	647.01
		12	17	0.53	440.83
		13	16	2.90	110.21
		14	15	0.30	419.48
		15	14	1.30	412.37
		16	13	2.09	323.51
		17	11	0.22	95.98
		18	11	2.46	117.32
		19	10	3.14	0
		20	9	1.56	88.87

SATURN (cont.)	R4	21	9	2.28	21.34
		22	9	0.68	216.48
		23	8	1.27	234.64
SATURN	R5	1	129	5.913	220.413
		2	32	0.69	7.11
		3	27	5.91	227.53
		4	20	4.95	433.71
		5	20	0.67	14.23
		6	14	2.67	206.19
		7	14	1.46	199.07
		8	13	4.59	426.60
		9	7	4.63	213.30
		10	5	3.61	639.90
		11	4	4.90	440.83
		12	3	4.07	647.01
		13	3	4.66	191.96
		14	3	0.49	323.51
		15	3	3.18	419.48
		16	2	3.70	88.87
		17	2	3.32	95.98
		18	2	0.56	117.32
URANUS	L0	1	548 129 294	0	0
		2	9 260 408	0.891 064 2	74.781 598 6
		3	1 504 248	3.627 192 6	1.484 472 7
		4	365 982	1.899 622	73.297 126
		5	272 328	3.358 237	149.563 197
		6	70 328	5.392 54	63.735 90
		7	68 893	6.092 92	76.266 07
		8	61 999	2.269 52	2.968 95
		9	61 951	2.850 99	11.045 70
		10	26 469	3.141 52	71.812 65
		11	25 711	6.113 80	454.909 37
		12	21 079	4.360 59	148.078 72
		13	17 819	1.744 37	36.648 56
		14	14 613	4.737 32	3.932 15
		15	11 163	5.826 82	224.344 80
		16	10 998	0.488 65	138.517 50
		17	9 527	2.955 2	35.164 1
		18	7 546	5.236 3	109.945 7
		19	4 220	3.233 3	70.849 4
		20	4 052	2.277 5	151.047 7
		21	3 490	5.483 1	146.594 3
		22	3 355	1.065 5	4.453 4
		23	3 144	4.752 0	77.750 5
		24	2 927	4.629 0	9.561 2
		25	2 922	5.352 4	85.827 3
		26	2 273	4.366 0	70.328 2
		27	2 149	0.607 5	38.133 0
		28	2 051	1.517 7	0.111 9
		29	1 992	4.924 4	277.035 0
		30	1 667	3.627 4	380.127 8
		31	1 533	2.585 9	52.690 2
		32	1 376	2.042 8	65.220 4
		33	1 372	4.196 4	111.430 2
		34	1 284	3.113 5	202.253 4
		35	1 282	0.542 7	222.860 3
		36	1 244	0.916 1	2.447 7
		37	1 221	0.199 0	108.461 2
		38	1 151	4.179 0	33.679 6
		39	1 150	0.933 4	3.181 4

URANUS (cont.)	L0	40	1 090	1.775 0	12.530 2
		41	1 072	0.235 6	62.251 4
		42	946	1.192	127.472
		43	708	5.183	213.299
		44	653	0.966	78.714
		45	628	0.182	984.600
		46	607	5.432	529.691
		47	559	3.358	0.521
		48	524	2.013	299.126
		49	483	2.106	0.963
		50	471	1.407	184.727
		51	467	0.415	145.110
		52	434	5.521	183.243
		53	405	5.987	8.077
		54	399	0.338	415.552
		55	396	5.870	351.817
		56	379	2.350	56.622
		57	310	5.833	145.631
		58	300	5.644	22.091
		59	294	5.839	39.618
		60	252	1.637	221.376
		61	249	4.746	225.829
		62	239	2.350	137.033
		63	224	0.516	84.343
		64	223	2.843	0.261
		65	220	1.922	67.668
		66	217	6.142	5.938
		67	216	4.778	340.771
		68	208	5.580	68.844
		69	202	1.297	0.048
		70	199	0.956	152.532
		71	194	1.888	456.394
		72	193	0.916	453.425
		73	187	1.319	0.160
		74	182	3.536	79.235
		75	173	1.539	160.609
		76	172	5.680	219.891
		77	170	3.677	5.417
		78	169	5.879	18.159
		79	165	1.424	106.977
		80	163	3.050	112.915
		81	158	0.738	54.175
		82	147	1.263	59.804
		83	143	1.300	35.425
		84	139	5.386	32.195
		85	139	4.260	909.819
		86	124	1.374	7.114
		87	110	2.027	554.070
		88	109	5.706	77.963
		89	104	5.028	0.751
		90	104	1.458	24.379
		91	103	0.681	14.978
URANUS	L1	1	7 502 543 122	0	0
		2	154 458	5.242 017	74.781 599
		3	24 456	1.712 56	1.484 47
		4	9 258	0.428 4	11.045 7
		5	8 266	1.502 2	63.735 9
		6	7 842	1.319 8	149.563 2
		7	3 899	0.464 8	3.932 2
		8	2 284	4.173 7	76.266 1
		9	1 927	0.530 1	2.968 9
		10	1 233	1.586 3	70.849 4

URANUS (cont.)	L1	11	791	5.436	3.181
		12	767	1.996	73.297
		13	482	2.984	85.827
		14	450	4.138	138.517
		15	446	3.723	224.345
		16	427	4.731	71.813
		17	354	2.583	148.079
		18	348	2.454	9.561
		19	317	5.579	52.690
		20	206	2.363	2.448
		21	189	4.202	56.622
		22	184	0.284	151.048
		23	180	5.684	12.530
		24	171	3.001	78.714
		25	158	2.909	0.963
		26	155	5.591	4.453
		27	154	4.652	35.164
		28	152	2.942	77.751
		29	143	2.590	62.251
		30	121	4.148	127.472
		31	116	3.732	65.220
		32	102	4.188	145.631
		33	102	6.034	0.112
		34	88	3.99	18.16
		35	88	6.16	202.25
		36	81	2.64	22.09
		37	72	6.05	70.33
		38	69	4.05	77.96
		39	59	3.70	67.67
		40	47	3.54	351.82
		41	44	5.91	7.11
		42	43	5.72	5.42
		43	39	4.92	222.86
		44	36	5.90	33.68
		45	36	3.29	8.08
		46	36	3.33	71.60
		47	35	5.08	38.13
		48	31	5.62	984.60
		49	31	5.50	59.80
		50	31	5.46	160.61
		51	30	1.66	447.80
		52	29	1.15	462.02
		53	29	4.52	84.34
		54	27	5.54	131.40
		55	27	6.15	299.13
		56	26	4.99	137.03
		57	25	5.74	380.13
URANUS	L2	1	53033	0	0
		2	2358	2.2601	74.7816
		3	769	4.526	11.046
		4	552	3.258	63.736
		5	542	2.276	3.932
		6	529	4.923	1.484
		7	258	3.691	3.181
		8	239	5.858	149.563
		9	182	6.218	70.849
		10	54	1.44	76.27
		11	49	6.03	56.62
		12	45	3.91	2.45
		13	45	0.81	85.83
		14	38	1.78	52.69
		15	37	4.46	2.97

URANUS (cont.)	L2	16	33	0.86	9.56
		17	29	5.10	73.30
		18	24	2.11	18.16
		19	22	5.99	138.52
		20	22	4.82	78.71
		21	21	2.40	77.96
		22	21	2.17	224.34
		23	17	2.54	145.63
		24	17	3.47	12.53
		25	12	0.02	22.09
		26	11	0.08	127.47
		27	10	5.16	71.60
		28	10	4.46	62.25
		29	9	4.26	7.11
		30	8	5.50	67.67
		31	7	1.25	5.42
		32	6	3.36	447.80
		33	6	5.45	65.22
		34	6	4.52	151.05
		35	6	5.73	462.02
URANUS	L3	1	121	0.024	74.782
		2	68	4.12	3.93
		3	53	2.39	11.05
		4	46	0	0
		5	45	2.04	3.18
		6	44	2.96	1.48
		7	25	4.89	63.74
		8	21	4.55	70.85
		9	20	2.31	149.56
		10	9	1.58	56.62
		11	4	0.23	18.16
		12	4	5.39	76.27
		13	4	0.95	77.96
		14	3	4.98	85.83
		15	3	4.13	52.69
		16	3	0.37	78.71
		17	2	0.86	145.63
		18	2	5.66	9.56
URANUS	L4	1	114	3.142	0
		2	6	4.58	74.78
		3	3	0.35	11.05
		4	1	3.42	56.62
URANUS	B0	1	1 346 278	2.618 778 1	74.781 598 6
		2	62 341	5.081 11	149.563 20
		3	61 601	3.141 59	0
		4	9 964	1.616 0	76.266 1
		5	9 926	0.576 3	73.297 1
		6	3 259	1.261 2	224.344 8
		7	2 972	2.243 7	1.484 5
		8	2 010	6.055 5	148.078 7
		9	1 522	0.279 6	63.735 9
		10	924	4.038	151.048
		11	761	6.140	71.813
		12	522	3.321	138.517
		13	463	0.743	85.827
		14	437	3.381	529.691
		15	435	0.341	77.751
		16	431	3.554	213.299
		17	420	5.213	11.046
		18	245	0.788	2.969

URANUS (cont.)	B0	19	233	2.257	222.860
		20	216	1.591	38.133
		21	180	3.725	299.126
		22	175	1.236	146.594
		23	174	1.937	380.128
		24	160	5.336	111.430
		25	144	5.962	35.164
		26	116	5.739	70.849
		27	106	0.941	70.328
		28	102	2.619	78.714
URANUS	B1	1	206 366	4.123 943	74.781 599
		2	8 563	0.338 2	149.563 2
		3	1 726	2.121 9	73.297 1
		4	1 374	0	0
		5	1 369	3.068 6	76.266 1
		6	451	3.777	1.484
		7	400	2.848	224.345
		8	307	1.255	148.079
		9	154	3.786	63.736
		10	112	5.573	151.048
		11	111	5.329	138.517
		12	83	3.59	71.81
		13	56	3.40	85.83
		14	54	1.70	77.75
		15	42	1.21	11.05
		16	41	4.45	78.71
		17	32	3.77	222.86
		18	30	2.56	2.97
		19	27	5.34	213.30
		20	26	0.42	380.13
URANUS	B2	1	9 212	5.800 4	74.781 6
		2	557	0	0
		3	286	2.177	149.563
		4	95	3.84	73.30
		5	45	4.88	76.27
		6	20	5.46	1.48
		7	15	0.88	138.52
		8	14	2.85	148.08
		9	14	5.07	63.74
		10	10	5.00	224.34
		11	8	6.27	78.71
URANUS	B3	1	268	1.251	74.782
		2	11	3.14	0
		3	6	4.01	149.56
		4	3	5.78	73.30
URANUS	B4	1	6	2.85	74.78
URANUS	R0	1	1 921 264 848	0	0
		2	88 784 984	5.603 775 27	74.781 598 57
		3	3 440 836	0.328 361 0	73.297 125 9
		4	2 055 653	1.782 951 7	149.563 197 1
		5	649 322	4.522 473	76.266 071
		6	602 248	3.860 038	63.735 898
		7	496 404	1.401 399	454.909 367
		8	338 526	1.580 027	138.517 497
		9	243 508	1.570 866	71.812 653
		10	190 522	1.998 094	1.484 473
		11	161 858	2.791 379	148.078 724
		12	143 706	1.383 686	11.045 700

URANUS (cont.)	R0	13	93 192	0.174 37	36.648 56
		14	89 806	3.661 05	109.945 69
		15	71 424	4.245 09	224.344 80
		16	46 677	1.399 77	35.164 09
		17	39 026	3.362 35	277.034 99
		18	39 010	1.669 71	70.849 45
		19	36 755	3.886 49	146.594 25
		20	30 349	0.701 00	151.047 67
		21	29 156	3.180 56	77.750 54
		22	25 786	3.785 38	85.827 30
		23	25 620	5.256 56	380.127 77
		24	22 637	0.725 19	529.690 97
		25	20 473	2.796 40	70.328 18
		26	20 472	1.555 89	202.253 40
		27	17 901	0.554 55	2.968 95
		28	15 503	5.354 05	38.133 04
		29	14 702	4.904 34	108.461 22
		30	12 897	2.621 54	111.430 16
		31	12 328	5.960 39	127.471 80
		32	11 959	1.750 44	984.600 33
		33	11 853	0.993 43	52.690 20
		34	11 696	3.298 26	3.932 15
		35	11 495	0.437 74	65.220 37
		36	10 793	1.421 05	213.299 10
		37	9 111	4.996 4	62.251 4
		38	8 421	5.253 5	222.860 3
		39	8 402	5.038 8	415.552 5
		40	7 449	0.794 9	351.816 6
		41	7 329	3.972 8	183.242 8
		42	6 046	5.679 6	78.713 8
		43	5 524	3.115 0	9.561 2
		44	5 445	5.105 8	145.109 8
		45	5 238	2.629 6	33.679 6
		46	4 079	3.220 6	340.770 9
		47	3 919	4.250 2	39.617 5
		48	3 802	6.109 9	184.727 3
		49	3 781	3.458 4	456.393 8
		50	3 687	2.487 2	453.424 9
		51	3 102	4.140 3	219.891 4
		52	2 963	0.829 8	56.622 4
		53	2 942	0.423 9	299.126 4
		54	2 940	2.146 4	137.033 0
		55	2 938	3.676 6	140.002 0
		56	2 865	0.310 0	12.530 2
		57	2 538	4.854 6	131.403 9
		58	2 364	0.442 5	554.070 0
		59	2 183	2.940 4	305.346 2
URANUS	R1	1	1 479 896	3.672 057 1	74.781 598 6
		2	71 212	6.226 01	63.735 90
		3	68 627	6.134 11	149.563 20
		4	24 060	3.141 59	0
		5	21 468	2.601 77	76.266 07
		6	20 857	5.246 25	11.045 70
		7	11 405	0.018 48	70.849 45
		8	7 497	0.423 6	73.297 1
		9	4 244	1.416 9	85.827 3
		10	3 927	3.155 1	71.812 7
		11	3 578	2.311 6	224.344 8
		12	3 506	2.583 5	138.517 5
		13	3 229	5.255 0	3.932 2
		14	3 060	0.153 2	1.484 5
		15	2 564	0.980 8	148.078 7

URANUS (cont.)	R1	16	2 429	3.994 4	52.690 2
		17	1 645	2.653 5	127.471 8
		18	1 584	1.430 5	78.713 8
		19	1 508	5.060 0	151.047 7
		20	1 490	2.675 6	56.622 4
		21	1 413	4.574 6	202.253 4
		22	1 403	1.369 9	77.750 5
		23	1 228	1.047 0	62.251 4
		24	1 033	0.264 6	131.403 9
		25	992	2.172	65.220
		26	862	5.055	351.817
		27	744	3.076	35.164
		28	687	2.499	77.963
		29	647	4.473	70.328
		30	624	0.863	9.561
		31	604	0.907	984.600
		32	575	3.231	447.796
		33	562	2.718	462.023
		34	530	5.917	213.299
		35	528	5.151	2.969
URANUS	R2	1	22 440	0.699 53	74.781 60
		2	4 727	1.699 0	63.735 9
		3	1 682	4.648 3	70.849 4
		4	1 650	3.096 6	11.045 7
		5	1 434	3.521 2	149.563 2
		6	770	0	0
		7	500	6.172	76.266
		8	461	0.767	3.932
		9	390	4.496	56.622
		10	390	5.527	85.827
		11	292	0.204	52.690
		12	287	3.534	73.297
		13	273	3.847	138.517
		14	220	1.964	131.404
		15	216	0.848	77.963
		16	205	3.248	78.714
		17	149	4.898	127.472
		18	129	2.081	3.181
URANUS	R3	1	1 164	4.734 5	74.781 6
		2	212	3.343	63.736
		3	196	2.980	70.849
		4	105	0.958	11.046
		5	73	1.00	149.56
		6	72	0.03	56.62
		7	55	2.59	3.93
		8	36	5.65	77.96
		9	34	3.82	76.27
		10	32	3.60	131.40
URANUS	R4	1	53	3.01	74.78
		2	10	1.91	56.62

NEPTUNE	L0	1	531 188 633	0	0
		2	1 798 476	2.901 012 7	38.133 035 6
		3	1 019 728	0.485 809 2	1.484 472 7
		4	124 532	4.830 081	36.648 563
		5	42 064	5.410 55	2.968 95
		6	37 715	6.092 22	35.164 09
		7	33 785	1.244 89	76.266 07
		8	16 483	0.000 08	491.557 93
		9	9 199	4.937 5	39.617 5
		10	8 994	0.274 6	175.166 1
		11	4 216	1.987 1	73.297 1
		12	3 365	1.035 9	33.679 6
		13	2 285	4.206 1	4.453 4
		14	1 434	2.783 4	74.781 6
		15	900	2.076	109.946
		16	745	3.190	71.813
		17	506	5.748	114.399
		18	400	0.350	1 021.249
		19	345	3.462	41.102
		20	340	3.304	77.751
		21	323	2.248	32.195
		22	306	0.497	0.521
		23	287	4.505	0.048
		24	282	2.246	146.594
		25	267	4.889	0.963
		26	252	5.782	388.465
		27	245	1.247	9.561
		28	233	2.505	137.033
		29	227	1.797	453.425
		30	170	3.324	108.461
		31	151	2.192	33.940
		32	150	2.997	5.938
		33	148	0.859	111.430
		34	119	3.677	2.448
		35	109	2.416	183.243
		36	103	0.041	0.261
		37	103	4.404	70.328
		38	102	5.705	0.112

NEPTUNE	L1	1	3 837 687 717	0	0
		2	16 604	4.863 19	1.484 47
		3	15 807	2.279 23	38.133 04
		4	3 335	3.682 0	76.266 1
		5	1 306	3.673 2	2.968 9
		6	605	1.505	35.164
		7	179	3.453	39.618
		8	107	2.451	4.453
		9	106	2.755	33.680
		10	73	5.49	36.65
		11	57	1.86	114.40
		12	57	5.22	0.52
		13	35	4.52	74.78
		14	32	5.90	77.75
		15	30	3.67	388.47
		16	29	5.17	9.56
		17	29	5.17	2.45
		18	26	5.25	168.05

NEPTUNE	L2	1	53 893	0	0
		2	296	1.855	1.484
		3	281	1.191	38.133
		4	270	5.721	76.266
		5	23	1.21	2.97
		6	9	4.43	35.16
		7	7	0.54	2.45
NEPTUNE	L3	1	31	0	0
		2	15	1.35	76.27
		3	12	6.04	1.48
		4	12	6.11	38.13
NEPTUNE	L4	1	114	3.142	0
NEPTUNE	B0	1	3 088 623	1.441 043 7	38.133 035 6
		2	27 780	5.912 72	76.266 07
		3	27 624	0	0
		4	15 448	3.508 77	39.617 51
		5	15 355	2.521 24	36.648 56
		6	2 000	1.510 0	74.781 6
		7	1 968	4.377 8	1.484 5
		8	1 015	3.215 6	35.164 1
		9	606	2.802	73.297
		10	595	2.129	41.102
		11	589	3.187	2.969
		12	402	4.169	114.399
		13	280	1.682	77.751
		14	262	3.767	213.299
		15	254	3.271	453.425
		16	206	4.257	529.691
		17	140	3.530	137.033
NEPTUNE	B1	1	227 279	3.807 931	38.133 036
		2	1 803	1.975 8	76.266 1
		3	1 433	3.141 6	0
		4	1 386	4.825 6	36.648 6
		5	1 073	6.080 5	39.617 5
		6	148	3.858	74.782
		7	136	0.478	1.484
		8	70	6.19	35.16
		9	52	5.05	73.30
		10	43	0.31	114.40
		11	37	4.89	41.10
		12	37	5.76	2.97
		13	26	5.22	213.30
NEPTUNE	B2	1	9 691	5.571 2	38.133 0
		2	79	3.63	76.27
		3	72	0.45	36.65
		4	59	3.14	0
		5	30	1.61	39.62
		6	6	5.61	74.78
NEPTUNE	B3	1	273	1.017	38.133
		2	2	0	0
		3	2	2.37	36.65
		4	2	5.33	76.27
NEPTUNE	B4	1	6	2.67	38.13

NEPTUNE	R0	1	3 007 013 206	0	0
		2	27 062 259	1.329 994 59	38.133 035 64
		3	1 691 764	3.251 861 4	36.648 562 9
		4	807 831	5.185 928	1.484 473
		5	537 761	4.521 139	35.164 090
		6	495 726	1.571 057	491.557 929
		7	274 572	1.845 523	175.166 060
		8	135 134	3.372 206	39.617 508
		9	121 802	5.797 544	76.266 071
		10	100 895	0.377 027	73.297 126
		11	69 792	3.796 17	2.968 95
		12	46 688	5.749 38	33.679 62
		13	24 594	0.508 02	109.945 69
		14	16 939	1.594 22	71.812 65
		15	14 230	1.077 86	74.781 60
		16	12 012	1.920 62	1 021.248 89
		17	8 395	0.678 2	146.594 3
		18	7 572	1.071 5	388.465 2
		19	5 721	2.590 6	4.453 4
		20	4 840	1.906 9	41.102 0
		21	4 483	2.905 7	529.691 0
		22	4 421	1.749 9	108.461 2
		23	4 354	0.679 9	32.195 1
		24	4 270	3.413 4	453.424 9
		25	3 381	0.848 1	183.242 8
		26	2 881	1.986 0	137.033 0
		27	2 879	3.674 2	350.332 1
		28	2 636	3.097 6	213.299 1
		29	2 530	5.798 4	490.073 5
		30	2 523	0.486 3	493.042 4
		31	2 306	2.809 6	70.328 2
		32	2 087	0.618 6	33.940 2
NEPTUNE	R1	1	236 339	0.704 980	38.133 036
		2	13 220	3.320 15	1.484 47
		3	8 622	6.216 3	35.164 1
		4	2 702	1.881 4	39.617 5
		5	2 155	2.094 3	2.968 9
		6	2 153	5.168 7	76.266 1
		7	1 603	0	0
		8	1 464	1.184 2	33.679 6
		9	1 136	3.918 9	36.648 6
		10	898	5.241	388.465
		11	790	0.533	168.053
		12	760	0.021	182.280
		13	607	1.077	1 021.249
		14	572	3.401	484.444
		15	561	2.887	498.671
NEPTUNE	R2	1	4 247	5.899 1	38.133 0
		2	218	0.346	1.484
		3	163	2.239	168.053
		4	156	4.594	182.280
		5	127	2.848	35.164
NEPTUNE	R3	1	166	4.552	38.133

Appendix III

Companion diskette

Separately available is a companion diskette for IBM PC's and compatibles (360K $5\frac{1}{4}$" or 720K $3\frac{1}{2}$"). Instructions on how to use the disk are given on the disk.

The disk contains source codes for all the necessary routines for developers of astronomical software, in one of three languages: Turbo Pascal 4.0 (or higher), QuickBasic 4.5, and C. It saves the programmer the time of writing the basic routines and entering thousands of numerical constants (*including the 2430 periodic terms from Appendix II*), allowing him to concentrate on the application and the presentation of the results. The disk is therefore useful for beginners as well as experienced programmers.

To illustrate how the various routines work together, the following sample programs are provided. They all use the routines in the library and cover all the important topics in the book.

1. ALMANAC

This program is an 'automatic almanac generator'. It calculates a.o. equinoxes and solstices, conjunctions and oppositions, and greatest elongations for a given year.

2. CALENDAR

This is a comprehensive general-purpose calendar program which also provides the rise and setting times of the Sun and the Moon.

3. EPHEM

Calculates an ephemeris for a comet, an asteroid, or any other solar system object from its orbital elements.

4. JSATS

Shows the positions of the four Galilean satellites of Jupiter *graphically*. Uses the low precision formulae.

5. MOON

Gives a complete positional and physical ephemeris of the Moon : ecliptic and equatorial coordinates, phase, rise and setting times, and both physical and optical libration. Calculating a lunar ephemeris is one of the more difficult subjects covered in this book.

6. PLANETS

Calculates a complete ephemeris of the Sun or any of the planets : heliocentric, geocentric and topocentric coordinates, rise and setting times, phase and magnitude, and a physical ephemeris where available. *This program includes all the planetary terms from Appendix II !*

7. STAR

Gives the apparent place of a star at any instant.

A math coprocessor can be useful for speeding up some of the calculations, but it is not needed.

The programs were written by Jeffrey Sax. The author would like to thank Christian Steyaert and other members of the Working Group Information Technology of the VVS (the Belgian astronomical association) for the testing and the many useful tips and suggestions.

See last of page this book for software order form

Index

The numbers refer to the pages

Some Other Books Published by Willmann-Bell, Inc.

Methods of Orbit Determination for the Micrcomputer, by D.L. Boulet, 6.00" by 9.00", 584 pages, hardbound, $24.95, ISBN 0-943396-34-4. This book describes how the principles of celestial mechanics may be applied to determine the orbits of planets, comets and earth satellites. Until recently, this exciting adventure with nature was beyond the reach of nearly all non-specialists. However the power of the microcomputer has swept away the drudgery of tedious calculations fraught with endless opportunities for careless error. With this book and a microcomputer the enthusiast may have the satisfaction of conquering problems which preoccupied astronomy for hundreds of years, and, in the process, gain a fresh appreciation for the genius and industry of the great mathematicians of the seventeenth, eighteenth, and nineteenth centuries.

This is a how-to-do-it book. Even though the derivations of many important relationships are described in some detail, the emphasis throughout is on practical applications. The reader need only accept the validity of the key equations and understand their symbology in order to use the computer programs to explore the power of the mathematical models. All the important principles have been reduced to complete computer programs written in simple BASIC that will execute directly on a Macintosh using Microsoft BASIC or (with the addition of a statement as line 1005 to reserve extra space in memory) an IBM-PC using BASICA or GWBASIC. For clarity, each program is preceded by an algorithm that describes the sequence of computation and ties it to the mathematics in the text. Further, the program is illustrated by at least one numerical example. Finally, the output from the examples is shown in the format produced by the computer routine. Magnetic media versions of the source code are also available for both IBM-PC and Macintosh computers at $19.95 each. Who will find this book of value? Amateur astronomer's who want to determine the orbits of planets, comets or Earth satellites. Teachers and students of basic calculus, physics, astronomy or computer programming for a source of material that illuminates and expands upon subjects covered in the classroom.

Fundamentals of Celestial Mechanics, by J.M.A. Danby, 6.00" by 9.00", 466 pages, hardbound, $24.95, ISBN 0-943396-20-4. This text assumes an understanding of calculus and elementary differential equations. Emphasis is on computations. Sample BASIC program listings (for a PC) are included. Covered are the problem of two bodies including the use of universal variables, several methods (including that of Laguerre) for solving Kepler's equation, and three methods for solving the two point boundary value problem. The chapter on the determination of orbits includes two versions of Gauss' method, the application of least squares and an introduction to recursive methods. The chapter on numerical methods includes three methods for the numerical integration of differential equations, one of which has full stepsize control. There are also chapters on perturbations, the three- and n-body problems plus much, much more.

Astronomical Formulae for Calculators, by Jean Meeus, 6.00" by 7.00", 201 pages, softbound, $14.95, ISBN 0-943396-22-0. Now in its 4th edition, it is the one source to which thousands turn when they need both equations and worked examples. Almost every other astronomical math book now in print credits Meeus as a reference but they do not provide you with his detailed formulae. *The Journal of The Royal Astronomical Society of Canada* said: "In just under 200 pages, he has covered most of the calculations one is likely to encounter in practical, computational astronomy. The writing is crisp and clear throughout and the examples are thoughtful and well-formatted."

Elements of Solar Eclipses 1951–2200, by Jean Meeus, 8.50" by 11.00", 112 pages, softbound, $19.95, ISBN 0-943396-21-2. This book contains Besselian elements for the 570 solar eclipses during the 250 years between 1951 and 2200. The elements were calculated using highly accurate modern theories of the Sun and Moon developed at the Bureau des Longitudes of Paris. Formulae are provided for the calculation of local circumstances, points on the central line or the northern and southern limits, etc. These algorithms can easily be programmed on a home computer and checked against numerical examples in this book.

Transits, Jean Meeus, 8.50" by 11.00", 75 pages. softbound, $14.95, ISBN 0-943396-26-3. Transits of Venus across face of the Sun rank among the rarest astronomical phenomena—only 81 occur during the 6,000 year period spanning −2000 to +4000. Two transits of Venus will occur early in the next century (2004 and 2012). Transits of Mercury are somewhat more frequent—117 occur during the 700 year period +1600 to +2300. Four Mercury transits will take place between 1993 and 2006: 1993, 1999, 2003 and 2006. This book presents elements, geocentric data for all transits of Venus from −2000 to +4000 and Mercury from +0160 to +2300. These elements allow the calculation of local circumstances and Jean Meeus has provided all necessary formulae and worked examples to do this. Also presented is a discussion (without elements) of transits seen from other planets.

Introduction to BASIC Astronomy With a PC, J.L. Lawrence, 8.5" by 11.00", 130 pages, softbound, $19.95, ISBN 0-943396-23-9. Introductory astronomy books usually follow two broad paths; descriptive or mathematical. This book is fundamentally a mathematical approach but with a difference. The author has written the text and IBM-PC computer programs to emphasize concepts rather than derivation of formulas. In order that you can begin immediately to learn the how and why of astronomy, the programs (BASICA) are provided on a diskette (IBM-PC, 5.25-inch DSDD) along with this book.

Lunar Tables and Programs from 4000 B.C to A.D. 8000 Michelle Chapront-Touzé and Jean Chapron, 8.50" by 11.00", 176 pages, softbound, $19.95, ISBN 0-943396-33-6. This book provides high precision time-dependent expansions of the longitude, latitude, and radius vector of the Moon, referred to the mean ecliptic and equinox of date. For historians and others who do not need this full precision, more compact procedures are given. Included are formulae for computing coordinates referred to other reference frames and for corrections of aberration. Microcomputer program listings in FORTRAN, BASIC and PASCAL are provided to implement the tables and formulae presented in the book. The best accuracy of the longitude for a date in ephemeris time for periods between −4000 to +8000 is as follows: −4000 to −2000: 0.80°, −2000° to −500: 0.37°, − 500 to +500: 0.136°, +500 to +1500: 0.0474°, +1500 to +1900: 0.0054°, +1900 to +2100: 0.0004°, +2100 to +2500: 0.0054°, +2500 to +3500: 0.0474°, +3500 to +4500: 0.136°, +4500 to +6000: 0.37°, +6000 to +8000: 0.80°.

Planetary Programs and Tables from −4000 to +2800 by Pierre Bretagnon and Jean-Louis Simon, 8.50" by 11.00", 150 pages, softbound, $19.95, ISBN 0-943396-08-5 Included in this book are formulae, tables and microcomputer program listings in FORTRAN and BASIC to compute the longitude of the Sun, the geocentric longitudes and latitudes of Mercury, Venus, Mars, Jupiter, Saturn, Uranus and Neptune with a precision that is always better than 0.01° over a period of time up to 12,000 years (Uranus and Neptune 1,200 years). The positions calculated using the procedures detailed in this book will be far more precise than is usually required by astronomers or historians working on the dating of historical events or documents. For the Sun through Mars each coordinate is represented by only one formula and Jupiter through Neptune for time-spans of five years by power series with seven coefficients.

Prices Subject To Change Without Notice

Turn Page
For This Book's
Optional Software Order Form

Please send me the following magnetic media versions of the programs and data files described on pages 423–24 of this book (Note: This software is sold only to purchasers of the book):

○ **QuickBasic 4.5 for the IBM-PC** at $24.95 each. $________
○ 5.25-inch 360K ○ 3.5-inch 720K

○ **Turbo Pascal 4.0 for the IBM-PC** at $24.95 each. $________
○ 5.25-inch 360K ○ 3.5-inch 720K

○ **"C" for the IBM-PC** at $24.95 each. $________
○ 5.25-inch 360K ○ 3.5-inch 720K

Handling[1] **1.00**

TOTAL $________

I wish to pay with:
○ **Check** ○ **Money Order**
○ **Visa** ○ **MasterCard** ○ **American Express**

Card No.________

Card expiration date________

Signature________

Name (Please Print)________

Street________

City, State, ZIP________

Willmann–Bell, Inc.
P.O. Box 35025
Richmond, Virginia, 23235
Voice (804) 320-7016 FAX (804) 272-5920

Prices Subject To Change Without Notice

This Book's Serial Number is 028996

[1]Foreign orders: shipping charges are additional. Write for proforma invoice which details your exact costs for various shipping options.